Environmental Crises:
Geographical Case Studies in Post-socialist Eurasia

Environmental Crises:

Geographical Case Studies in Post-socialist Eurasia

Tatyana Saiko

Routledge
Taylor & Francis Group

LONDON AND NEW YORK

First published 2001 by Pearson Education Limited

Published 2013 by Routledge
2 Park Square, Milton Park, Abingdon, Oxon OX14 4RN
711 Third Avenue, New York, NY 10017, USA

Routledge is an imprint of the Taylor & Francis Group, an informa business

ISBN 13: 978-0-582-35695-5 (pbk)

British Library Cataloguing-in-Publication Data
A catalogue record for this book can be obtained from the British Library

Library of Congress Cataloging-in-Publication Data
Saiko, Tatyana.
    Environmental crises : geographical case studies in post-socialist Eurasia /
Tatyana Saiko.
        p. cm.
    Includes bibliographical references and index.
    ISBN 0-582-35695-4 (alk. paper)
    1. Environmental degradation--Russia (Federation) 2. Russia
(Federation)--Environmental conditions. I. Title.

    GE160.R9 S25 2000
    363.7'02'0947--dc21
                                                    00-041652

Typeset in Palatino 10/12pt by 30.

This book is dedicated to my dear late father,
Alexey Yevgenyevich Furman,
who was a philosopher at Moscow State University

Contents

Acknowledgements   x
List of figures   xii
List of tables   xiv
List of plates   xvi

Introduction   1
1 *The conceptual framework*   9
   1.1 Introduction   9
   1.2 The concept of geographical zonality   11
   1.3 The concept of environmental degeneration   20

2 *The Arctic North: intensive industrial development in
   the Russian tundra*   31
   2.1 Introduction   31
   2.2 Physical geographical setting   32
   2.3 Historical perspective and economic development   43
   2.4 Ecological consequences and recent changes   49
   2.5 Socio-economic and health issues   62
   2.6 Prospects for the future   64

3 *Lake Baikal: a unique ecosystem under threat in the
   Siberian taiga*   66
   3.1 Introduction   66
   3.2 Physical geographical setting   68
   3.3 Historical perspective and economic development   79
   3.4 Ecological consequences and recent changes   81
   3.5 Socio-economic and health issues   94
   3.6 Prospects for the future   94

CONTENTS

4 The Moscow region: complex problems of Europe's
   major conurbation   98
   4.1 Introduction   98
   4.2 Physical geographical setting   100
   4.3 History of urban development   102
   4.4 Ecological consequences and recent changes   104
   4.5 Health and demographic issues   117
   4.6 Prospects for the future   118

5 Chernobyl: the world's worst nuclear accident   120
   5.1 Introduction   120
   5.2 Physical geographical setting   121
   5.3 The causes of the accident   124
   5.4 Ecological consequences and recent changes   127
   5.5 Health and socio-economic issues   134
   5.6 Prospects for the future   137

6 The 'Black Triangle' of Central Europe: air pollution comes
   under control?   140
   6.1 Introduction   140
   6.2 Physical geographical setting   141
   6.3 Historical perspective and economic development   144
   6.4 Ecological consequences and recent changes   147
   6.5 Socio-economic and health issues   154
   6.6 Prospects for the future   154

7 Kalmykia: the first anthropogenic desert in Europe   159
   7.1 Introduction   159
   7.2 Physical geographical setting   161
   7.3 Historical perspective and agricultural development 165
   7.4 Ecological consequences and recent changes   172
   7.5 Socio-economic implications and health issues   181
   7.6 Prospects for the future   182

8 The Volga basin: a massive degradation of ecosystems   185
   8.1 Introduction   185
   8.2 Physical geographical setting   187
   8.3 Historical perspective and economic development   193
   8.4 Ecological consequences and recent changes   196
   8.5 Socio-economic and health issues   212
   8.6 Prospects for the future   214

9   *The Caspian Sea region: inter-state problems of a*
    shared water body   217
    9.1   Introduction   217
    9.2   Physical geographical setting   219
    9.3   Historical perspective and economic development   222
    9.4   Ecological consequences and recent changes   224
    9.5   Socio-economic and political issues   237
    9.6   Prospects for the future   239

10 *Desiccation of the Aral Sea: the hidden costs of irrigation*   242
    10.1 Introduction   242
    10.2 Physical geographical setting   244
    10.3 Historical perspective and economic development   250
    10.4 Ecological consequences and recent changes   254
    10.5 Socio-economic and health issues   262
    10.6 Prospects for the future   268

CONCLUSION   273
References   286
Index   307

Acknowledgements

Writing a book on a multidisciplinary topic in a foreign language was a challenging and complex task. I would like to extend gratitude to the many people who gave me invaluable assistance in all stages of my research and writing of this book. Firstly, I would like to thank all my colleagues in the Department of Geographical Sciences at the University of Plymouth for advice and encouragement, especially to Professor Mark Blacksell, the Head. I am grateful to all my former Russian colleagues and friends who supplied me with relevant materials and assistance.

My sincere thanks go to Professor Igor Zonn from UNEPCOM who generously provided various valuable materials useful for different case studies. I am very grateful to my friend Vera Sidorova, of the Institute of Geography, Russian Academy of Sciences, for supplying me with hundreds of newspaper cuttings and publications over the last several years. I wish to thank my former colleagues from the Centre for International Projects and particularly Tatyana Butylina, the Deputy Director, for providing a few Russian state-of-the-environment reports and material on the Caspian region.

I would further like to extend thanks to my former colleagues at Moscow State University, to Dr Valentin Burov, Deputy Dean of the Faculty of Geography; Dr Elena Milanova from the Department of the Global Physical Geography and Geoecology; Dr Tatyana Krasovskaya and Professor Alexandr Yevseev for their assistance and authored books on the ecological problems of the Arctic North. My thanks go to Professor German Kust, the Deputy Director of the Soil Sciences Institute, for material on the Aral Sea region.

I am also grateful to the staff of the Institute of Geography, Russian Academy of Sciences, namely Dr Dmitriy Lyuri for his authored book on environmental crises; Professor Arkadiy Tishkov for the first report on the conservation of biodiversity in Russia, and Yuriy Reteyum for very useful publications on the state of the environment in the FSU.

I would also like to thank Larisa Svintsova for her cartographic assistance in compiling some maps for this book. I am very grateful to the friendly cartographic team of the Department of Geographical Sciences, University of Plymouth, and particularly Tim Absalom and Amanda Richardson for compiling a few excellent 'families of maps' for this book. My sincere thanks go to Pauline Framington and David Antwis who helped me to sort out my urgent computing problems. I am very thankful to Kate Hopewell for typing my tables and other assistance. As language presented one of the major challenges in writing this book, I would very much like to thank John Sallnow for all his linguistic and editorial advice and help. I am also grateful to my anonymous referees for their invaluable comments. And finally, my thanks and love go to my son Maxim, whose optimism and sense of humour encouraged me through this most difficult period of life.

I express my gratitude to the 'Novosti' Press Agency of the Russian Federation for their permission to reproduce plates 4, 5, 7, 10–17, 19 and 20 published in this book.

Tatyana Saiko
Department of Geographical Sciences
University of Plymouth

The publishers wish to thank the following for permission to reproduce the material:

Figure 3.2 from Moldau, B., 'Czech Republic' in Klarer, J. and Moldau, B. (eds), *The Environmental Challenge for Central European Economics in Transition*, pp. 107–129 (page 114) © John Wiley and Sons Ltd.

Map from *Water, Air and Soil Pollution*, Vol. 110, 1999, pp. 35–58 (p. 38) from article by Burrough, U., Perk, M.V.D., Holbard, B.J., Prister, B.S. Sansome, U. and Voitsekhovitch, O.V., 'Environmental mobility of radio-caesium in the Pripyat catchment, Ukraine/Belarus': Figure 2.1 with kind permission from Kluwer Academic Publishers.

List of figures

Figure 1.1 Conceptual scheme of interrelationships between   10
            human and natural factors which cause environmental
            degeneration
Figure 1.2 A graph of geographical zonality   16
Figure 1.3 Natural landscape zones in northern Eurasia and the   21
            location of case studies
Figure 1.4 Integral resilience potential of landscapes in the former   24
            Soviet Union
Figure 2.1 Industrial pollution by sulphur dioxide and degradation   56
            of natural rangelands on the Kola peninsula, 1990–1995
Figure 2.2 Pollution of soil by heavy metals and location of sites   57
            of radioactive waste storage, processing and disposal
            on the Kola peninsula, 1990–1995
Figure 2.3 The present ecological situation on the Yamal peninsula   61
            and in the Norilsk region
Figure 3.1 The scheme of interrelationships between anthropogenic   69
            factors and physical environment within the Baikal basin
Figure 3.2 The present ecological situation in the Lake Baikal region   78
Figure 4.1 Distribution of industrial centres and administrative   99
            division of Moscow city
Figure 4.2 Atmospheric pollutant emissions from stationary   106
            sources and transport, 1992–1997/1998
Figure 4.3 The present ecological situation in Moscow city   115
Figure 4.4 The present ecological situation in Moscow oblast   116
Figure 4.5 Demographic changes in Moscow city, 1970–1997   117
Figure 5.1 Distribution of organic peat soils in the Pripyat   123
            catchment
Figure 5.2 Radioactive contamination of the Ukraine and the   128
            adjacent regions of Belarus with caesium-137 on 10
            May 1986

Figure 5.3   Intensity of β radiation within the 'exclusion zone',   132
             1994–1995
Figure 6.1   Acid rains in Central Europe, 1988   149
Figure 6.2   Impact of pollution on forests in Central Europe   152
Figure 6.3   Reduction in sulphur dioxide pollution in West, Central 155
             and Eastern Europe, and the European part of the FSU
Figure 7.1   The present ecological situation in Kalmykia   160
Figure 7.2   Desertification in the 'Black lands' of Kalmykia   176
Figure 8.1   The present ecological situation in the Volga   186
             drainage basin
Figure 8.2   A scheme of the Volga 'river-reservoir system'   200
Figure 9.1   Political divisions in the Caspian Sea region, exploited 218
             oilfields and marine pollution with hydrocarbons
Figure 9.2   Change in the level of the Caspian Sea during   219
             this century
Figure 9.3   A scheme of environmental problems faced by littoral   225
             states in the Caspian Sea region
Figure 10.1 Political divisions in the Aral Sea region and the main 243
             areas of irrigated lands
Figure 10.2 Dynamics of desertification processes in the Circum-   256
             Aral region, 1960–1996
Figure 10.3 Change in the structure of sown agricultural lands in   266
             Central Asia excluding Kazakstan, 1913–1996
Figure 10.4 Per capita gross domestic product in the countries of   270
             Central Asia and Russia, 1989–1996

List of tables

Table 1.1   Climatic, soil and vegetation features of the major   14
             landscape zones in northern Eurasia
Table 1.2   Selected geographical characteristics of the major   18
             landscape zones in northern Eurasia
Table 1.3   Classification of stages in environmental degeneration   27
Table 1.4   Selected criteria for the evaluation of the ecological   29
             situation
Table 2.1   The legacy of economic development and the ecological   46
             change in the Arctic North
Table 2.2   Total airborne pollutants emitted from stationary sources 50
             by selected Russian cities, 1992–1998
Table 2.3   The main pollutants in 'Norilsk Nickel' emissions,   51
             1996–1997
Table 3.1   Discharge of waste waters within the Baikal basin,   82
             1995–1996
Table 3.2   Cellulose production and pollutants discharged by   84
             Baikalsk PPM, 1994–1996
Table 3.3   The legacy of economic development, ecological change   92
             and environmental protection in Baikal region
Table 3.4   The structure of financial sources of the Federal   96
             programme on the protection of Baikal, 1995–1996
Table 4.1   Changes in concentrations of pollutants in the   104
             atmosphere over Moscow city, 1990–1997
Table 4.2   Environmental pressures and health in Moscow city by   110
             district
Table 4.3   Changes in population numbers in the Moscow region,   113
             1975–1998
Table 5.1   Areas contaminated with caesium-137 by the Chernobyl   134
             accident by country 1993–1995
Table 5.2   Population settled in contaminated areas by country,   135
             1995–1996

Table 5.3   Radioactive contamination of agricultural lands in   136
Belarus and the Ukraine

Table 6.1   Total emissions of SOx and NOx in selected countries of   148
Europe, 1980–1990

Table 6.2   Mean concentrations of heavy metals in mosses in the   150
countries of 'Black Triangle', 1991–1992

Table 6.3   Forest damage by air pollution in the Czech Republic   151

Table 7.1   Change in animal pressure on pastures in Kalmykia   167
during the 20th century

Table 7.2   The legacy of agricultural development and ecological   170
change in Kalmykia

Table 7.3   The present status of desertification in Kalmykia   180

Table 8.1   Freshwater use in selected regions of the Volga basin,   198
1985–1997

Table 8.2   Pollutants discharged into the Volga basin, 1994–1997   201

Table 8.3   Decline in fish catch in selected Volga reservoirs,   210
1989–1997

Table 9.1   Oil production in four littoral countries of the Caspian   224
basin, 1990–1997

Table 9.2   Discharge of polluted waste waters by Caspian littoral   227
countries, 1991–1997

Table 10.1   Cotton production in Central Asia in 1997 and in 1980   252
compared to 1913, 1940 and 1961–1965

Table 10.2   Per capita water withdrawal in Central Asian republics   254
and Russia, 1985–1994

Table 10.3   Environmental changes in the Aral Sea basin, 1960–1995   257

Table 10.4   Per capita grain production in Central Asian republics,   267
1913–1997

List of plates

Plate 1   A typical tundra landscape on the slope of the Khibiny   37
          mountains with dwarf birches in the foreground
Plate 2   Coniferous trees are the first victims of air pollution: a fir   40
          tree on the Kola peninsula with a 'flag-pole' top
Plate 3   Mining in the former Soviet Union was always lethal for   45
          nature: a giant quarry in 'Apatity'
Plate 4   A rusting memory of the Yamal's explorers – a technogenic   54
          desert in the tundra
Plate 5   A serene view of the 'Sacred Sea'   67
Plate 6   The typical light-coniferous taiga landscape   75
Plate 7   The unique freshwater nerpa seals bathing in Baikal   77
Plate 8   The industrial landscape dominates this view in the city of 105
          Moscow
Plate 9   The role of traffic in urban air pollution has dramatically 108
          increased over the last decade: the Russian Foreign
          Ministry on the Moscow Garden Ring
Plate 10 Chernobyl: two weeks after the explosion   125
Plate 11 One of the genetic effects of Chernobyl: a radiomorphose   130
          in the plant's stem
Plate 12 Poland possesses Europe's largest reserves of coal   144
Plate 13 A unique antelope of Europe: young saiga in the steppes   166
          of Kalmykia
Plate 14 One of the former Soviet giants, the Volzhskaya dam at   195
          Volgograd, has cut off almost 90% of sturgeon's spawning
          grounds
Plate 15 Construction of this Astrakhan gas condensate complex   206
          near the Volga delta had a dire impact on its ecosystem
Plate 16 Sturgeon versus oil: the main dilemma of the Caspian Sea   222
Plate 17 Fire on the offshore rig near Baku, Azerbaijan: a major   231
          threat to the Caspian comes from its wealth of oil
Plate 18 A typical landscape of the Karakum sand desert   248
Plate 19 Schoolchildren of Uzbekistan harvesting the 'white gold'   253
Plate 20 Will the sea come back to Aralsk?   264

Introduction

The collapse of the socialist system has revealed the grave environmental inheritance of the former regimes. The ecological problems in the former Soviet Union (FSU) and post-socialist countries of Central Eastern Europe (CEE) are acute and diverse. In the early 1990s Altshuler *et al.* stated, 'Practically the whole USSR's territory can be classified as an ecological disaster zone, with very few exceptions. Not only Lake Baikal, not only Chernobyl, not only the Aral Sea or Kalmykia, not only the Volga basin, but many other areas as well can be regarded as ecological disaster areas' (1992, 200).

At the beginning of the 1990s around 300 areas with a very acute environmental situation were recorded in the former Soviet Union. In 1991, 13 regions in Russia were officially designated as 'ecological disaster areas' (Gosudarstvenniy... 1992). Not much has changed for the better during the post-socialist period. More recently, it was estimated, that the total crisis area amounted to 3.7 million sq km or 16% of Russia's whole territory (Okruzhayushchaya... 1996). The state of some regions was found to be particularly critical.

This dramatic situation has very deep-rooted, complex and diverse causes stemming from the essence of the socialist system. Although all of these crisis areas have been 'inherited' from the previous political system, to a certain extent their ecological state has been complicated by some recent changes. The last decade has also brought about new challenges to these independent states, which are associated with their transition to the free market economy. In addition, a recent financial and economic crisis, which hit the Russian Federation in August 1998, also affected some other countries of the Commonwealth of Independent States (CIS), and has reduced the potential for environmental enhancement. At the same time, it appears that those countries of Central Europe that took an early and radical start on the way to economic reform and democracy have already benefited in terms of improvement of their ecological situation.

It should be realized, however, that along with grave ecological problems many of these countries enjoy a major environmental advantage.

The latter is related to the existence of large areas with pristine natural landscapes, almost untouched by human activities, and a unique system of nature reserves. Unfortunately, at the present time these areas also experience an increasing threat from the new profit-oriented industrial and agricultural development.

The central objective of this book is to attempt to develop a holistic understanding of complex environmental problems within the territory of the former 'socialist bloc' countries of Europe and Asia. This is achieved through integration of the knowledge obtained from physical geography and environmental science. At the same time relevant socio-economic and political issues are taken into account. A 'landscape approach' and historical geographical study of human impacts are used to achieve this. The main assumption is that various landscapes have different natural abilities for self-rehabilitation after human disturbance.

An historical geographical context is the basis for the assessment of cumulative changes in the environment, which often occur unnoticed until a certain critical threshold is reached, at which stage the situation becomes a 'crisis' or, ultimately, a 'catastrophe'. One of the aims of this book is to examine different types and degrees of environmental degeneration occurring in various physical geographical settings. To achieve this, a case study analysis is carried out and, where possible, a similar structure of discussion is attempted.

The author also aims to introduce a wide variety of physical geographical landscapes where these ecological disasters occur, which range from the Arctic tundra and the Siberian taiga to the Central Asian deserts. These landscapes are discussed within the concept of geographical zonality with a view to examining their complexity and the factors responsible for natural vulnerability and/or resilience to human impacts. The range of the latter includes the impacts of mining, industrial and urban activities; technical accidents; over-cultivation and overgrazing; irrigation development and forestry or a combination of the above factors. This book is intended for undergraduates and experts in environmental sciences and human and physical geography.

The standpoint adopted to address these questions is that a landscape is in a state of an evolving interrelationship between its individual components, i.e. relief, climate, hydrology, vegetation, soils, and any external anthropogenic disturbance will inevitably affect the whole set of them. The magnitude of changes will depend on the adaptability, resilience and rehabilitative capacity of each zonal type of the landscape. The classification of environmental changes is based on the stages of ecological deterioration, which involves not only implications for the environment but also for human health and socio-economic conditions. Examples for discussion of different types of environmental problems are drawn from the specific areas within the former socialist countries, which have currently reached or are reaching the state of environmental crisis or catastrophe.

The author of this book does not attempt to examine 'the geography of environmental crises' which would need the inclusion of many more adversely affected regions of post-socialist Eurasia. It is, rather, an assessment of selected geographical case studies located in different landscape zones that also differ in terms of the character and magnitude of impact of various anthropogenic factors. The focus here is on the varying natural vulnerability and self-rehabilitative capacity of diverse landscapes defined in Chapter 1.

Three areas, considered in the context of the Arctic North (Chapter 2), are found within the tundra, forested tundra and northern taiga landscape zones. Lake Baikal (Chapter 3) lies within the Siberian taiga or boreal coniferous forest; furthermore, its drainage basin features mountainous relief, and, therefore, has a clear pattern of 'altitudinal zonality'. The Moscow region (Chapter 4) is located in the mixed forest zone that is found only within Europe. The regions of the Ukraine and Belarus affected by the Chernobyl accident (Chapter 5) also include distinctive forested steppe and marshland landscapes along with mixed and deciduous forests.

The 'Black Triangle' of Poland, former Czechoslovakia and GDR lies within the belt of broad-leaved deciduous forests. Kalmykia (Chapter 7) in the Russian Federation features a range of steppe and semi-desert conditions, while the Aral Sea (Chapter 10) drainage basin lies mainly within the desert landscape zone. At the same time an extensive drainage basin of the Volga, Europe's longest river, covers a whole range of landscapes which alternate from the north to the south: from the European taiga to the deserts of the Circum-Caspian lowland (Chapter 8). The landlocked Caspian region (Chapter 9) is predominantly arid and features a variety of landscapes. In addition, four major aquatic systems appear in these case studies: that of Lake Baikal, the Volga, the Aral Sea and the Caspian Sea.

Each of the above landscape types has a varying vulnerability to human impact and a different self-rehabilitation capacity. The most fragile and sensitive of these are the tundra and desert landscapes; at the other end of the scale, the most stable and resistant appear to be the deciduous forest and forested steppe zones.

The range of operating anthropogenic factors and associated environmental problems is also extremely wide. Three studied areas of the Arctic North are affected by an extended legacy of intensive industrial development that includes mining, metallurgy and, more recently, oil and gas extraction. Consequently, the main environmental problems here include severe sulphur dioxide pollution; contamination of soils and vegetation with heavy metals; oil pollution; degradation of forests and reindeer pastures. These are further aggravated by major socio-economic implications for the indigenous and migrant population.

In the Baikal case study, the focus is on the unique nature of this lake and on the need to adopt a 'drainage basin' approach to its study. The latter appears essential because all adverse impacts within this whole territory would ultimately affect the water quality in the lake and its endemic aquatic life. The range of anthropogenic factors operating in this case comprises timber and cellulose production, mining and energy generating industries, settlements and transport, all of which directly or indirectly contribute to Baikal's pollution.

The Moscow region faces a completely different set of problems that are typical of major urban agglomerations. With probably the highest population density among the world capitals, Moscow has an extremely high industrial pressure within its city limits. Excessive water consumption patterns inherited from the Soviet period contribute to large-scale water pollution. New environmental challenges during the transition to a 'free market economy' are related to the increasing urban encroachment into the 'green belt' around Moscow, increasing consumerism and an associated exponential increase in transport pollution and waste generation.

A completely different type of environmental problem is associated with radioactive contamination. This is examined in the Chernobyl case study. Along with ecological consequences, health issues become very important in this context. The focus here is on the universal nature of the impact on the environment and the importance of long-term implications. The 'Black Triangle' in Central Eastern Europe has an ill reputation for extensive air pollution. This case study examines the main causes and repercussions of these problems and underscores the recent positive developments in environmental enhancement in this region.

The Russian Republic of Kalmykia is relatively unknown to Western scholars although it is unique in many ways. During the Soviet period excessive animal pressures were applied to sensitive dry steppe ecosystems. These, combined with the effect of cultivating fragile pastures, have resulted in the creation of Europe's first anthropogenic desert. Another type of unsound agricultural development, i.e. mismanagement of irrigation, is associated with the so-called 'cotton independence' policy pursued by the Soviet government in the 1950s and 1960s. These were responsible for the desiccation of the Aral Sea and tremendous environmental problems in the adjacent region. Socio-economic and health issues in this context assume major importance as well. This case study is also one of the two examples, along with the Caspian region, where international co-operation is essential to achieve any improvement in the environmental sphere.

The Volga drainage basin comprises over half of European Russia and has the longest history of industrial and agricultural development in the country. Centuries of increasing anthropogenic pressures have substantially transformed the natural landscapes of this vast territory. However, the most adverse changes date back to the policies of the Soviet period, namely 'electrification' and 'industrialization', which completely ignored the possible ecological consequences. As a result of these attitudes, massive degradation of ecosystems has become a grave reality throughout the Volga region.

Since the Volga flows into the Caspian Sea, the ecological problems of the latter can be primarily attributed to the pollution from this mighty river. Furthermore, a recently emerging threat to this sea comes from the already-started offshore oil production. It will acquire further magnitude in view of its expected substantial increase in the next century. The geopolitical interests of many countries are focused on this region. The case study covering this area looks at the growing contradiction between the economic and environmental interests. The world's single commercial stock of sturgeon species located here has already experienced a catastrophic decline in population numbers over previous decades. Other issues examined by this case study are the various implications of the rise in the sea level, the importance of international co-operation and the need to adopt an agreed legal regime for the Caspian Sea.

The conceptual framework of this book is discussed in Chapter 1. It includes the concept of geographical zonality that underlies the differences in landscape types, which are apparently also responsible for their varying sensitivity to human impacts. Another applied concept is that of environmental degeneration which comprises not only the process of adverse ecological change, but also the associated socio-economic and health implications. The character of different stages in environmental degradation is examined, and selected criteria for their identification are outlined. The complexity of such an assessment is emphasized.

To allow a certain degree of comparability between the different case studies, the author has attempted to use a similar structure within each chapter wherever possible. Therefore, each case study outlines the physical geographical setting, including geographical location and relief; climate and hydrology; vegetation and soil cover. The above issues are important for the evaluation of the natural vulnerability of landscape(s), which is examined in a final subsection. The degree of detail differs in various case studies.

The application of an historical geographical approach helps to achieve an in-depth understanding of the causes and changes in anthropogenic impact both during the Soviet/socialist period and in the more recent period of transition to the market economy. Thus each case study gives an overview of the past and contemporary economic development or, in the case of Chernobyl, of the legacy of events related to this technical accident. The main focus of this book is on the ecological consequences of economic development and recent changes, which are discussed in detail within appropriate subsections. The current overall ecological situation is summed up in the last subsection and, where possible from the available data, the analysis of dynamic changes to the past ecological status is also given (e.g. in Chapter 7, on Kalmykia).

Each chapter further looks at the existing socio-economic and health implications of adverse ecological changes, the importance and scope of which differ in the various contexts of case studies. A review is included of the recent major developments in environmental protection, if any, and socio-economic aspects that may affect the situation in future. Along with these issues, the last section attempts to give an outline of a possible future scenario.

The issues raised in this book have received different degrees of coverage in the scientific publications. Many urgent environmental problems have been examined by the geographical English language literature over the last decade (e.g. Bradshaw 1997; Carter and Turnock 1993, 1996; Feshbach and Friendly 1992; Fodor 1994; Klarer and Moldan 1997; Manser 1993; Mnatsakanian 1992; Peterson 1993; Pryde 1991; Shaw 1999; Stewart 1992; Symons 1990; Tickle and Welsh 1998; Vari and Tamaz 1993; Wolfson 1994). Health issues related to the FSU were discussed in the work of O. Bridges and J. Bridges (1996). Regional in-depth monographs have been published on the Aral Sea problems (Glantz 1999; Kobori and Glantz 1998; Letolle and Mainguet 1993; Micklin and Williams 1996), the Caspian Sea region (Glantz and Zonn 1997) and Chernobyl (Marples 1996; Medvedev 1990; Savchenko 1995).

Various ecological problems associated with Arctic development have been discussed in numerous articles in international journals (cited in Chapter 2). The latter have often raised the issue of extreme vulnerability of landscapes found in these landscapes. However, in publications related to many other ecological disaster areas, with an exception of Chernobyl, this aspect has normally found only limited coverage. Insufficient information still exists in the English language publications on such territories as Kalmykia, the Moscow region and the Volga basin. Therefore, this book also aims to fill the existing gaps in Western knowledge of the current environmental status of these regions.

In the early 1990s Altshuler *et al.* noted that 'The seventy-three year history is a history of systematic misinformation on the environmental situation in the USSR' (Altshuler *et al.* 1992, 201). After 1986, when censorship on environmental information was lifted, and particularly during the post-Soviet period, the overall amount of statistical and other data on the ecological situation in post-socialist Eurasia has substantially increased. 1989 onwards has seen the publication of annual State Reports on the Status of the Environment in the Russian Federation (Gosudarstvenniy... 1992–1998a–c).

Recently the overall reliability of such information published within the Commonwealth of Independent States has been questioned (e.g. Shaw and Oldfield 1998). It is worth noting that the author of this book happened to work for 12 years in the organization responsible for the collation of material for the Russian State Reports over the last few years, namely, the Centre for International Projects. To the best of my knowledge, in the case of the Russian Federation, this lack of reliability during the post-Soviet period can be primarily attributed to substantially weakened monitoring capabilities throughout Russia due to financial difficulties. This issue is discussed in the text, for example in the context of the Volga basin case study. In addition, in Russia as well as in most other countries of the former Soviet Union, one of the main problems with reliability of data is the problem of bribing poorly paid officials where an enterprise or organization needs to withhold information for commercial reasons. Furthermore, provision of ecological data is, sometimes, the responsibility of the polluter e.g. in the case of oil spills, and, as discussed in the chapter on the Arctic North, the exact dimensions of the disaster are often underestimated.

A joint publication on the state of the environment in the 12 member-states of the Commonwealth of Independent States (CIS) was published in 1996 (Sostoyaniye... 1996). However, as noted in the context of the Aral Sea case study, after 1993–1994 some politically sensitive information, such as that related to water withdrawals and losses, is not provided by individual countries of the CIS. Information on the state of air and water quality and environment protection forms a section in the annual statistical yearbook of CIS countries (Sodruzhestvo... 1997; Statisticheskiy... 1996). In addition, a few high-quality detailed thematic atlases were published recently on Lake Baikal (Baikal 1993), the Chernobyl region (Atlas... 1996, 1998) and on the environment and health in Russia (Atlas... 1995). Two ecological maps have been published in the Russian Federation recently, one at a scale of 1:4,000,000 (see Poyasnitel'naya... 1996) and another at a scale of 1:8,000,000 (Ekologichaskaya... 1999); the latter map was used in compiling several ecological situation maps for this book.

The evaluation of all the above sources, along with the author's own research on selected ecological disaster areas, allows us to develop an understanding of the underlying causes of the continuing environmental degeneration in the post-socialist world. However, only limited data exists on the actual critical anthropogenic loads and thresholds of change within the wide variety of natural landscapes. Even more limited attempts have been made to compare such information. It appears that extensive long-term studies are needed in the future to assess these values and provide a valid quantitative basis for any comparative studies. Clearly, the sensitivity of a specific natural environment is a dynamic parameter that changes both in time and space even within a single type of landscape. Nevertheless, the differences in vulnerability of different landscapes should be accounted for in environmental management policies and in adopting permissible norms of anthropogenic impacts, separately for each case.

Any attempt to integrate and synthesize the existing multi-disciplinary material on such a wide range of geographical study areas within the scope of one book is a major challenge and it is fraught with many obvious limitations. These are, for example, the lack of comparability between the levels of detail combined with the problem of description; furthermore, there are individual aspects of each case study that are unique and cannot be compared. In addition, the availability and quality of data was not always adequate. This text should, therefore, be considered as an attempt towards synthesis in this area of scientific research.

The material included in this book has been presented and systematically revised within a course taught for several years to undergraduate students entitled 'Environmental Crises: Case Studies in Europe and Asia' by the author in the Department of Geographical Sciences at the University of Plymouth, United Kingdom.

CHAPTER 1

The conceptual framework

We live in an exceptional time in the history of our biosphere ... when the geological role of mankind becomes dominant ...

(V. Vernadskiy 1969, 270)

1.1 Introduction

The conceptual basis which forms the framework for this book comprises two main concepts: the global regularity of 'geographical zonality' and the concept of 'environmental crises'. Both are equally important for understanding the scope and magnitude of environmental problems in a given area. The state of the environment in different regions of the world is governed by the two sets of factors: natural and human. Natural factors comprise the specific features of a local physical environment which are responsible for its stability and/or vulnerability and the potential ability to resist external disturbance. At the same time human factors determine the nature, character, magnitude and duration of anthropogenic impacts. It is important that the same kind and magnitude of anthropogenic impact can produce different degrees of influence or even effects in various types of physical environment.

The most important issue is the character of the interrelationships between mankind and the environment. Therefore, it is essential to define what the 'environment' actually is and to examine whether humans are part of it. The 'environment' is probably the most widely used term in contemporary science but there is no consensus on what it means. For example, the *Dictionary of Environmental Science* defines it as 'the sum total of external influences acting on an organism' (Lawrence *et al*. 1998, 136). In this definition the 'organism' is clearly not incorporated in the notion of the 'environment' but rather opposed to it.

The *Encyclopedic Dictionary of Physical Geography* does not give a definition for the 'environment' but in discussing 'man-environment relations' states that 'Mankind tends to regard the environment as a set of resources and a receptacle for wastes' (Goudie 1997, 314). In this case, humans are also considered not as an innate component of the environment but as its external consumers and users. Again, in a recent dictionary of geography the environment is defined as 'that which surrounds, the sum total of the surroundings within which an organism, or group, or an object, exists' (Clark 1998, 132).

The approach taken in the context of this book considers humans as an inherent part of the environment. It appears that such a position is vital for the better understanding of the role of humans in the biosphere and for achieving a better control of their actions regarding the physical environment. This comprehension of an extremely complex 'humans–nature'

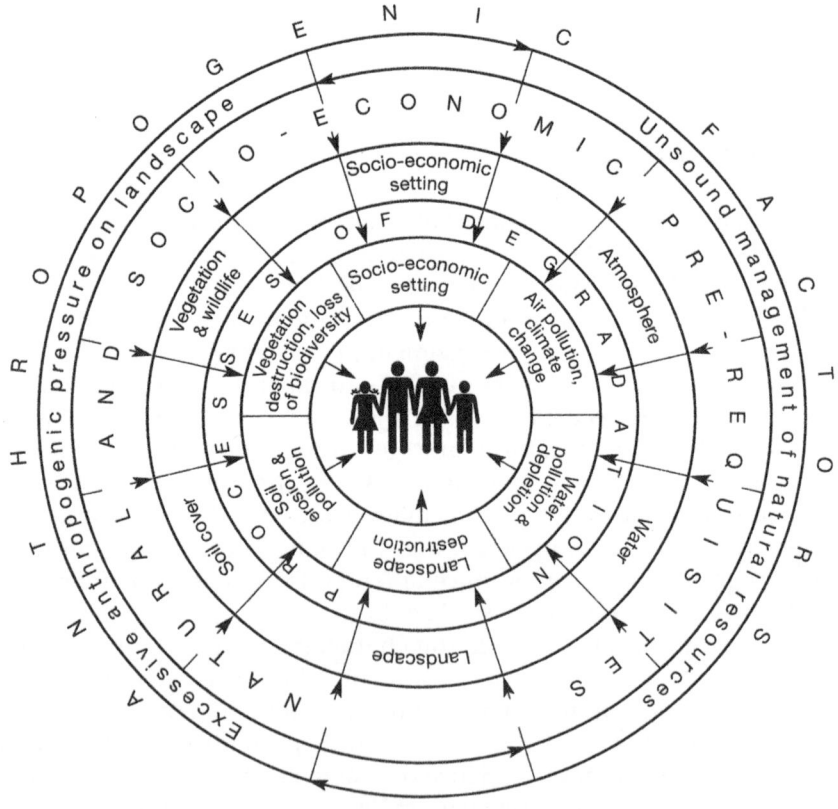

Figure 1.1   Conceptual scheme of interrelationships between human and natural factors which cause environmental degeneration

relationship has been increasingly used in the geographical science of the Soviet and post-Soviet periods (Gorshkov 1998; Danilov-Danilyan and Kotlyakov 1993; Milanova and Ryabchikov 1987; Protasov and Molchanov 1995). Some Western scholars have also shared this approach. For example, Mote stated, 'Environment is a comprehensive term that includes both human and physical factors, the equivalent of the French notion of "milieu". Accordingly, environmental quality bespeaks the quality of both the physical and human environment' (Mote 1992, 52).

An attempt to illustrate the conceptual basis adopted by the author of this book is given in Figure 1.1. All human factors in this concept are grouped into excessive anthropogenic pressure on the landscape, the definition of which will be introduced later in this chapter, and the environmentally unsound management of natural resources. These factors interrelate with each other and affect the state of the landscape as a whole and of its components, i.e. the atmosphere or climate, surface and ground water, soil cover, vegetation and wildlife. In turn, their properties, along with the socio-economic setting, represent the natural and socio-economic prerequisites of potential degeneration. In other words, they determine the degree of the landscape's or geosystem's resilience to external disturbance. The processes of degradation or degeneration operate individually in each landscape component and in the landscape as a whole, and all these changes have direct or indirect implications for humans who are the focus of this complex and intertwined system of relationships.

1.2 The concept of geographical zonality

The area of the former Soviet Union (FSU) amounts to an impressive one-sixth sixth of the total land surface of the world. As noted by L. Symons, 'In the geologically recent formative period following the Pleistocene period, the low and subdued nature of the relief over the vast areas of the Russo-Siberian plains enabled climate to use a "broader brush" in the creation of continuous, fairly homogeneous latitudinal "natural zones" of soil and vegetation than anywhere else in the world' (Symons 1990, 43). These so-called 'natural' or 'geographical' or 'landscape' zones alternate from the north to the south in a specific pattern. The alternation of different types of landscapes ranging from the Arctic polar deserts to the 'true' hot deserts of Central Asia was a unique feature of the FSU. Russia with an area of over 17 million sq km has the greatest share of this diversity.

It should be noted that in the Western geographical media, the concept of a 'landscape' basically differs from the one developed and widely used by Russian geographers. For example, in the *Dictionary of Environmental Science* the definition of a landscape reads: 'a distinct region, e.g. coastal or volcanic, composed of individual landforms' (Lawrence *et al.* 1998, 225). This term originated in Germany as

'Landschaft', implying both an area of land and an aesthetic response to physical environment. The approach most often taken in relation to landscape is historic, cultural or aesthetic.

In a recent work, Muir examined the evolution of attitudes to the notion of landscape in the UK and Western geography in general. He has revealed that in 1939 Hartshorne had rejected a landscape as the central organizing concept in geography in favour of geography as a science of region and space. This had affected further attitudes: 'The perspectives of more than half a century ago still affect the geographical treatment of landscape and endure in ambiguity between objective and subjective attitudes towards landscape, while in late twentieth century geography, the subjective approaches are ascendant ... Although landscape has returned to favour in geography, the study of landscape is fragmented and unbalanced' (Muir 1998, 263, 269).

The concept of a 'landscape' used in the Dobris Assessment of Europe's environment is not defined but the examined range of landscapes included natural, cultural and rural types. It was, however, noted in the above context that 'The interrelationship between nature and people varies from place to place, due to differences in physical conditions, such as topography, climate, geology, soils and biotic factors and the type of human use or occupancy that range from minimal to intensive' (Europe's Environment 1995a, 172).

The reason why landscape studies have 'lost' favour in Western European geography seems to be attributed to two factors. Firstly, the main part of Western Europe lies within one geographical zone, namely the broad-leaved deciduous forest. Secondly, there are virtually no pristine landscapes left within this territory, unlike in the former Soviet Union. As the above source stated, 'There are practically no areas in Europe that can be considered "natural" in the sense that there is no human influence whatsoever, and few where there is no human presence' (ibid.).

In contrast to the above attitudes and approaches, 'landscape' was and remains the focal concept in Russian physical geography. There are dozens of definitions of a landscape. The closest to our concept is the one suggested by Perel'man (1975): 'Landscape is a complex dynamic system of the Earth's surface, in which living organisms and inorganic matter penetrate into each other, are very closely linked and deeply interdependent'. According to his definition, 'landscape is the fundamental unit of the environment'. A hierarchy of landscapes is distinguished: they range from elementary to global, the latter represented by the biosphere. Most commonly the term 'landscape' is used as an equivalent of a 'geographical zone'. However, it can be still applied to a micro natural complex e.g. that evolved on a slope of a local river valley. Along with zonal distribution, some landscapes called 'intrazonal' can be found in various types of location e.g. alluvial floodplains. Another definition of a landscape as a 'fairly homogeneous complex of interrelated and interacting physical geographical components i.e. climate, relief, hydrology, soils and vegetation cover that have a common history of evolution and structure' specifies the main genetic and structural parts of this unit.

The concept of an 'ecosystem' which, by Odum's definition, 'includes all of the organisms in a given area interacting with their physical environment so that a flow of energy leads to ... exchange of materials between living and non-living parts of the system' (Goudie 1997, 167) has a certain similarity with the notion of a landscape. However, the ecosystem's concept is normally focused on the biotic component. The most commonly used equivalent of a 'landscape' in the Russian geographical literature is a 'geosystem' (e.g. in Armand 1975; Gorshkov 1998). The link between the two was noted by the former author: 'Ecosystems are geosystems in which biocomponents play a significant role'.

The observed sequence of alternating landscape zones is a reflection of the global effect of the regularity that is known as a *geographical zonality*. It was initially revealed and examined at a regional level by a well-known Russian soil scientist, V. Dokuchaev (1846–1903) in the 19th century in his study of chernozem soils. His work entitled 'Natural soil zones' was first published in 1900 and republished in 1948 as 'Natural zones and classification of soils'. V. Vernadskiy (1863–1945), an outstanding Russian scientist, has further developed Dokuchaev's ideas. This was essential for the elaboration of a theoretical concept of the 'biosphere'. Vernadskiy's work entitled 'Biosphere' was published in 1929 and only many decades later was this concept recognized and widely accepted by Western scholars.

Geographical zonality is a global geographical regularity. It is expressed in the consecutive, regular succession of geographical belts and zones due to the specific character of distribution of solar energy and moisture on the Earth. A 'geographical' or 'natural' or 'landscape zone' features common thermal conditions and a moisture regime, which together determine a specific nature of hydrological conditions, geochemical processes, soil formation and character of vegetation cover (Table 1.1). The total amount of solar radiation increases from the poles to the lower latitudes. At the same time, distribution of moisture has a more complex pattern due to the nature of global atmospheric circulation. It tends to increase from the poles to mid-latitudes, then decrease in the direction toward the tropics and increase again closer to the equator. A study of the impact of climatic factors on geographical zonality was made by M. Budyko, a Russian climatologist. A graph of global geographical zonality is shown in Figure 1.2, which illustrates that the global sequence of landscapes is as follows: tundra – forest – steppe – semi-desert – desert. Variation is greatest within the forest zone due to its wide diversity of radiation conditions and favourable moisture.

Table 1.1 Climatic, soil and vegetation features of the major landscape zones in northern Eurasia

Climatic belt, landscape zone	Heat and moisture regime	Radiation index of dryness*	Soil types	Water regime of soils	Vegetation type
POLAR BELT					
Polar desert	extremely cold excessive moisture	0–0.2	permanent ice frozen ground	permanently frozen	lack of vegetation
Arctic tundra	extremely cold excessive moisture	0–0.2	cryogenic tundra soils	frozen for much of year excessive moisture during brief thawing period	mosses, lichens, rare grasses
SUBPOLAR BELT					
Tundra	very cold winter brief cool summer excessive moisture	0.2–0.3	gley tundra soils boggy tundra soil cryogenic soils	water stagnation above permafrost during thawing period	tundra vegetation: mosses, lichens, shrubs, dwarf small-leaved trees
Forested tundra	very cold winter cool summer excessive moisture	0.3–0.4	gley tundra soils gley podzolic soil cryogenic soils	waterlogging above permafrost; brief leaching period	tundra vegetation with with patches of small-leaved and coniferous forest
TEMPERATE BELT					
Taiga: boreal coniferous forest	very cold/cold winter warm summer increased moisture	0.4–0.8	podzolic taiga soils peaty-podzolic soil	seasonally waterlogged periodically leached long frozen	light or dark-coniferous forests with small-leaved deciduous component
Mixed forest	cold winter warm summer increased moisture	0.6–0.8	peaty-podzolic soils grey forest soil	leaching regime frozen in winter	combination of coniferous and broad-leaved deciduous forest with small-leaved component

Table 1.1 Continued

Broad-leaved deciduous forest	mild winter warm summer optimal moisture	0.8–1.0	grey forest soil brown forest soil	seasonal leaching regime leaching with filtration frozen in winter	broad-leaved forests with small-leaved and coniferous component
Forested steppe	mild winter warm summer optimal moisture	0.8–1.0	brown forest soil podzolized chernozem	periodic leaching regime frozen in winter	combination of broad-leaved forests with grasslands composed of mixed grasses
Steppe	mild winter hot summer insufficient moisture	1–2	chernozems (black earths) chestnut soils in dry steppe	periodic lack of leaching briefly frozen in winter	grasslands composed of gramineous or xerophytic plants (in dry steppe) with woodland in river valleys
Semi-desert	cool winter hot dry summer lack of moisture	2–3	chestnut soils brown desert soil saline soils	periodic leaching in pulse regime briefly frozen	xerophytic shrubs and grasses with ephemeral plant component
Desert	cool winter very hot summer extreme lack of moisture	more than 3	grey desert soil saline soils: 'solonchak' and 'solonets'	pulse leaching regime severe lack of moisture increased mineralisation of soil water	sparse xerophytic and halophytic shrubs and grasses including ephemeral plants

*Source: Grigoriev and Budyko (1956)

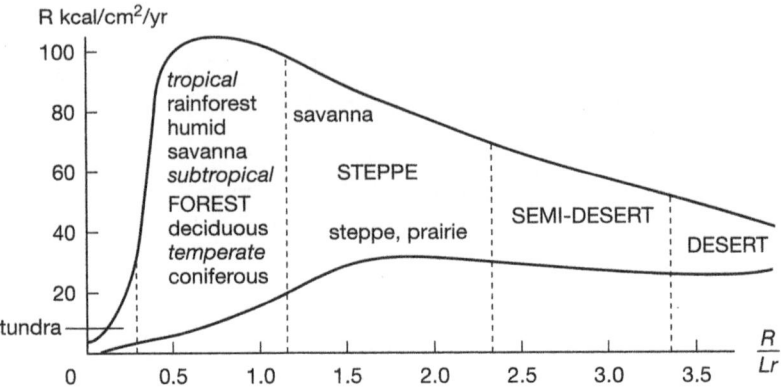

Figure 1.2 A graph of geographical zonality (after Grigoriev and Budyko 1956): R/Lr – radiation index of dryness; R – radiation balance; L – latent evaporation heat in kcal/yr; r – precipitation

Budyko has developed a quantitative basis for the assessment of various landscape zones included in the above graph. In 1955 he published a global map of the distribution of the radiation index of dryness. This index is defined by the ratio between the radiation balance (R) and the moisture balance calculated as the value of latent heat of evaporation (L) multiplied by the amount of precipitation (r). In essence, it defines the relationship between heat and moisture. The wettest conditions are found in the high latitudes where potential evaporation is extremely small. As is clear from Figure 1.2, the tundra appears to be the wettest landscape on Earth.

The values of the radiation index of dryness for each landscape zone in northern Eurasia are given in Table 1.1. This shows that the optimum balance between heat and moisture that corresponds to the index value of 1.0 is observed approximately along the border between the broad-leaved deciduous forest and forested steppe zones. In general, the forest zone features values from 0.3 to 1.0; steppe, 1–2; semi-desert, 2–3 and desert, >3. This regularity is equally reflected in the soil changes. In the global scope, with an increase in the dryness index, soil types change in the following sequence: (a) tundra soils; (b) podzols, brown forest soils, yellow and red earths and laterite soils (wide variations are due to a great amplitude in the value of R); (c) black earths or chernozems of steppes and black soils of savannahs; (d) chestnut soils; (e) grey desert soils. Furthermore, in different latitudinal belts the same index values of dryness correlate to geographical zones that are similar in many significant features. Grigoriev and Budyko called this global regularity a 'periodic law of geographical zonality' (Budyko 1977; Grigoriev and Budyko 1956).

The role of climate in 'shaping' landscapes is paramount. Climate affects the formation of the hydrological system within a territory, its soil geochemistry and vegetation growth. Table 1.1 illustrates the impact of various combinations of heat and moisture on soil formation and vegetation cover in the different landscape zones of northern Eurasia. For example, in European Russia, to the north of the forested steppe zone, moisture is sufficient but heat is inadequate and the soils are leached and acidic. To the south of the 'equilibrium' or 'balance' line, there is enough heat but inadequate moisture, so leaching gives way to accumulation. Soils, therefore, become richer in humus content. At the same time, in the more arid and hot regions of semi-desert and desert landscapes, lack of moisture and intensive heat bring salts closer to the surface and create saline types of soils.

Vegetation change between different zones is also spectacular. While in the tundra landscape, mosses and lichens alternate with shrubs and dwarf trees, the taiga has a vast continuous coniferous forest cover. These give way to mixed forests and then deciduous forests in the European part of Eurasia, and directly to deciduous forests in the Asiatic part. Further south, after a transitional zone of forested steppe, there are the infinite open spaces of the steppes. A continuous cover of rich and diverse grassland in this landscape zone gives way to the more sparse vegetation of semi-deserts. The latter are replaced by deserts comprising drought and salt-resistant shrubs and grasses. The boundaries of natural zones are gradual and they roughly coincide with the borders of the vegetation zones. These are called 'biomes' in the context of biogeography.

Table 1.2 provides further detail on the geographical characteristics of each landscape zone. This data includes radiation balance, annual precipitation, annual biological production, the value of phytomass, plant biodiversity, duration of primary and secondary successions. It is evident from this table, that, for example, the value of annual biological production is the minimum in the Arctic or polar climatic belt and it reaches its peak within the forested steppe and steppe landscapes. At the same time, the overall value of phytomass is greatest in the broad-leaved deciduous forest zone and it drastically declines in the steppes. Plant biological diversity reaches its maximum in the forested steppes where both trees and grass species are represented in a great variety. The duration of primary succession in the Arctic and Subarctic ecosystems takes thousands of years. It is interesting that it lasts for 800–1000 years in the light coniferous taiga of Siberia, but is estimated 'only' at 150–200 years in the dark coniferous taiga found mostly within the European part of Eurasia.

Table 1.2 Selected geographical characteristics of the major landscape zones in northern Eurasia

Landscape zone	Radiation balance kcal/sq cm/yr	Annual precipitation mm	Annual biological production t/ha/yr	Phytomass t/ha	Plant biodiversity species / 100 sq km	Duration of primary succession years	Duration of secondary succession
POLAR BELT							
Polar desert	0–10	100–150	0.1–0.3	0.5–2.0	20–30	3000–3500	ND
Arctic tundra	10–15	150–250	1–3	5–10	70–100	1000–3000	400–500
SUBARCTIC BELT							
Subarctic tundra	15–20	250–400	2–4	10–40	150–200	1000–1500	30–80
Forested tundra	20–25	400–500	3–5	30–100	250–300	1000–1200	ND
TEMPERATE BELT							
Taiga: light coniferous	25–30	400–700	2–8	60–180	400–450	800–1200	350–400
Taiga: dark coniferous	25–35	600–750	4–12	100–350	400–700	150–200	120–150
Mixed forest	30–40	600–800	12–20	300–400	600–700	ND	ND
Deciduous broad-leaved forest	35–45	600–1000	12–25	300–500	700–800	300–500	100–200
Forested steppe	40–42	400–800	18–20	100–300	900–1100	ND	ND
Steppe	40–45	500–700	18–25	15–30	800–900	100–150	35–40
Dry steppe	40–49	300–400	6–15	8–15	400–700	ND	ND
Semi-desert	42–49	200–350	4–8	5–10	300–400	ND	ND
Desert	40–55	100–200	2–5	5–10	200–300	100–300	50–300

Compiled by author from sources: Bazilevich et al. (1986); Mil'kov and Gvozdetskiy (1986); Gvozdetskiy and Mikhailov (1987); Sokhraneniye... (1997); Tishkov (1993)

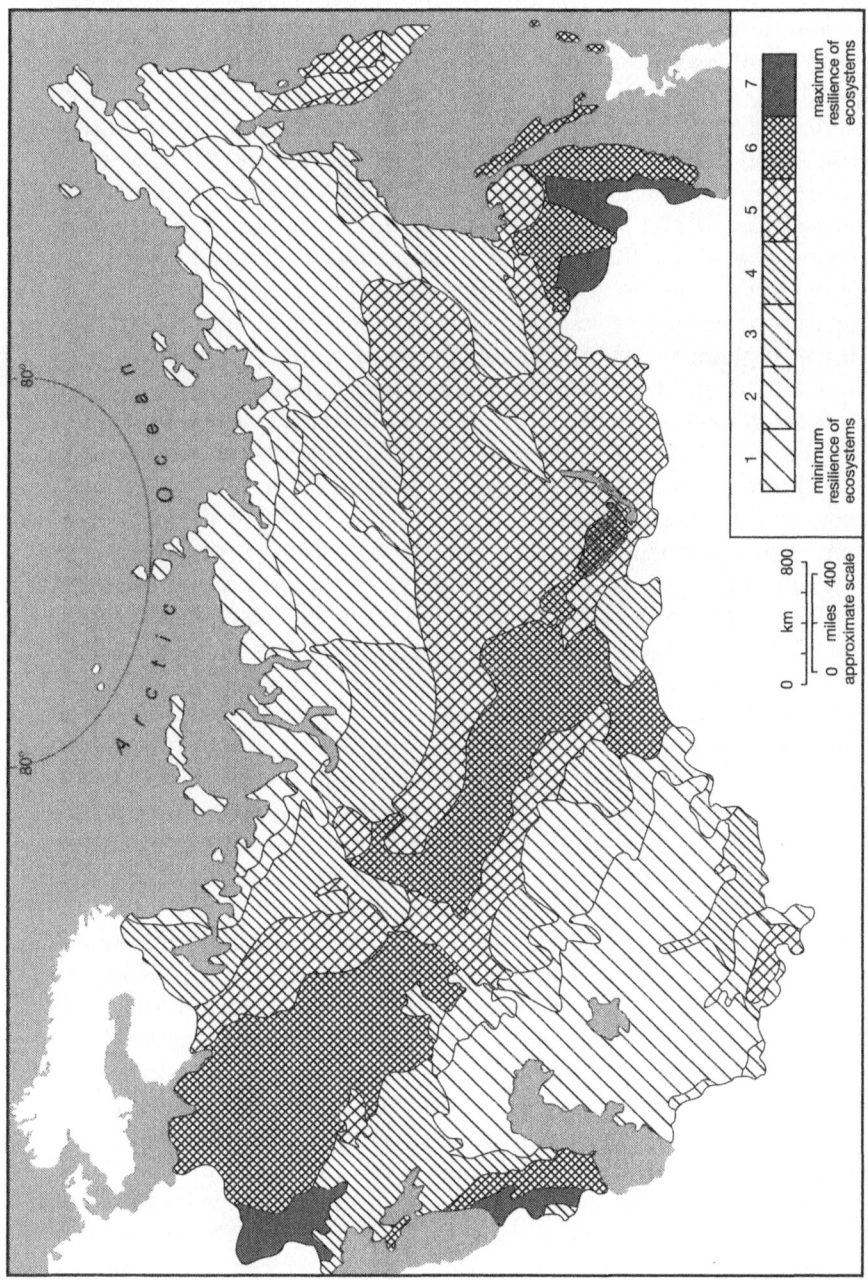

Figure 1.4 Integral resilience potential of landscapes in the former Soviet Union (after Atlas... 1995)

Table 1.2 also reveals the potential ability of different types of landscapes for self-rehabilitation after anthropogenic disturbance. For example, if human impact terminates, normally steppes could recover faster than forests: the duration of secondary succession in steppes on fallow is 35–40 years compared to 100–200 years in oak groves after logging and 120–150 years in the dark coniferous taiga (Sokhraneniye... 1997). According to Tishkov (1993), the existence of a powerful organic horizon in the steppe soils provides the needed 'reserve' for a relatively fast plant regeneration if the steppes are left idle. Another peculiar phenomenon is a contrast between the duration of primary and secondary succession after fire in the Subarctic tundra zone. This could probably be attributed to the absence of trees in this geographical zone.

Apparently, secondary successions feature a more simplified structure and lower biological production and diversity compared to those in the original landscapes. For example, in the taiga landscapes secondary successions often have a greater share of small-leaved deciduous species. It should be noted that for regions with large areas of disturbed or degraded territories and transformed aboriginal biota, self-rehabilitation of ecosystems proves difficult.

Many of these landscapes have been modified to various degrees by human activities. Milanova has developed a classification of present-day landscapes (Milanova and Kushlin 1991), on the basis of which, the 'World Map of the Present Status of Landscapes' was elaborated at the Department of Global Physical Geography and Geoecology, Faculty of Geography, Moscow State University. In this map, landscapes are divided into two groups: 'natural' or 'primary' and 'natural-anthropogenic'. The latter group is further subdivided into 'secondary' or 'derivative', 'anthropogenic' and 'technogenic' dependent on the degree of transformation. In order to have a better understanding of anthropogenic modifications in various geographical zones it is essential to study the original features of natural landscapes.

The effect of geographical zonality which otherwise would create a strictly latitudinal pattern of landscapes is complicated in the FSU by the global distribution of land and water and by the orographic pattern. Therefore, along with *latitudinal zonality* most clearly represented in the two major plains, namely the East European or Russian and the West Siberian, there is a similar phenomenon called *altitudinal* or *vertical zonality*. It operates in mountainous regions and is expressed by a succession of consecutive natural landscape zones. They alternate uphill with altitude in a similar pattern to latitudinal sequence, normally starting from the 'zonal type' of landscape dominant in the foothills. The actual range of landscapes actually depends on the height of the mountain system, climate and other factors. Sometimes one or two zones from the 'latitudinal zonal range' may be missing.

The distribution of various geographical zones within northern Eurasia and the location of case studies considered in this book are shown in Figure 1.3. Each of these landscapes is described in detail in the context of each case study. The nature of the tundra, forested tundra and northern taiga landscapes is examined in the context of the 'Arctic North' (Chapter 2). Both light and dark coniferous forests are found in a clear pattern of altitudinal zonality along the coasts of Lake Baikal (Chapter 3). The nature of the mixed forest landscapes is considered within the Moscow region and Chernobyl case studies (Chapters 4, 5). The latter also describes the forested steppe zone. The steppe and semi-desert landscapes are found in the Republic of Kalmykia (Chapter 7), while deserts are widely spread in the Aral Sea drainage basin (Chapter 10). A comparative overview of a variety of landscapes stretching across the European Russia is presented in the context of the extensive Volga basin (Chapter 8).

1.3 The concept of environmental degeneration

The philosophical basis of the contents and the role of environmental degeneration or degradation are governed by the principles of dialectics. It is an integral part of the overall concept of the development or evolution of material systems. According to Furman and Livanova, 'Evolution is a complex integral process inherent to material systems. It includes a number of elements: conservation of systems and their variability; dissipation of matter and energy and their utilisation; progress of the system as a whole and cycles of elements; change at the same level of complexity; regress and degradation, etc.' (Furman and Livanova 1978, 50).

This concept implies that the evolution of self-developing material systems is overall a progressive and universal process. In general terms, it involves a change from simple to complex and from the lowest to the highest. However, it proceeds in a multifaceted manner through cycles; through a transition to its opposite; through processes of origination and destruction. This kind of complex change ultimately results in the typical spiral character of the overall development. Each circle reverts to a higher-level point before evolving further. This, in the sense of landscape evolution, is expressed in accumulation of matter and energy.

As the above authors stated, 'Development always includes cycles. Thus evolution of organic life is impossible without a permanent action of cycles of oxygen, nitrogen and other elements. Origination and destruction – are two sides of a cycle. The latter does not exist outside development' (Furman and Livanova 1978, 4). And, 'Cycles serve as the main decisive condition of the biological process, as the form of its implementation, as its internal mechanism' (ibid., 120).

Degradation, in its turn, involves a dissipation of matter and energy when a material system loses the required conditions for self-development

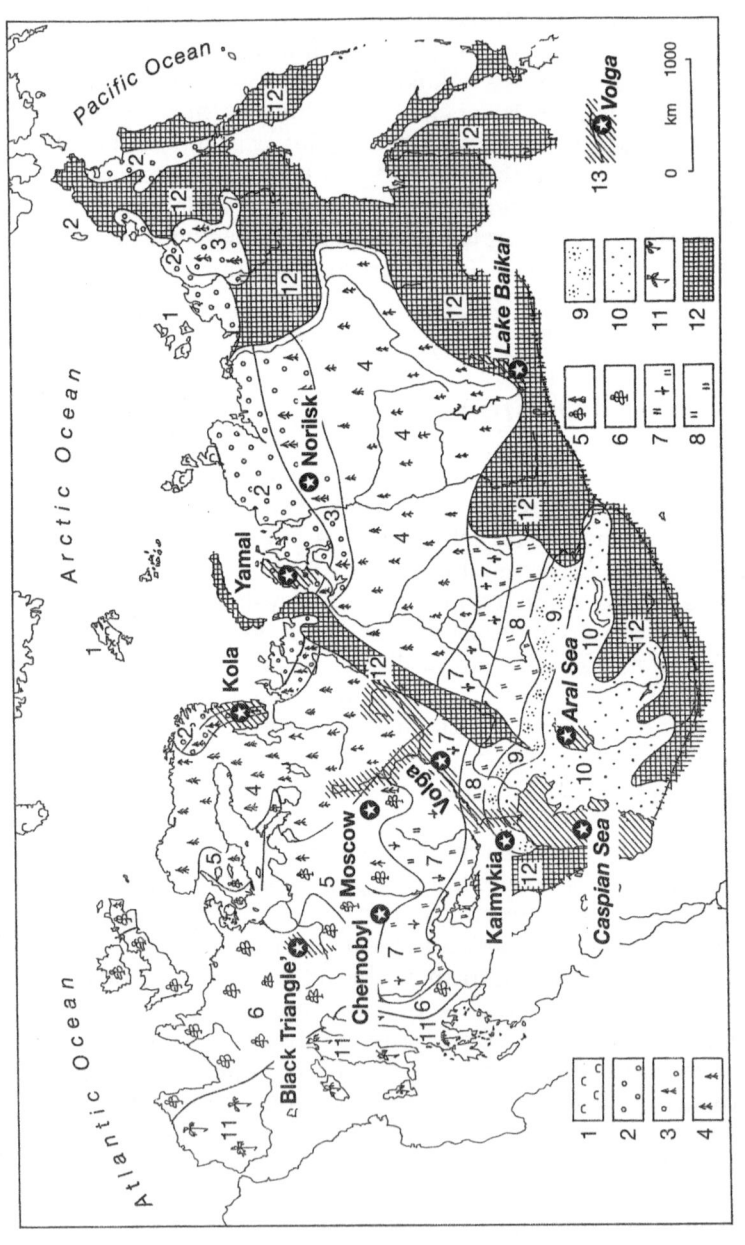

Figure 1.3   Natural landscape zones in northern Eurasia and the loction of case studies: 1 – Arctic desert; 2 – tundra; 3 – forested tundra; 4 – taiga; 5 – mixed forest; 6 – broad-leaved deciduous forest; 7 – forested steppe; 8 – steppe; 9 – semi-desert; 10 – desert; 11 – Mediterranean landscapes; 12 – mountain landscapes; 13 – approximate location of case studies (compiled by T. Saiko and L. Svintsova from sources: Fiziko-Geograficheskoye... 1967; Gvozdetskiy and Mikhailov 1987; Mil'kov and Gvozdetskiy 1986; Romanova 1997)

and becomes unstable. Every degrading material system is a part of a higher-level progressive development. 'Progress of the whole is achieved not only at the expense of progress of its individual components but also at the expense of changes at the same level and regressive changes' (ibid., 115). Therefore, cycles are essential components of the overall evolutionary movement along with degradation. Both represent the internal driving forces of any development process.

Therefore, from the philosophical point of view, 'degrading' or 'entropic' material systems appear as an essential element of an overall progressive process of natural evolution. However, humans, who by Vladimir Vernadskiy's definition in the epigraph to this chapter have become a 'geological force', have dramatically increased the share of degradation processes in the environment. These are particularly destructive under political systems that have practically boundless control over people. Such was the case in a totalitarian type of regime where large-scale suppression of the individual's interests was accompanied by a similar violence over nature.

In the context of this book, environmental change comprises alterations in the ecological situation, socio-economic conditions and the state of human health. It would be useful to define some terms used in the discussion of ecological degeneration. A degree to which an individual landscape is vulnerable, is determined by its capacity of resistance or resilience to external natural and *anthropogenic* disturbances, the latter being those caused by humans. *Disturbance* is a factor that displaces a component of the natural landscape, such as relief or landform, water body or system, soil, vegetation and animal life beyond its normal limits of variation. *Resistance* is the ability of a component or a landscape as a whole to withstand disturbance without changing its initial state. *Resilience* is the ability of a component or landscape to return towards its original state once a change has occurred.

Resistance of landscapes is to a large degree determined by the character of the interrelationship between its integral parts, which can weaken or accelerate the effect of external impacts. Each component performs its own function. Biological components, particularly the vegetation cover, play a special role in the processes of self-regulation and reproduction of the natural environment. They are probably the most mobile elements in a landscape and are the first to transform under any external impact. At the same time, vegetation is the main stabilizing force of nature, which counteracts any violation of the ecological balance.

As Sochava (1974) noted, the more diverse a biotic community is, the more effective is its stabilizing role in supporting the ecological balance of geosystems. Plasticity of the vegetation cover is most important, because it can significantly alleviate fluctuations in hydrothermal conditions and weaken the destructive processes in a landscape. Vegetation is 'programmed' to develop the maximum biomass, which determines its potential for inertia and sensitivity to external disturbances.

The integral resilience potential of various landscapes found within the former Soviet Union is shown in Figure 1.4. If a comparison is made between this map and the map of geographical zones, it becomes clear that the most vulnerable landscapes are those located in the Arctic latitudes and in the hot dry lands of Central Asia. At the same time, mixed and deciduous forests along with the northern parts of forested steppe zone have a much greater potential for restoration. This fact gives further evidence that the global regularity of geographical zonality is universal and has a wide range of various geographical implications. The causes of natural vulnerability in each type of landscape are examined in detail in the context of the appropriate case studies.

After anthropogenic disturbance, *functional* recovery may occur when a landscape has been permanently or severely altered but has recovered to a point where it remains a stable landscape that functions normally. A *complete* recovery is achieved when a landscape has regained its original status and has restored a system of relationships that existed between individual components before disturbance. It would be useful to note that degeneration in one component of a landscape inevitably initiates a chain reaction of degradation in the landscape as a whole. However these processes are complicated due to the nature of destabilized environments studied by Zaletayev (1989). He defined the current stage in the evolution of the biosphere as a 'phase of "ecologically destabilised environment" – super dynamic, with multiple disturbances in its functional mechanisms; with failing natural rhythms and accelerated successions ...' (Zaletayev 1989, 5). The author further noted that in such destabilized conditions 'along with natural strive of system's components for restoration of the equilibrium to provide its integrity, a mechanism of imbalance of biological links emerges which can persist and even self-amplify the effect' (ibid.).

The adverse anthropogenic changes in individual natural components, or in a landscape as a whole, accumulate gradually until they enter a 'crisis' stage. The latter is very close to a certain critical 'threshold', after passing which they attain a new quality. Again, this process finds an explanation in the philosophical laws and categories. In accordance with the principles of dialectics, gradual and accumulative change(s) in any material system are ultimately followed by a substantial transformation with a transition of this system into a new quality. As Furman stated, 'the general law of the transition of quantitative changes into a new quality operates continuously, but, at the same time, in the evolution of systems there is a period of gradual changes and a period of a leap' (Furman and Livanova 1978, 10).

After the leap takes place, in material systems, a reversal cannot be achieved under the continuing set of conditions. An example of this process in the material world can be taken from a simple operation of heating water: when boiling point is reached, water turns into vapour thus obtaining a completely new different quality. Similarly, after the *resistance threshold*, which is different in each landscape zone, has been passed, alterations in the landscape are irreversible and it is no longer able to return to its original state. It should be pointed out that dissimilarly to the above example, in reality it is often difficult to determine the quantitative value of a threshold in a given landscape zone or ecosystem.

In Western geographical research, historically the issues of human awareness and perception of problems have been the focus of attention for a much longer period than in 'socialist' science. In assessing ecological status, these thresholds are sometimes evaluated on the basis of perceptions of environmental changes rather than of actual physical alterations. For example, an important concept of *creeping environmental problems* (CEP) has been developed by Michael Glantz (Glantz 1999, 1994). CEP are defined as 'long-term, low-grade, incremental but cumulative environmental problems' (Glantz 1999, vii). The threshold values in this approach are associated with awareness of different stages in environmental changes, again associated with perceptions of a problem, crisis and the need for action. These perceptions are closely linked to various physical changes e.g. in the levels of the salinity in the Aral Sea and associated drastic reduction in commercial fisheries (Aladin 1999 in Glantz 1999). This approach gives an additional 'human dimension' to the environmental problems. It would be useful to make further research into the interrelationship between 'physical thresholds' and 'awareness thresholds'.

In Russian geography, research in this field has traditionally been focused on the assessment of various ecological and socio-economic parameters that characterize different stages in environmental degeneration. For example, the Institute of Geography of the Russian Academy of Sciences has developed a comprehensive approach to the assessment of crisis ecological and economic situations (Glazovskiy *et al.* 1991). It is based on the evaluation of adverse changes in the landscape, economy and population health. In this concept, a *crisis* in the landscape system consists in approaching the irreversibility threshold while an overall *catastrophic* ecological and economic situation involves irreversible processes of degradation.

According to these authors, *ecological disaster zones* (*catastrophes*) are 'territories (aquatic systems) where permissible levels of natural resource use have been exceeded to such a degree that, even if stabilization of anthropogenic pressure is achieved, degradation of natural life-supporting systems occurs, economic efficiency continues to decline, the population's health deteriorates further and social tensions grow; major (international) investments are needed for saving the humans' (Glazovskiy *et al.* 1991, 13).

The above concept was further developed in the work of Lyuri (1997) from the same institute, entitled 'Development of resource use and ecological crises'. This author recognized the multifaceted nature of the ecological crisis and specified three aspects in the analysis of such problems: natural resource, geoecological and socio-economic. His comprehensive quantitative approach is mainly based on the natural resource capacity of the environment. According to this author, 'an ecological crisis is a stage in the evolution of "society-nature" system when a highly-efficient increase in the rates of resource use and consumption achieved at the expense of their depletion, changes in the direction of the resource-ecology balance' (Lyuri 1997, 63). It appears, however, that account should be made not only of the natural resource value of nature for humans, but also of the overall 'health' of a landscape even if it is not currently in use. An important point made by the above author is that a 'crisis' could be 'useful' for humans because it makes them find the way out and move to a new level of resource utilization and attitude.

An attempt to further develop the classification of different stages in environmental degeneration, partially based on the approach of Glazovskiy *et al.* (1991), is presented in Table 1.3. This table adds a social and demographic dimension to the criteria suggested by these authors, suggests different contents for some stages and also provides a detailed description of each category, particularly those related to degeneration of the landscape. We assume that the disruption of the equilibrium in a landscape or ecosystem marks the beginning of a crisis stage, while the irreversibility threshold forms the boundary between a crisis and a catastrophe.

Within the conceptual framework of this book, *environmental crisis* involves an ecological crisis in a landscape *and* major social problems, signs of disruption in the economic system and deterioration of health in a few population groups. *Ecological crisis*, in its turn, implies persistent adverse change, e.g. severe pollution and/or decline in biological productivity and diversity and/or degradation of soil cover. All changes in this stage occur close to the irreversibility threshold. Therefore, when this threshold is approached and the crisis becomes acute, a catastrophe can still be avoided if the necessary urgent measures are taken.

Summing up the situation as regards environmental awareness by the end of the Soviet period at state level, Alshuler *et al.* noted that the official report on the status of the environment in the USSR in 1988 (Sostoyaniye... 1990) lacked 'clear definitions of such terms as ..."zone of ecological disaster", ... "catastrophic state"'(Altshuler *et al.* 1992, 203). This gap in the Russian Federation was filled in December 1991 by the adoption of the 'Law on the Protection of the Environment' which defined both terms. It was further detailed in the document 'Criteria for the assessment of the ecological situation in the zones of "urgent ecological situation" [equivalent to ecological crisis – T.S.] and zones of "ecological disaster" [ecological catastrophe]' adopted by the Minpriroda [Russian Ministry for the Protection of Nature and Natural Resources] on 30 November 1992.

The latter document provided recommendations for the evaluation of the state of the environment in the worst-affected regions of Russia. According to the definitions included in these documents, the zones of

Table 1.3 Classification of stages in environmental degeneration

Overall situation	Landscape	Socio-economic conditions	Human health
Satisfactory state	Transformed but functions normally	No major social problems; economy functions normally	Satisfactory state of health
Strained situation	Signs of deterioration in individual landscape components	Limited socio-economic problems in different areas	Signs of health deterioration: increase in illness frequencies
Critical situation	Moderate to severe degeneration in individual landscape components; disruption of equilibrium	Signs of serious social problems; decline in economic production or efficiency in certain branches of economy	Threat of serious deterioration in population health due to increase in illness frequencies
Environmental crisis	Ecological crisis: persistent adverse change i.e. severe pollution or decline in biological production and diversity; degradation of soil cover	Major social problems in certain areas; moderate economic losses; signs of disruption in the existing economic system	Deterioration in health in a few population or age groups; growth of rates of illness frequencies; increase in death rates
Environmental catastrophe	Ecological catastrophe: irreversible changes in landscape structure and very severe destruction of individual components or landscape transition to a lower energy/biomass type or genetic damage	Large-scale or acute social problems; negative demographic changes; major economic losses; disruption of the economic system	Substantial overall deterioration of population health; increase in death rates and decline in birth rates or evident adverse genetic changes

ecological crises are 'the territories, where economic and other human activity has brought about persistent adverse changes in the environment which threaten human health, the state of natural ecosystems and genetic funds of plants and animals' (Clause 58). The zones of ecological catastrophe are 'the territories, where economic and other human activity has brought about deep irreversible changes in the environment which had resulted in substantial deterioration of human health, disruption of natural equilibrium, destruction of natural ecosystems and degradation of flora and fauna' (Clause 59). It should be noted that such zones attain higher priority in the governmental allocation of funds for environmental improvement.

Hundreds of criteria related to ecology and health were elaborated within the framework of the former Minprirody RF (Ministry of Environmental Protection and Natural Resources of Russia). For example, Protasov and Molchanov (1995) recommended a set of various criteria for the assessment of the crisis and catastrophic stages. A number of criteria from these recommendations, amended and supplemented by further subdivision of values into the 'critical' and 'strained' categories, are included in Table 1.4. It is clear that most of these criteria need further consideration for application to different landscapes. For example, the same degree of productivity decline in the typical steppe and desert pastures can mean a crisis stage for the former and a catastrophe for the latter. The zonal principle is applied only in the 'loss of biodiversity' category.

It could be noted that in the applied concept a subjective view is of course inevitable if the issue of 'irreversibility' is thoroughly examined. This is because it depends both on the physical parameters of estimated adverse changes and the potential contemporary ability to deal with them. The latter clearly changes with time and depends on the current 'know-how', financial possibilities and the will to achieve environmental enhancement. It should be reiterated that in the real world it is quite difficult to distinguish between different stages of environmental degeneration. As Professor S. Bogolyubov has noted, 'in some cases multiple excess of a few parameters can "outweigh" normal values in many others' (*Zelyoniy Mir* 23, 1998, 6).

In addition, adverse anthropogenic changes in the landscape are not always accompanied by similar levels of change in the economic situation or in human health. For example, in Kalmykia a catastrophic degradation of rangelands did not involve major health implications of a similar magnitude, that is, classed as 'trends of extinction' in the classification of Glazovskiy *et al.* or as 'evident adverse genetic changes' as in Table 1.3. The latter are however evident, for example, in the study of Chernobyl after-effects. In this case, it appears that the Kalmyk situation should be classified as an 'ecological catastrophe' rather than an 'environmental catastrophe'.

When a landscape reaches its critical level or threshold of irreversibility, in other words the state at which it loses its self-supporting properties, the whole geosystem can result in a transition to the lower-level landscape e.g. from dry steppe to semi-desert and from semi-desert to desert in case of Kalmykia, or from forested tundra to tundra in West Siberia. A transition

Table 1.4 Selected criteria for the evaluation of the ecological situation

Index	Ecological catastrophe	Ecological crisis	Critical situation	Strained situation	Satisfactory state
Proportion of degraded territory, % of total	>75	50–75	25–50	5–25	<5
Annual rates of degradation of terrestrial ecosystems, %	> 4	2–4	1–2	0.5–1.0	< 0.5
Proportion of very severely degraded territory	>30	20–30	10–20	5–10	<5
Air pollution with most toxic substances, excess of average daily value of MPC, times	3	2–3	1–2	0.5–1.0	–
Average annual sulphur dioxide pollution of vegetation, MPC, times (MPC=<0.02 mg/cubic m)	>10	5–10	2–5	1.2	<1
Groundwater pollution with nitrates, phenols, heavy metals, surfactants, oil, MPC, times	>50	10–50	5–10	1.5	3–5
Groundwater pollution with pesticides, benzapyrene, MPC, times	>3	2–3	1–2	0.5–1.0	<1
Percentage vegetation cover of rangelands, % of normal	< 25	25–50	50–75	75–90	>80
Destruction of coniferous species by technogenic emissions, % of needles	>50	30–50	15–30	5–15	<5
Productivity of natural fodder vegetation in pastures, % of potential	<25	25–50	50–75	75–90	80
Loss in biodiversity, % of normal for zonal landscape	> 50	30–50	15–30	5–15	<5
Annual rates of shifting sands encroachment, % of territory	>4	2–4	1–2	0.5–1.0	<0.5
Annual rates of increase in area of saline soils, % territory	>5	2–5	1–2	0.5–1.0	<0.5
Annual rates of decline in the soil organic matter, %	>7	3–7	2–3	0.5–2.0	<0.5
Density of radioactive contamination with caesium-137, Ci/sq km	>40	15–40	5–15	1–5	<1
Status of aquatic fauna	Loss of valuable or endemic species; total loss of commercial fish stocks	Decline in the share of valuable or rare species; >50% decline of commercial fish stocks	25–50% decline in commercial fish stocks	Up to 25% decline in commercial fish stocks	Normal state

could also occur from natural to modified or from modified to anthropogenic landscape. An example of the latter is the cultivation of black earths or chernozems in European Russia that resulted in the creation of a single steppe farmland with remains of forests only along the river valleys.

Within each geographical zone in the former Soviet Union the proportion of transformed landscapes varies. For example, in the Russian Federation, according to the National Report on the Conservation of Biological Diversity (Sokhraneniye... 1997), 0.06% of the Arctic desert and tundra zone have been completely transformed, mainly by the extracting industries. In the taiga the share of these landscapes varies from 0.84% in the northern part to 1.8% in the middle and 10.2% in the southern sub-zone. The main anthropogenic factors responsible for this were logging practices, fires, extracting industries, air pollution and cultivation. Approximately one third of the mixed and deciduous forest zones have been completely transformed by cultivation, urban development and hydroelectric construction. Furthermore, over 40% of forested steppes and steppes have been totally modified by cultivation, grazing of domestic animals, water erosion, hydroelectric construction and urban development. Finally, 21% of semi-deserts and deserts have been completely transformed by animal husbandry, irrigation and land salinization.

However, an important peculiarity of the Russian Federation is the existence of pristine and relatively untouched areas, which in total account for over 14% and 60% of the country area particularly in its northern and eastern parts. During the Soviet period a unique system of protected areas was established throughout the USSR which was described by Pryde (1991). It included a whole range of 'zapovedniki' or strict nature reserves (SNR) in which any kind of human activities was forbidden. By the end of 1991 when the Soviet Union was dissolved, the total number of 'zapovedniki' within the Russian Federation only was 75 and they covered an area of almost 20 million hectares. Along with 'zapovedniki', 1519 game reserves and 17 national parks were established by 1992. By 1 January 1998 the number of zapovedniki had reached 98, which included 20 biosphere reserves (Sokhraneniye... 1997).

These zapovedniki were established in a great variety of different types of landscapes throughout Russia, thus enabling scientists to study the complex interrelationship between their various components in 'pristine' condition. These natural 'models' are used for investigation into the changes associated with human impacts. Unfortunately, during the post-Soviet period these areas experienced an increasing pressure and persistent violation of the protected regime.

It should be noted finally that all environmental factors and changes are extremely dynamic. Therefore, it is also essential to consider the duration of human influence, which can serve as an index of accumulated changes and help to develop an adequate abatement strategy. Hence a historical geographical perspective should always be applied when examining the state of the environment in any specific region.

CHAPTER 2

The Arctic North: intensive industrial development in the Russian tundra

The Arctic – is one of the most important regions in the biosphere which comprises unique natural complexes extremely vulnerable and unstable to anthropogenic impact. Their degradation will have unpredictable consequences on the global scale.

(A. Yablokov 1996, 12)

2.1 Introduction

The focus of this chapter is on the fragile nature of the Russian Arctic, which is on the threshold of a large-scale ecological catastrophe. The Arctic North of Russia is an extensive territory within the Arctic Circle, which stretches along the whole of the Russian Federation both in Europe and Asia. The region is exceptionally rich in various mineral and energy resources and during the 20th century has attracted explorers and developers of its immense wealth. Consequently, this austere but extremely sensitive environment has been dramatically affected by intensive industrial development during the Soviet period.

The delicate 'tundra' landscapes of the Arctic North, which for the most part are underlaid by a permanently frozen ground called 'permafrost', have the lowest resilience to human activities. At the same time, the national minorities of the indigenous population are very poorly adaptable to any major changes. People of European origin who settled in the Arctic North more recently in the course of its industrial development think of themselves as living away from the 'continent' or 'mainland'. Life in the harsh conditions of the Arctic puts both groups of people on the verge of everyday struggle for survival. At the present time, both people and the environment appear to be even more vulnerable and worst hit by the realities of the transitional period.

This chapter examines the natural causes of such extreme vulnerability and investigates a range of contemporary anthropogenic factors which operate in this frail environment at an ever expanding scale. A multitude of environmental consequences of Arctic development is assessed by examining the effects of mining, non-ferrous metallurgy and oil and gas industries. Examples will be drawn from three regions: the Kola peninsula in the European part of the Russian Federation, the Yamal peninsula in West Siberia and Norilsk city in East Siberia. The location of these areas is shown in Figure 1.3. Administratively, the Kola peninsula lies within Murmansk oblast [province], the Yamal peninsula forms part of Yamal-Nenets autonomous okrug [district] within Tyumen oblast, while Norilsk is part of Taimyr autonomous okrug within Krasnoyarsk krai [territory].

The environmental situation in all three areas is almost desperate. As Yevseev has stated, 'A distinguishing feature of the Russian North is formation of severely transformed technogenic territories or "impact zones". Only here, disruption of the dynamic equilibrium in the environment has caused an emergence of acute ecological situations and catastrophes' (Yevseev 1996, 47). Already by the end of the Soviet period, anthropogenic badlands have been formed near Monchegorsk on the Kola peninsula and near Norilsk, the world's major producer of nickel. The latter city is responsible for over 70% of all Russian pollutant emissions from non-ferrous industry. Furthermore, extensive areas of reindeer pastures have been irreversibly damaged by prospecting for oil and gas and more recently, gas production in the Yamal.

Over the last two decades, many authors have been attracted and fascinated by the unique physical environment of the tundra zone and the peculiar nature of its indigenous peoples. All these issues, along with the analysis of sometimes contradictory Soviet economic development and its diverse ecological implications, have been considered earlier in the works of Agranat (1998, 1992), Aleksandrova (1980), Bradshaw (1995), Chaturvedi (1996), Chernov (1985), Fondahl (1995), Golubchikov (1996), Yevseev and Krasovskaya (1998, 1996), Osherenko (1995), Yablokov (1996) and other authors.

2.2 Physical geographical setting

It is useful to note that there are different concepts of the 'North', which do not always coincide with the concept of the 'Arctic North'. Thus, immense areas in the Russian Far East are also ascribed to the 'northern territories'. The reasons for this are (a) the harsh climate over much of Asiatic Russia and (b) the inaccessibility of these remote areas from the industrially developed European Russia due to poor transport infrastructure. Bradshaw (1995) noted that the North's concept is very wide in scope and incorporates approximately 40% of Russia. Some Russian authors (Kotlakov and Agranat 1995) attribute an even larger proportion, of 65% of the Russian Federation's territory, to this definition.

The concept of the Russian 'Arctic North' used in this chapter is closest to the perception of 'Krainiy Sever' [Extreme North] found in Soviet and Russian scientific literature. In the north, the region is washed by the Arctic Ocean. Its southern border corresponds to the northern boundary of the world's largest uninterrupted forest stand, called *taiga*. Geographically, the Russian Arctic incorporates the *tundra* and, in some parts of the country, it also includes the *forested tundra* landscape zone.

2.2.1 Geographical location and relief

The Kola Peninsula is located in the north-east of the European part of the Russian Federation and lies almost entirely above the Arctic Circle. It is also called the 'Kola North' by Russian geographers. The peninsula extends for 390 km from south to north and for about 550 km from west to east. The total area is 98,500 sq km, which exceeds the size of Scotland and is three times greater than Belgium.

The relief of the Kola is very diverse. The most common land forms are low mountain ridges, flat-topped mountainous massifs and lowland platform plains. The highest mountain massif, the Khibiny mountains, are located in the central part of the peninsula with the highest point (Chasnachorr Mountain) at 1191 m above sea level. The age of the most ancient glaciation is estimated at about 3.5 million years ago (Krasovskaya and Yevseev 1990) while the period of the end of deglaciation is dated 8000 before the present time (Vaschalova 1986). Therefore glacial land forms including various types of moraines and drumlins are common on the peninsula.

The Yamal peninsula is located in the extreme north of the West Siberian lowland plain, and in the south-west is bordered by the Ural Mountains, which separate Europe from Asia. The total area of the Yamal peninsula is 122,000 sq km, which is three times the area of the Netherlands. It stretches into the Kara Sea for several hundred kilometres and has a width of up to 240 km. The peninsula has a featureless lowland plain topography generally not exceeding 80–90 m. In the southern and inner parts of the peninsula, relief is more elevated and dissected. Landforms are predominantly glacial or cryogenic by origin and are composed of sandy-clay marine and fluvio-glacial sediments. Microrelief is formed of such permafrost landforms as thermokarstic depressions, peat mounds and polygons. In places devoid of vegetation, bare rock, massives of sand barchans and blowout depressions can be seen.

Norilsk is located in the north-west of the elevated Central Siberian Plateau 300 km north of the Arctic Circle. With population of 156,300 people in 1997 (Chislennost'... 1998), it is the world's second largest polar city after Murmansk (387,400) but it is located further north inside the Arctic Circle than the latter, and has the most extreme climatic conditions compared to any other city. American geographers call it 'the Russian urban Arctic laboratory'. The city borders the Norilsk mountains in the north-western part of the Central Siberian Plateau.

2.2.2 Climate and hydrology

Owing to the geographical position of the **Kola** peninsula and the influence of the warm North Atlantic current, it has a relatively mild climate compared to Asiatic Arctic. Intensive cyclonic activity with heavy snowfalls and blizzards alternating with frequent thaws are typical for the winter period. The average depth of snow cover is up to 70 cm on the plains and 150 cm in the mountains, although in places it may reach 2 to 4 m. Average January temperatures vary from –8 degrees C in the northern part to –13 degrees C in the centre of the peninsula. Summers are short and cool with average July temperatures ranging from +8 degrees to +14 degrees C. Mean annual precipitation is relatively high and varies between 500–600 mm in the central part and 1000 mm in the mountains. A distinctive feature of the Arctic North is the existence of a prolonged period of darkness or a 'polar night' which in the Kola peninsula normally lasts from December to the end of January and a 'polar day' for the months of June and July.

The whole territory of the **Yamal** peninsula lies within the Arctic climatic belt. Therefore, its climate is extreme continental and is much more severe than that of the Kola peninsula. Winter starts in October and lasts until mid-May. In some years with very cool summers half of the snow cover remains unthawed. Winds are quite strong and thick clouds cover the sky for most of the year, particularly during a short cool summer. The annual amount of precipitation does not exceed 140–200 mm (Gvozdetskiy and Mikhailov 1987) and occurs mainly in summer. Evaporation is also insignificant due to the predominance of freezing temperatures, which can reach the absolute minimum of –51 degrees C to –56 degrees C. These features make this type of landscape one of the wettest in the world. The frost-free period lasts from 25 days in the north to 60 days in the south of the Yamal. The polar night lasts from November to the end of January suppressing growth of vegetation and has a profound impact on the functioning of tundra ecosystems.

The climate in the **Norilsk** region is also classified as extreme continental. But being located further to the east and into the heartland of the Siberian continent, Norilsk features a more severe climate than the Kola or Yamal peninsulas. Winter here lasts for nine months and the average January temperatures reach –28 degrees C. The absolute minimum temperatures on some days may fall down to as low as –60 degrees C. Temperature inversions are quite common, particularly during the winter period, when stale air dominates in the polar latitudes. At the same time summers are warmer than in the western tundra and the average July temperature amounts to +11 degrees to +12 degrees C. The total amount of precipitation is also greater, reaching 300–350 mm.

There are some specific climatic factors contributing to the vulnera-bility of the tundra landscapes. A peculiar polar phenomena called an 'Arctic haze' involves the creation of a temperature inversion in the low atmos-pheric layers. It means that air temperature at the height of about 500 m is a few degrees higher than at the ground level, thus increasing the effect of air pollution. This phenomena is generally responsible for the accumulation of pollutants in the polar latitudes. In the central part of the Kola peninsula, a typical example of this inversion can be evidenced in the Monchegorsk region, where the main polluting 'Severonickel' industrial complex is located. According to Yevseev (1998), in winter, temperature inversions here are recorded within an atmospheric layer of up to 1 km in 79% of cases, and in summer, within a layer of up to 2 km in 92% of cases. Light winds with average speed of up to 5 m/sec are quite common as well as calm periods. Both features impede the turbulent mixing of air masses and contribute to the concentration of toxic chemicals in the lower atmosphere.

The removal of pollutants from the atmosphere is associated with pre-cipitation, which in the polar latitudes is represented primarily by snow. Thus, another factor contributing to intensive accumulation of techno-genic pollutants in the landscape is the existence of a long cold period with snow cover, which even under the mildest conditions of the Kola peninsula lasts from October to June. It determines the character of the biological and geochemical cycles preventing fast removal of pollutants from the landscape.

Hydrology in the Arctic region had acquired some specific features. Generally speaking, because of the typically excessive moisture, the sur-face runoff in the tundra is significant, and in summer the rivers have rather high water levels. Thousands of shallow lakes of moraine and thermokarst origin are located in the Arctic. Some of the world's mighty rivers, such as the Ob' and Yenisey, flow into the Arctic Ocean crossing the tundra zone. The tundra is also featured by the wide distribution of *permafrost*, permanently frozen ground, which is the relic of the Glacial period. Permafrost underlies an extensive area of approximately 11 mil-lion sq km in the Russian Federation, which includes much of the Arctic North zone, reaching depths of up to 500 m and even more. The upper level of the ground, which thaws during the summer period, is called an 'active layer'.

The hydrological network on the **Kola** peninsula is dense, well devel-oped and consists of about 100,000 lakes and 21,000 rivers. Many of those are young and, like the landforms, bear the evidence from the Quaternary Glaciation. Rivers are predominantly recharged by thawing snow and are frozen for up to seven months in the year. Bogs and swamps occupy about 37% of the total area of the peninsula (Kremenetsky *et al.* 1997). Geographical location and climate explain an interesting feature of the

Kola North – the absence of continuous permafrost. Sporadic permafrost, however, occurs in the Khibiny mountains and in some eastern parts of the peninsula.

The **Yamal** peninsula has an extensive hydrological system with numerous lakes and meandering shallow rivers. In the east, Yamal is bordered by the estuary of Siberia's longest river, the Ob'. The central part of the peninsula is boggy. River erosion is limited and develops only during a short summer period, when thawing of 20 to 90 cm of the upper ground is accompanied by the processes of solifluction and thermokarst. Rivers are normally rather short and in winter the river channels freeze right to the bottom. The whole territory of the Yamal peninsula is underlain by an extensive permafrost layer reaching a depth of up to 300–600 m. In summer thawing creates a 'quagmire' and an overall waterlogging of the territory, which makes the tundra terrain almost impassable. Mechanical disturbances present a major threat to the tundra environment as they can result in development of various cryogenic processes, solifluction, creeping and landslides.

The hydrological network of Central Siberia is well developed and dense. **Norilsk** city is located on the southern border of the Taimyr peninsula between the Norilsk mountains and the Pyasina river, which flows into the Kara Sea. The river originates in the shallow Pyasino lake and stays frozen from late September to June. Permafrost is well developed and has a particularly great depth in the whole of Central Siberia, and particularly in its north, where Norilsk is situated. Only 30 to 60 cm of soil thaw during the short summer period.

The high vulnerability of the tundra landscapes is also due to a number of specific features of permafrost. According to Solomatin (1992), cryogeosystems are characterized by significantly lower energy expenditures needed to disrupt their balance and a much greater time needed for 'relaxation' or return to the initial state, compared with geosystems in the temperate and lower latitudes outside the 'cryolithozone'. The stability of the ecological balance can become lower if 'alien' elements, for example of technogenic origin, enter the system. In this situation the soil and vegetation cover, lower atmospheric layer, snow cover and soil moisture all serve as the 'buffer' between the permafrost and external disturbance. Mechanical disturbances are the second most destructive after vegetation removal, but they tend to remain in the landscape for an even longer period. Economic development of this region has an outstanding impact on the energy-related stability of the permafrost.

Any technogenic impact is almost inevitably accompanied by a partial disturbance or complete destruction of the snow, vegetation and soil cover, change in their moisture level as well as of the underlying ground. This further leads to an increase in the thawing depth or freezing of wet sediments, development of dangerous hazardous processes, i.e. thermokarst, thermal erosion, solifluction, swelling, etc. and results in irreversible deformations of extensive areas in the Arctic North.

2.2.3 Vegetation and soil cover

The vegetation cover in all three areas under study is quite diverse, according to their zonal types of landscape and some local peculiarities. The range of landscapes found within this region includes the tundra, forested tundra and the taiga (tayga) natural zones. Each zone from north to south is further subdivided into a number of subzones associated with differences in vegetation. For instance, the tundra zone is further subdivided into 'Arctic tundra'; 'moss and lichen' or 'typical' tundra and the 'shrub' or 'southern tundra'. All three subzones are represented within the Yamal peninsula, while the Norilsk oblast is located within the forested tundra zone. Only the northern taiga subzone is found within the Arctic North region. As mentioned in Chapter 1, the most definitive characteristic of each landscape zone is its specific type of vegetation.

The **tundra** zone on the **Kola** peninsula covers relatively large areas in the northern and eastern parts of the peninsula and also the summits of the lower mountain ridges called 'tundras' as well as lower areas within the Khibiny mountains (Plate 1). Overall, this zone occupies about 20% of the territory of the Kola (Makarova *et al.* 1997). The tundra belt is only 20 to 30 km wide in the west and almost 100 km wide in the central and eastern parts. Due to cold climate and an excessive moisture regime, plant growth is inhibited. The main role in the vegetation cover

Plate 1   A typical tundra landscape on the slope of the Khibiny mountains with dwarf birches in the foreground

of the tundra is played by mosses and lichens, which form continuous mats or separate patches alternating with extensive areas or patches of bare ground, particularly in the northern tundra. Flowering plants consist mainly of sedges (*Carex spp.*) and cotton grasses (*Eriophorum spp.*), but also of grasses, dwarf shrubs such as willows (*Salix spp.*), birches (e.g. *Betula nana*) reaching the height of 1 m and alders, and of berry-bearing shrubs (e.g. foxberries, bilberries, wortleberries), juniper, etc.

The vegetation cover of the **Yamal** peninsula is thin and quite poor. 'Spotted' or 'polygonal' lichen tundras, which are even called 'semi-deserts' by some scholars (Golubchikov 1996), have developed along the northern coastline. Small polygons of 5 to 7 m in diameter are separated by frost cracks filled with ice wedges. Scanty vegetation is represented by mosses and sedges (*Carex spp.*). Thawing in summer often creates a boggy terrain with flat mounds. The central part is taken up by a 'typical tundra' subzone. Normally, mosses dominate on clayey divides, while sandy and gravel sediments are covered by lichen (*Alectoria jubata*) tundras. Less space is occupied with precious lichens, favoured by reindeer (*Cladonia rangiferina* and *Cetraria sp.*). The 'shrub tundra' subzone has developed in the southern part of the peninsula. It is most commonly represented by dwarf birches and lichens, also found on the Kola peninsula with some willow species, grasses and mosses on clays.

The **forested tundra** zone called 'lesotundra' in Russian in the **Kola** peninsula is manifested by 'redkolesye' [sparse open woodlands] and mountainous elfin woodlands, predominantly composed of birch species (*Betula callosa, B. tortuosa*). The height of birches does not exceed 1.5–5 m (Makarova *et al.* 1997). Crooked tree trunks and bent crowns are quite common due to the heavy snow cover and unfavourable soil conditions. Underwood is mainly composed of dwarf birch and juniper. Lichens, mostly the reindeer moss (*Cladonia rangiferina*), cover as much as 40–70% of the soil cover (Krasovskaya 1998). The forested tundra zone stretches from north-west to south-east in a rather narrow belt 20 to 100 km wide.

Vegetation composition of forested tundra landscapes in the **Norilsk** region is different from that of the Kola peninsula. Larch species start to dominate in the taiga (*Larix gmelini* and *L. dahurica*), and their height does not exceed two to three metres. They are rather sparsely spread between areas of the tundra vegetation, composed, similarly, of dwarf birches (*Betula nana*), which reach a lesser height of 50–60 cm, and normally have their crowns bent low to the ground.

The **taiga** is a coniferous forest zone described in detail in Chapter 3. The **Kola** peninsula is unique in being the only region in Russia where the taiga landscape zone is found above the Arctic Circle. Forests occupy 23% of the total area of the peninsula (Makarova *et al.* 1998). Only the northern taiga subzone is present. It is characterized by very low productivity ranging from 25 to 60 cubic m of wood per hectare. Along with sparsely growing conifer species, represented by pine (*Pinus friesiana*) in the west and south of the peninsula, or fir (*Picea fennica*) in the north and east, this subzone is composed of some small-leaved deciduous species. The overstory of conifers reaching 20 m is supplemented by birch (*Betula subpolaris*) of 10–15 m height. Due to prevailing strong winds and heavy snow cover, fir trees sometimes have a curious 'flag-pole' shape – in which branches get thicker on the leeward side (Plate 2). It is also evident from the plate that coniferous trees are more vulnerable to air pollution compared to deciduous species. Fir trees can also have an abundant lower layer of branches, which in winter is protected by a thick snow cover.

The underwood is composed of young trees of the main stand, rowan tree, bird cherry, and shrubs, including willows and Siberian juniper. The ground layer is composed of the following combinations: mosses and lichens; mosses and small shrubs; grasses and berry-bearing shrubs like blueberries, foxberries and bilberries. Trees grow very slowly but can live until 200–300 years of age (Krasovskaya and Yevseev 1990).

Altitudinal or *vertical zonality* is represented both on the Kola peninsula and the Norilsk region. In the Khibiny mountains it is expressed in a succession of taiga vegetation up to the altitude of 370 m; forested tundra in a narrow band of up to 400 m; typical tundra from 400 to 900 m and arctic or polar deserts at altitudes exceeding 900 m above sea level. Vegetation in the landscapes around Norilsk is very diverse: the slopes of river valleys are covered by dense larch and fir forests up to the altitudes of 300–350 m; further up the slope they are replaced by the sparse larch forest followed by a band of shrubs (*Alnaster fruticocus, Betula exilis* and *Salix*). At altitudes of over 700 m the vegetation cover becomes sparse, and the highest mountain tops are partially covered by patches of lichens, mosses interspersed with bare ground.

Generally, Arctic latitudes are characterized by a slow biological cycle, and very low values of the phytomass and biological production. In the Arctic tundra, the value of the total phytomass varies from five to ten tonnes per hectare; it increases to 10–40 tonnes in the Subarctic tundra. The annual biological production is also very low at one to three tonnes per hectare in the Arctic and two to four tonnes in the Subarctic tundra (Table 1.2). Tundra features a relatively simple ecosystem structure and low biological diversity. The latter, expressed by the numbers of plant species per 100 sq km, amounts to 70–100 in the Arctic and 250–300 species in the Subarctic or typical tundra zone. It can be compared to the broad-leaved deciduous forest with a complex multi-tier structure and biodiversity reaching 700–800 species per 100 sq km (Tishkov 1993). In the taiga zone all three above-mentioned parameters further increase (Chapter 3) which makes it more stable to human impact. It should also be emphasized that, compared to all other geographical zones, the Arctic landscapes are characterized by the longest duration of primary successions estimated at more than a thousand years.

Plate 2 Coniferous trees are the first victims of air pollution: a fir tree on the Kola peninsula with a 'flag-pole' top

The role of vegetation in stabilizing a cryolithic environment cannot be overestimated, particularly on the slopes, where it prevents development of thermal and water erosion. This vegetation cover serves as a thermal isolator limiting thawing in summer and freezing in winter. It has been found (Belopukhova *et al.* 1976) that the maximum thermal resistance is provided by powerful green moss and lichen covers. However, the resistance and rehabilitative ability of the Arctic vegetation is very low due to the overall weakened biological cycle.

Each of the three areas under study have certain similarities, but also some major differences in the **soil cover**. Due to unfavourable climatic conditions described above, soils are poorly developed, acidic and have a shallow profile and low humus content. Active gleyic processes have formed gley soils (Table 1.1) which are widely spread in the tundra. To the south of the tundra zone, podzolic processes start to dominate under better drainage conditions, and the gley podzolic and podzolic type of soils is more typical for the northern taiga zone.

The soil cover on the **Kola** peninsula is mottled due to the combination of plains and mountains and presence of all three zonal types of landscape. Typical mountainous soils have short profiles ranging from a few millimetres near the tops of the mountains to a few dozens of centimetres in the inter-mountain depressions. In the tundra zone on the upper slopes of the mountains, the podzolic horizon is normally absent, while it is clearly developed down slope within the taiga zone. High erosion risks in soils on steep mountainous slopes contribute to the overall high degree of landscape vulnerability to human impact. They are particularly hazardous in areas of extensive ore-extraction activities, for example within the main mining areas of Apatity. Podzolic soils are more widely developed on plains due to the reasons explained above.

The development of the soil cover on the **Yamal** peninsula is to a great extent determined by the wide distribution of permafrost. The main features of cryogenic soil formation are discussed by Golubchikov (1992). They are distinguished by the following: (1) prevalence of shallow profile soils of the initial stage of formation; (2) substantial horizontal (side) movement of soil masses including over the screen of permafrost; (3) clearly expressed complex character, diversity and mosaic pattern of the soil cover; (4) weak metamorphism of the soil; (5) slowed transformation of organic matter and formation of peat horizons; (6) limited vertical transfer of the soil formation products from the profile and their accumulation in the humus form or as carbonates.

Wide distribution of permafrost in the Yamal determines the specific nature of these soils. The depth of the subsurface peat cover can vary from 10 to 50 cm. The following rough humus horizon is rather shallow and is underlain by a bluish gleyic layer. A permanent water-confining stratum under the seasonally thawing shallow active layer enables 'sliding' of the soil/ground mass in the horizontal direction due to excessive humidity of soils. This creates favourable conditions for horizontal migration of pollutants rather than vertical. In this way they become 'trapped' and accumulate within the upper soil horizon.

When the active layer freezes after the summer period, strong cryogenic soil movements create a very complicated, poorly differentiated profile of over-saturated soils with quick ground horizons. Under the action of solifluction, soil horizons keep bending and deforming. As has been stated by Golubchikov: 'The Yamal is a region of an extraordinary vulnerability of nature. In essence, it is an extensive block of ice, which had run aground and got covered by a thin mire of tundra. There is more underground ice under it, than the ground itself. Nowhere else in the world there is a space, so vulnerable to contemporary machinery' (Golubchikov 1996, 288).

In the territory surrounding **Norilsk**, plain areas have developed a similar gleyic tundra soils. However, better drainage conditions on elevated areas have created a different type of soil. Owing to the typically warmer summers and relatively dry air, the mountain soils of Central Siberian forested tundra around Norilsk have a greater humus horizon reaching 15–20 cm.

The potential for the removal of pollutants from soils in the tundra is very low due to the predominance of freezing temperatures throughout the year. Micro-organisms in the tundra soils are rather low in numbers and, due to the cold climate, microbiological activity is weakened. This factor along with the lack of free oxygen in water and low capacity of the biological cycle in the tundra determine the slow pattern of phytomass decomposition. In addition a prolonged period of darkness leads to the limitation of active processes of photosynthesis and, consequently, self-rehabilitation in soils to only a few months. Normally, organic remains become decomposed and mineralized only within 20 to 50 years (Solomatin 1992). Other estimates suggest that the maximum possible rate of decomposition for Sphagnum mosses in the tundra landscapes is even greater and exceeds 100 years (Yevseev and Krasovskaya 1996).

Decomposition of 'alien' substances is also retarded, which makes their adverse impact more prolonged and extended in magnitude. Overall, the areas underlaid by permafrost are particularly vulnerable to the destructive impact of machinery and equipment, which, in most cases in the FSU, was not suited to the natural conditions of the fragile Arctic landscapes.

2.2.4 Overall vulnerability of landscapes

The causes of the overall high vulnerability of Arctic and Subarctic landscapes to natural and anthropogenic disturbances had been examined previously by the author of this book (Saiko 1998c). As discussed earlier, the following orographic, climatic, hydrologic, biological and other factors contribute to the overall vulnerability of the Arctic North landscapes:

(a) high soil erosion risks on mountainous landscapes;
(b) predominantly freezing temperatures during the year which slow down and adversely affect all natural processes in the landscape;
(c) temperature inversions in the low atmospheric layers in the polar latitudes;
(d) existence of a long-lasting snow cover;
(e) prolonged freezing of water bodies and their subsequently low potential assimilative capacity;
(f) lack of free oxygen in water;
(g) wide distribution of permafrost and subsequent cryogenic processes;
(h) regime of excessive moisture which suppresses biological activity and soil formation;
(i) prolonged period of darkness which limits the duration of active photosynthesis;
(j) slow metabolism of natural processes and biological activity;
(k) low total phytomass, biological production and plant biodiversity in the tundra zones;
(l) simple structure of the tundra ecosystems;
(m) weakened microbiological activity;
(n) slow plant decomposition rates;
(o) low potential ability for removal of pollutants from soils.

All the above-mentioned features of the landscape components are responsible for an overall very poor ability of landscapes to assimilate or remove pollutants and to rehabilitate after an external anthropogenic impact, even if the magnitude of such impact is not substantial. In essence, they represent natural prerequisites for the aggravation or acceleration of the incurred damage (Figure 1.1). To summarize, the Arctic North landscapes are distinguished by an extreme sensitivity and low resilience to human impacts.

2.3 Historical perspective and economic development

Human settlement of the Arctic latitudes started later than in the more favourable temperate zones of European Russia. It is generally agreed by scientists that the Russian Arctic has been experiencing human influence since the domestication of reindeer and the start of their breeding about 2000 years ago (Yablokov 1996). However, the environmental implications of these human impacts were relatively insignificant until the 20th century. This territory is very rich in various natural resources and contains one-third of the prospected world reserves of nickel and platinum, and half of palladium. It is also rich in gas, oil, coal, gold and diamonds.

In general, the role of the 'North' for providing many basic natural resources for economic development of the FSU was paramount during the Soviet period, and it has further increased with the evolution of Russia as a sovereign state after the disintegration of the USSR. Recent estimates suggest that it contains about 13% of the world's oil, 42% gas reserves and 15% of the global pristine lands (Zaidfudim 1998). Currently its territory accounts for 80% of all Russia's oil, 90% gas, 90% nickel and gold production, as well as a great share of mineral fertilizers, timber and other raw materials (Kotlyakov and Agranat 1995).

The **Kola** peninsula is one of the most industrially developed regions of the Russian North. Intensive economic development started in the 1930s with the discovery of rich mineral deposits of the apatite-nepheline ores. Murmansk oblast is important for its mining industry which covers both the extraction and processing of mineral ores. This oblast developed a major mineral producer in the Soviet period and continues to be of great value in the Russian Federation (Luzin *et al.* 1994). Apatite-nepheline ores are mined in the Khibiny mountains (Plate 3) and copper and nickel ores are located in the Pechenga region. Modern extraction processes at Kirovsk have contributed to a creation of a major integrated industrial complex. Nickel is mined and smelted at the 'Severonickel' complex in Monchegorsk starting from 1939 and 'Pechenganickel' complex at Nikel and Zapolyarny, while iron ores are mined at Olenegorsk and Kovdor. The production of non-ferrous minerals, notably cobalt, copper and nickel, forms the basis of the industrial complex of the region.

The **Yamal** peninsula was historically used by Nentsi people for grazing their reindeer. Natural reindeer tundra rangelands account for 62.5% of its area. Exploration of oil and gas deposits in the peninsula started only some 30 years ago. Immense reserves of gas have been discovered in Bovanenkovo in the central part of the peninsula. Other gas deposits include the rich Harasaway and Kruzenshtern deposits near the western coast and Northern and Southern Tambeisk deposits in the east of the peninsula. Additionally, nearby offshore oil and gas fields in the Kara Sea include the Leningradskoye and Russanovskoye deposits. On the other side of the Obskaya Guba bay a very rich Yamburg deposit of gas, which was discovered in 1969, is already in operation. As Kamyshev has stated: 'This region [North of West Siberia] concentrates gigantic reserves of oil and gas, therefore, the main volume of global investments into the economy of Russia in the future will be associated with the development of oil and gas industry in the peninsula part of Tyumen oblast (Yamal, Tazovskiy and Gydan peninsulas)' (Kamyshev 1999, 5). Although pressure from environmentalists initially caused the postponement of a large-scale development of the Yamal in 1988 (Stewart 1992a), 11 years later in 1997 the industrial extraction of gas on the Yamal peninsula started. An ambitious project has been already launched to construct an extensive gas pipeline, the 'Yamal-West', to transport gas to West Europe

Plate 3 Mining in the former Soviet Union was always lethal for nature: a giant quarry in 'Apatity'

Table 2.1 The legacy of economic development and ecological change in the Arctic North

Period	Legacy of economic development	Ecological situation	Stage in ecological degeneration
Before 1929 (early industrial stage)	Transhumance reindeer pastoralism, slow urban development, ore prospecting in the Kola; nomadic pastoralism, fishing, hunting in Yamal; limited copper smelting near Norilsk	Limited local damage to natural landscapes, sustained self-rehabilitation after anthropogenic and natural impacts in Kola and Yamal; mining related disturbance near Norilsk	Satisfactory state in Kola & Norilsk; Normal state in Yamal
1929–1941 (start of industrialization)	Policy of industrialization initiated by first 5-year plan; start of apatite-nepheline mining & urbanization in Kola; copper & nickel mining in Norilsk, then Kola; subsistence economy but creation of first collective farms and attempts at sedentarization in Yamal	Gradual local increase in industrial impacts; beginning of air and water pollution problems in Kola & Norilsk; ecological equilibrium remains in Yamal	Strained situation in Kola & Norilsk; Satisfactory state in Yamal
1942–1955 (war period and post-war recovery)	Intensified mining & production of non-ferrous metals and steel for defence needs, decline in 1946, post-war recovery: 2–3 times production increase, transition from forced labour to incentive policies; creation of large state farms, forced sedentarization in Yamal	Increased industrial impact in mining areas; persistent air pollution problems in Norilsk and Kola close to smelter centres; increase in water pollution; disruption of ecological equilibrium; beginning of damage to rangelands in Yamal	Critical situation in Kola & Norilsk; Satisfactory state in Yamal

Table 2.1 Continued

1956–1969 (intensive industrialization)	Rapid acceleration of heavy industrial development without regard for environment; Increase in mining production by 3–4 times during period; oil and gas prospecting in Yamal: discovery of large gas deposits by end of period	Intensification of impacts: increase in air pollution; severe pollution of adjacent small rivers; local soil contamination with heavy metals; damage to forests in Kola & around Norilsk; local oil pollution, permafrost disturbance, rangeland deterioration in Yamal	Ecological crisis in Kola & Norilsk; Critical situation in Yamal
1970–1990 (economic stagnation)	Retardation of overall economic production but persistent priority to heavy industry during 'Cold War' period; permanently high production of iron, non-ferrous metals and fertilizers in Kola, non-ferrous metals in Norilsk; expansion of prospecting for oil & gas in Yamal	Accumulation of ecological changes unabated in Kola and Yamal: large-scale forest damage around smelters, severe air pollution, large-scale water pollution & soil contamination in Kola and Norilsk; degradation of rangelands, pollution, solid waste accumulation in Yamal	Ecological crisis with local catastrophes in Kola & Norilsk; Ecological crisis in Yamal
1991 up to present time (transition to free market economy)	Reorganization of industrial production; shift from state to mainly mixed form of ownership: 70% in Norilsk, 63% Kola and 97% Yamal: creation of 'Norilsk Nickel' association; increase in production of non-ferrous metals along with decline in other industries; start of industrial gas extraction in Yamal; financial crisis; lack of ecological funds	Awareness of gravity of ecological situation which is similar to that in previous period; limited environmental improvement; increase in air pollution	Similar situation

via the whole territory of European Russia. Gas condensate will be initially transported along six branches of this pipeline, including a section of 80 km laid on the bottom of the Baidaratskaya bay to a new port terminal. From there icebreaker tankers will be transporting gas onwards. This venture is ecologically dangerous and unique in world practice of pipeline construction. There are also plans to build a railway line, Obskaya–Bovanenkovo–Harasaway.

The economic development of the **Norilsk** region has a longer record, as already in the 1860s copper was being smelted near the contemporary location of the city. Geological surveys started in 1919 and in 1935 a decision was made to establish the Norilsk mining and metallurgical works. These were constructed mainly by the prisoners of 'Noril'lag' (a local branch of 'Gulag'), who then worked at the complex until 1955. Production of nickel started in 1942, and at present the 'Norilsk' integrated metallurgical complex also produces copper, cobalt, selenium, platinum, palladium, gold, silver and other metals. Nickel in the region is found in sulphide ores, which are the main source of sulphur dioxide pollution. Mining is carried out from mines and in open quarries, as ores are found at depths from 150 m to 1.5 km.

The intensive economic development of the Arctic North, which has been responsible for the degeneration of the environment, is primarily associated with the intensification of industrial production during the 'Cold War' period (Table 2.1). It should be noted that reliable official data on the production of non-ferrous metals is still lacking. However, it can be assumed that overall industrial production indices, for example rates of production and total production output in the mining industry, could be used to get an approximation of changes in the rates of industrial development in the Arctic North. Estimates for the regions under study have become available only recently, with the publication of statistical yearbooks of the Russian Federation were published (Rossiyskiy... 1994–1999). According to these sources, in Murmansk oblast and Taimyr autonomous okrug the index of industrial production in the 1970s–1980s amounted to only 103–106% and 98–106% of the 1990 level (Rossiyskiy... 1996). This is evidence of the gradual decline in industrial production which started already in the 1970s.

Following structural changes during the post-Soviet period, the 'Norilskiy', 'Severonickel' and 'Pechenganickel' integrated mining and metallurgical complexes joined to form the share-holding association called 'Norilsk Nickel'. This association is one of a few successful examples of economic development during the period of transition to the market economy. The reason for its success, apart from this amalgamation, is the high price for non-ferrous metals on the world market. At the same time, this period in Russia featured economic and production decline in most other industries except gas production and some other raw resource producing industries.

2.4 Ecological consequences and recent changes

Even up to the beginning of this century human impact on the ecology of the region was relatively insignificant. As a result of a prolonged technogenic impact after the Second World War, the naturally low self-restoration and assimilation capacity of the environment of the Russian Arctic in the three study areas has been greatly exceeded. At present, the list of urgent ecological problems associated with intensive industrial development in the Russian Arctic is lengthy and varied. They include: severe air pollution by metallurgic smelter combines, cement plants, mining and oil industry; pollution of rivers and lakes by oil products; water and soil contamination by heavy metals and other hazardous substances; radioactive contamination of the environment in places of former nuclear testing or current waste dumping; mechanical disturbance of soils and destruction of vegetation by geological exploration and resource extraction, etc.

2.4.1 Air pollution

Pollution of the atmosphere is a major single type of environmental degradation. Chemical and dust pollution of the atmosphere are among the most hazardous environmental aspects of mining, non-ferrous metallurgical industry and oil production due to their consequent negative impact on human health. Furthermore, it is estimated that due to air pollution the costs of the capital repair of equipment at the local plants double every few years (Popov 1987).

Annually about 6.5 million tonnes of pollutants are emitted from stationary sources into the atmosphere over the Russian Arctic, which amounts to 18% of all Russian emissions. At the same time, the share of sulphur oxides and hydrocarbon emissions account for 30% of all similar emissions in the Russian Federation. In addition, emissions of nitrogen oxides, carbon dioxide and solid particles are estimated at 15%, 10% and 8% respectively (Myach 1996). The range of emitted pollutants involve over 30 substances, including such highly toxic contaminants as benzopyrene, lead, vanadium, chromium and mercury.

The **metallurgical industries** are represented by integrated industrial complexes on the Kola peninsula and in Norilsk. In the 1980s Norilsk was responsible for over 70% of all Russian pollutant emissions from non-ferrous industry. It is worth noting that this industry had also produced 98.9% of all Russian atmospheric discharges of nickel, 98.1% of copper, 93.5% of sulphuric acid and 92.7% of fluoride compounds. More recently, in 1997 its share reduced to 60.3% (Gosudarstvenniy... 1998b) but during the 1990s it was still responsible for more than 70% of the sulphur dioxide emissions in the Russian Arctic (Myach 1996). As Table 2.2 illustrates, by the overall amount of pollutants from stationary sources Norilsk is the 'record-holder' in the Russian Federation. Blais *et al.* (1999)

Table 2.2 Total airborne pollutants emitted from stationary sources by selected Russian cities, 1992–1998

City	Total airborne pollutants (000 tonnes)						
	1992	1993	1994	1995	1996	1997	1998
Angarsk	319	289	244	233	210	168	164
Bratsk	154	125	106	97	91	84	88
Cherepovets	521	438	387	416	438	429	390
Lipetsk	509	402	391	386	361	348	339
Moscow	251	219	196	174	171	151	131
Norilsk	2208	1946	1941	2041	2115	2185	2140
Samara	102	101	83	80	74	64	59

Source of data: Rossiya... (1999)

have noted that the smelters in Norilsk are now the largest source of atmospheric sulphur dioxide in the world, far ahead of such contenders as Sudbury, Canada, with 250,000 tonnes per year.

In Norilsk, the main polluter is the Norilskiy integrated mining and metallurgical works (NIMMW). In some years during the Soviet period, the total amount of pollutants discharged by NIMMW into the atmosphere reached 22.5 million tonnes per year (Myach 1996). During the period from 1991 to 1994 the total amount of discharged pollutants had a consequent trend of reduction, owing to the general economic and production decline of the transition period. However, since 1995 a steady growth of pollution load, associated with the increase in the production of copper, nickel and cobalt at the NIMMW and Severonickel complexes was recorded. At the Norilskiy complex the total discharge of pollutants had increased from 1.9 million tonnes in 1994 to 2.1 in 1996 and 2.2 in 1997 (Table 2.2). The multiyear average concentration of this pollutant in the atmosphere over Norilsk exceeds 2 MPC (maximum permissible concentration), which is 10 times higher than the all-Russian level (Gosudarstvenniy... 1998b).

In 1997 the total share of the Norilsk Nickel association, which also includes enterprises on the Kola peninsula, amounted to over one-third of the whole atmospheric pollutant discharges, including 81.5% of sulphuric anhydride, released by the Russian non-ferrous metallurgy. Some indices of the total pollution load emitted by the Norilsk Nickel association in 1996 and 1997 are presented in Table 2.3. It can be calculated by the included data that the share of sulphur anhydrite in the total pollution load in 1997 amounted to 96%. In 1997 the total amount of emitted sulphur dioxide increased by 137,000 tonnes compared to 1996, while only 37.2% of all pollutants were intercepted in both years. It is also evident from the above data that a large amount of nickel, copper and cobalt are emitted into the atmosphere in the airborne wastes.

Table 2.3 The main pollutants in 'Norilsk Nickel' emissions, 1996–1997

Pollutant index	Units	Amount of pollutants		1997 as % of 1996
		1996	1997	
Total emissions	000 tonnes	2485.0	2601.0	104.7
Including:				
Solid particles	000 tonnes	41.3	43.7	105.8
Nickel	tonnes	2806.0	3101.0	110.5
Copper	tonnes	3027.0	2902.0	95.8
Cobalt	tonnes	106.0	110.0	103.4
Liquid and gaseous substances	000 tonnes	2443.2	2557.7	104.7
Sulphuric anhydrite	000 tonnes	2361.0	2498.0	105.8
Polluted waste waters	mln cubic m	151.1	145.8	96.4

Source of data: Gosudarstvenniy... (1998b)

In 1995 the total amount of harmful substances released from stationary sources into the atmosphere over the Murmansk oblast amounted to 543,000 tonnes, which implied a 16% increase compared with 1994 (Murmanskaya... 1996). In 1996 it was slightly less at 505,000 tonnes (Murmanskaya... 1997). The Severonickel smelter works in Monchegorsk and the Pechenganickel complex in Nikel are the two main sources of pollution on the Kola peninsula, responsible for 86% of the total load of sulphur dioxide emitted in the Murmansk oblast. In 1997 both combines emitted 415,700 tonnes of sulphur dioxide, which was 28.8% more than in 1994 (Gosudarstvenniy... 1998b).

In 1983, at the peak of production, emissions of sulphur compounds at the Severonickel integrated metallurgical works reached 270,000 tonnes. In 1997 the total amount of these pollutants had decreased to approximately 130,000 because of production decline, but also due to some recent technological improvements which currently enable the combine to intercept and reproduce 70% of emitted sulphur into sulphuric acid. Emissions of dust have also been reduced by 30–35% and nickel by almost 50%. Before modernization the annual emissions of dust reached in some years 6000–7000 tonnes; copper, 2500 tonnes and nickel 4300 tonnes (Yevseev 1998). However, if comparison is made between the situation in 1996 and 1997, the data reveal a substantial growth (by 24.5%) of the total amount of discharged pollutants which is related to the increase in the production of nickel and copper by 155% and 130% accordingly (Gosudarstvenniy... 1998b). This fact supports the point previously made by the author of this book (Saiko 1998a), that the current *transitional environmental improvement* is only temporary and the situation will be reversed when the economic recession is over, unless more effective and resource-saving technologies replace the currently prevalent but out-dated polluting ones.

As was indicated in the 'Report on the State of the Environment in the Russian Federation in 1997', by 1993 the two complexes on the Kola peninsula had successfully reached the target set by the 'Convention on the Sulphur Dioxide Transboundary Pollution', and reduced the amount of sulphur dioxide emissions by 30% compared to 1980. In 1997 this value reached a new low of 33.4% (Gosudarstvenniy... 1998b). It was esti-mated that, at present, the average level of atmospheric pollution by sulphur dioxide near the Severonickel and Pechenganickel complexes, generally, *does not exceed* the maximum permissible concentration (MPC) (Myach 1996). It is also a stated fact that, at all three combines, the vol-umes of each of the current individual pollutant discharges do not exceed the officially permitted values (Gosudarstvenniy... 1998b). However, in spite of the above, and despite the recorded overall reduc-tion in pollutant emissions, the impact of atmospheric pollution on other components of the environment and the landscape as a whole is too obvi-ous to ignore, as changes have been, in many cases, irreversible.

This situation, in fact, casts doubt on the 'universal nature' of some generally adopted MPC values, and stresses their inadequacy in the cases of such sensitive environments as the Arctic regions. It appears that the current state of the environment is, to a large extent, 'inherited' from the Soviet past. It is also clear that the current decline in the pollution load is not sufficient for any substantial improvement of the environmen-tal situation. This fact, in turn, reveals that the changes have already been so dramatic because the relatively low resistance threshold in this fragile landscape had been greatly exceeded.

Additionally, large-scale **mining** activities associated with industries other than metallurgical have also greatly contributed to the aggravation of the ecological situation in the Russian Arctic. Compared to the environ-mental impact of metallurgical enterprises, mechanical disturbance of the ground and dust pollution become more significant than chemical pollu-tion. This normally takes the form of transportation of toxic dust from quarries, tailings and dumps. A major polluter on the Kola peninsula is the Apatity combine, which extracts and enriches the apatite–nepheline ores and produces fertilizers. It has been recorded that maximum concen-trations of dust containing various heavy metals can reach 200–270 grams per cubic m. Strontium, contained in dust, may reach the maximum con-centrations of 170 ng per cubic m, which is 100 times greater than the background values. These toxic substances reach the environment affect-ing vegetation, soils, surface and ground waters. Annually 30 million tonnes of waste rock is being produced by the Apatity industrial complex (Plate 3). As a result of over 60 years of its operation, the impact area of pollutants has reached approximately 3000 sq km (Myach 1996).

Additional input of sulphur dioxide is due to **transboundary pollution**, which is estimated at almost 60% of the total volume of sulphur compounds. Annual transboundary deposition of sulphur in the Kola peninsula amounts to 1.26 tonnes per sq km (Bryukhanov and Zaveryaeva 1988). Other pollutants arising from transboundary sources include nitrogen oxides, ammonium compounds, heavy metals and radionuclides. They reach the Russian Arctic not only from the European countries and southern Russia, but also from Kazakstan and Central Asia. Of course, the share of the Murmansk oblast in the transboundary atmospheric pollution of Norway and Finland should also not be underestimated.

Oil and gas industries are concentrated in the 'heartland' of West Siberia, namely the Tyumen oblast, which is accountable for 80% of all hydrocarbon discharges, while most natural gas is produced in the northern part of this oblast. Both oil and gas industries are also responsible for a major environmental impact. The burning of associated gases in oil production and exploration significantly contributes to the acceleration of the overall 'greenhouse' effect. It is estimated that annually up to 19 billion cubic m of associate gases is being burnt in West Siberia, releasing hydrocarbons, heavy metals, solid particles, carbon dioxide and nitrogen oxide into the environment and contributing to thermal pollution. Oil leaks from production sites and pipelines is another major cause of pollution. Gas production and processing industries are also the source of carbon and nitrogen oxides and hydrocarbons, which contain up to 90% of methane (Myach 1996).

2.4.2 Water pollution

Water pollution is also a major problem in the Arctic North. In 1997 all small rivers of the **Kola** peninsula located in the vicinity of Monchegorsk, Nikel and Zapolyarny towns were qualified as 'extremely dirty' and 'very dirty' (Gosudarstvenniy... 1998a). In total, 303 million cubic m of polluted wastes were discharged into the surface water bodies of the Murmansk oblast in 1995 (Murmanskaya... 1996). Two years later, the volume increased to 343 million, which included 87.5 million cubic m or over one quarter of completely untreated waste waters (Gosudarstvenniy... 1998c).

In 1993 the level of pollution in small rivers within the impact zones of the Norilsk Nickel enterprises was extremely high. The 'chronic pollution' zone included the Kolos-yoki, Nama-yoki, Luotti-yoki and Nyuduai rivers and also the Imandra lake near Monchegorsk. The worst polluted was the Nyuduai, flowing from Nyud'yavr lake into which Severonickel's waste waters are discharged. The pollutants included nickel and copper compounds, sulphates, chlorides and suspended matter. In 1993 the mean annual concentration of nickel amounted to 225 MPC while the maximum reached 465 MPC (Gosudarstvenniy... 1994).

Similar to the situation with air pollution, a system of water protection measures was recently implemented at the Severonickel and Pechenganickel metallurgical works. Although these succeeded in decreasing the total amount of nickel in waste waters, the overall water quality in the local rivers has not improved. In 1997 the worst polluted river remained the Nyuduai. Rivers flowing through Zapolyarny town and Murmansk city are also among the dirtiest in the Kola.

Water pollution in West Siberia remains a serious problem. Oil and gas prospecting and production has had multiple dramatic impacts on the environment of its Arctic North. Numerous tributaries of the Ob' river flow through the major oil fields of Russia. In 1997 only 25% of all polluted wastes in the Tyumen oblast were adequately treated. Although the **Yamal** peninsula is an area of more recent natural gas extraction, there is already extensive evidence of prospecting works, as surveying workers normally leave behind tonnes of waste material, polluted water and land and their rusting equipment in the tundra (Plate 4).

In the Yamal-Nenets autonomous okrug half of all drilling grounds are heavily polluted with hydrocarbons. Transportation of oil and gas provides additional pollution load for the fragile tundra environment. Ultimately, all types of industrial waste and oil products which enter the environment due to technical accidents flow into the Ob' river. For example, in 1997 as a result of one technical accident on the pipeline near

Plate 4   A rusting memory of the Yamal's explorers – a technogenic desert in the tundra

Surgut 500 tonnes of oil was spilt, 17 tonnes of which reached the Ob' river (Gosudarstvenniy... 1998a). Average concentrations of oil products in Ob's waters vary from 12 to 19 MPC: phenols from 5 to 19, copper from 2 to 25 and iron from 7 to 10 MPC (Gosudarstvenniy ... 1998a). It was recently evaluated that the total amount of oil products which enter the Ob' river is equal to half of the similar discharges into all navigable rivers of the world taken together. Another fact that came to attention lately is that the bottom sediments of the Obskaya Guba bay have accumulated up to 10 grams of oil per 100 grams of ground (Golubchikov 1996).

In 1997 the **Norilsk** industrial complex was responsible for 22.2% of the total volume of pollutants discharged into the surface waters by the non-ferrous metallurgical industry. The share of the other two combines amounted to 12.1% (Gosudarstvenniy ... 1998b). A similar trend of a slight decrease in the total amount of discharged waste water is evident from Table 2.2. In 1997 the reduction compared to the previous year was 3.6%. Lakes in the Norilsk region are contaminated with nickel, copper and mercury (Blais *et al.* 1999).

2.4.3 Soil contamination and disturbance of permafrost

Along with water pollution, contamination of soils presents a major threat. Acidification of water in lakes and soil pollution from sulphur deposition on the Kola peninsula is illustrated by Figure 2.1 and discussed later in the context of vegetation destruction.

Three to ten million tonnes of **oil** is being spilt annually in approximately 300 major and 11,000 minor technical accidents on pipelines and wells in the main oil-producing regions of Russia (Yablokov 1996). In West Siberia, major accidents which are officially recorded involve oil spills exceeding 10,000 tonnes each. In 1993 420,000 tonnes of oil was spilt in the aftermath of a pipeline rupture not far from Sos'vinskiy nature reserve at Nyagan' station (Yevseev 1996). This event by far overshadows the well-known Usinsk disaster in the European Russia which occurred a year later and which has been described by Sagers (1994).

More recently, in total about 40,000 technical accidents involving oil leaks were registered in West Siberia between 1995 and 1997. It was revealed recently that the total area of land polluted with a layer of oil in excess of five cm thickness within West Siberian oil and gas fields is greater than 200,000 hectares (Gosudarstvenniy... 1996). It should be also stressed that many enterprises in the oil and gas industries often try to withhold information about technical accidents and attempt to underplay their ecological consequences.

It has also been suggested that only half of all gas produced for export in West Siberia reaches the western border of Russia. The rest is lost due to numerous leaks from pipelines, technical accidents and through fuel

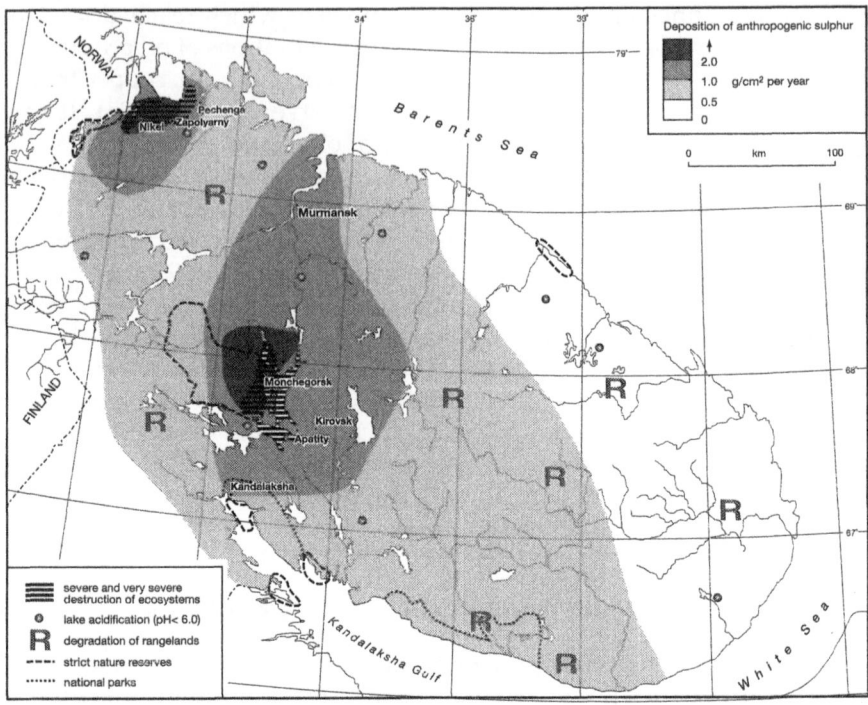

Figure 2.1 Industrial pollution by sulphur dioxide and degradation of natural rangelands on the Kola peninsula, 1990–1995 (compiled from Murmansk... 1996; Ekologicheskaya... 1999)

requirements for its pumping (Wolfson 1994). The toxic effect of oil products increases due to the wide distribution of impenetrable permafrost layer, which causes pollutants to be absorbed by the snow and ice cover, and thus remain in the environment for centuries or probably even longer. Likewise, the remains of mammoths found in these latitudes in permafrost are virtually unscathed.

Contamination of soils by heavy metals and radioactivity presents another serious problem. Heavy metals reach the environment via the discharge of untreated effluents, industrial waste and emission of aerosol particles. It is estimated that the total surface affected by the spread of heavy metals and sulphur dioxide pollutants amounts to 80% of the territory of the Kola peninsula (Makarova 1988). In 1993 contamination of lands around Monchegorsk was classified as 'extremely dangerous pollution with toxic substances' (Gosudarstvenniy... 1994).

A major problem of the Arctic North regions is radioactive contamination of the environment due to the impact of nuclear testing, the Chernobyl accident and other sources. It has become known that the former Soviet Union carried out 130 nuclear tests in the Arctic regions (*Zelyoniy Mir* No. 10–11, 1999) mainly on the Novaya Zemlya island located to the north-east of the Kola peninsula.

Another 'inheritance' of the Soviet period, when the Kola peninsula was a major strategic military base for the USSR, is the problem of radioactive waste disposal. Some of these wastes are from the Kol'skaya nuclear power plant, others are from decommissioned nuclear submarines, discarded rocket fuel, waste fuel from the icebreaker fleet, etc. (Figure 2.2). By the present time, over one million curies of solid and liquid radioactive wastes have been accumulated in the Murmansk oblast (Gosudarstvenniy... 1988c). It is worth emphasizing that this region has not until now developed a system for processing solid radioactive wastes.

Economic development in the tundra is often accompanied by large-scale damage to the soil cover and **permafrost** environment. It is estimated that subsurface ice in the upper 50–100 m layer occupies up to 70% of the total volume of ground, which amounts to about 20,000 cubic km of ice (Vil'chek *et al.* 1996a). The thawing of permafrost below

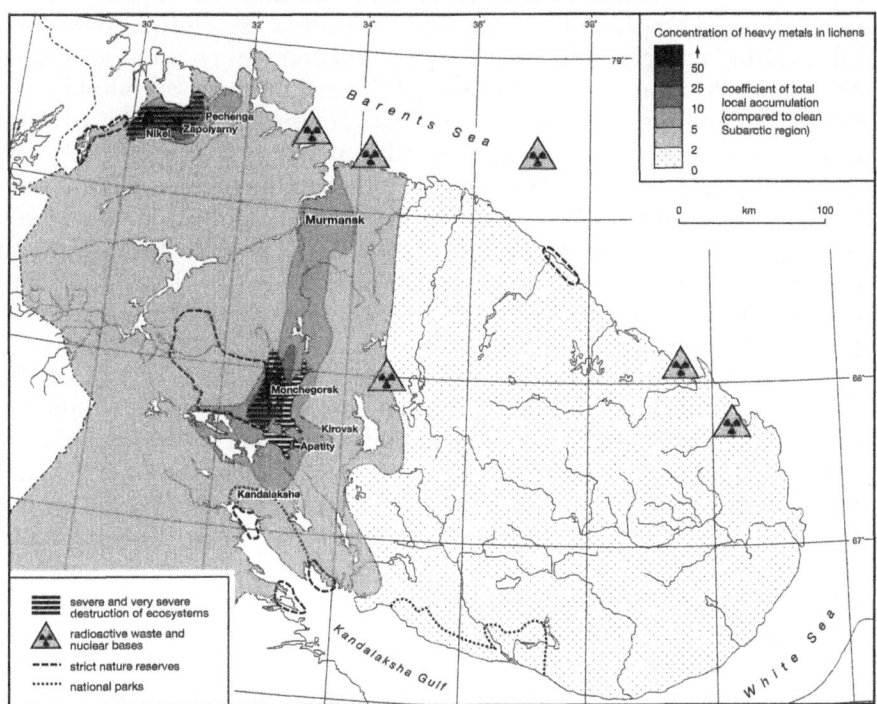

Figure 2.2 Pollution of soil by heavy metals and location of sites of radioactive waste storage, processing and disposal on the Kola peninsula, 1990–1995 (compiled from Murmansk... 1996; Ekologicheskaya... 1999)

disturbed vegetation results in the formation of thermokarst lakes and waterlogging. The western and central parts of the Yamal peninsula are affected by the processes of fast solifluction, which is evidenced by complete denudation of vegetation on slopes. In some areas, these processes have affected up to 60–70% of the slope's area (Voskresenskiy 1992). To the south-east of Yamal, the rates of thermokarstic processes along the route of the Yagel'naya-Yamburg railway road have increased by ten times since construction started (Kamyshev 1999).

2.4.4 Degradation of natural vegetation and pastures

It has been mentioned above that vegetation in the Arctic is extremely vulnerable to human influence. It has been estimated that although soil and vegetation cover have been completely destroyed within an area of only 3–8% of the Russian Arctic region, the area affected by vegetation removal or damage accounts for almost 25–50% (Yablokov 1996). Areas around the worst polluting metallurgical enterprises have been most affected. The impact of air pollution is normally expressed by the extent of surface deposition of various pollutants. Two maps of sulphur deposition and concentration of heavy metals in lichens on the **Kola** peninsula are included in Figures 2.1 and 2.2. They illustrate that both indices have a strong correlation with the location of the main industrial complexes in the central and north-western parts of the peninsula. The highest values of the mean annual anthropogenic sulphur deposition exceeding two grammes per sq m are evident around Nikel, to the north of Zapolyarny and around Monchegorsk. Clearly, air and soil pollution have been responsible for the degradation of rangelands in the central and southern parts of the Kola peninsula shown in Figure 2.1. Acidification of lakes presented in the same map is also at least partially due to the atmospheric sulphur deposition.

Concentrations of heavy metals in lichens (Figure 2.2) reflect a similar impact of the industrial activity associated with mining and metallurgical enterprises. The·values of the coefficient of total local accumulation are highest in the areas situated closest to the Pechenganickel and Severonickel complexes, where they are 10 to more than 50 times greater than the clean Subarctic region. Estimates of heavy metal contents in mosses near Monchegorsk have been made by Yevseev and Krasovskaya (1996, 1998). Concentrations of copper in green mosses (*Brios sp.*) reached 1516 mg/kg and nickel 1189 mg/kg.

Yevseev and Krasovskaya also found that because snow can retain and accumulate pollutants during a prolonged winter, by the end of the winter period copper contents in snow cover could be even higher at 2154 mg/kg. Excessive concentrations of heavy metals cover approximately half of the territory of the Kola peninsula, while radioactive contamination is associated with the location of the main nuclear bases, particularly around Murmansk and the Kola nuclear power plant. The combined effect of air pollution and soil contamination with sulphur compounds and heavy metals has created extensive areas of severe and very severe destruction of ecosystems, stretching in the vicinity of the above towns as far north as the Norwegian state border and as far south as Apatity.

The adverse impact of the Monchegorsk industrial complex is evident within a radius of at least 40 km from the plant. Closest to the plant there is a zone of complete destruction of northern taiga vegetation and severely degraded soils, which has even been called an 'anthropogenic desert' (Yevseev and Krasovskaya 1996) or 'industrial desert' (Roginko 1992). Here, existing only in depressions, can be seen sparse low willows and birches with discoloured leaves. Continuous ground cover is lacking and mosses and soils are heavily contaminated with sulphur and heavy metals. Species of deciduous trees and shrubs are less vulnerable to air pollution due to their ability to shed leaves every year (Plate 2); in addition, shrubs are protected by thick snow cover during the winter season. Thus it is not surprising that sparse small-leaved shrubs can be seen quite close to the Monchegorsk plant.

The first stage of destruction of forest involves discolouration of needles which become brown. While normally needles live for 11–12 years, under the heavy impact of industrial emissions they only survive for two years. It has been estimated that during the period of the most intensive production, average movement rates of the forest destruction boundary in the territory adjacent to Monchegorsk amounted to one km per year in the direction of the prevailing winds (Yevseev 1998). Expansion of the affected area around Monchegorsk has taken place at a rate of 20 km in 30 years (Yevseev and Krasovskaya 1996).

Half of all the forests have been destroyed to different degrees including the dead forest on one-fifth of the latter's area. The 'industrial desert' spreads over an area of 10,000 ha coniferous forests; there are no living trees and composition of moss and lichen cover has changed. Over an area of 40,000 ha, tree growth is depressed. It can be noted that forest rehabilitation within the worst affected sections takes up to 500–800 years, while in the less destroyed coniferous forests it would take 150–300 years. Moss and lichen cover can become restored in 20–30 and sometimes 80 years (Sokhraneniye... 1997).

Reindeer husbandry is the main traditional occupation of the indigenous people of the Arctic North. Human impact has greatly devastated the natural rangelands of reindeer in the tundra. During 1965–1990 the total area of reindeer pastures shrank by 70 million ha. During the early 1990s their area decreased by 20–30%, while reserves of valuable reindeer lichens shrank by 2–3 times. The total area of reindeer pastures in 1996 amounted to 328.1 million ha or 19% of the whole Russian territory. By 1 January 1996 the degraded area reached 230.6 million ha including 32% of severely and 46.6% of moderately degraded land (Gosudarstvenniy... 1996).

The main causes of the destruction of vegetation and soil cover in the tundra are exploration and extraction of mineral deposits, uncontrolled passage of track-vehicles and heavy machinery, excessive animal pressures on pastures and geological surveying. According to Roginko (1992) 6 million ha of pastures in the Yamal-Nenets Autonomous Okrug have been destroyed over the two previous decades. Large areas of rangelands in **Yamal** are degraded and polluted. Our calculations from data presented by Vil'chek (1996) show that over 50,000 ha of rangelands have been completely or partially destroyed by technogenic impact. Also, approximately 13 million ha of hunting grounds and reindeer pastures have been lost in the Taimyr autonomous okrug due to the multi-year impact of the Norilsk metallurgical complex.

In the Norilsk region the combined area of 'anthropogenic desert' and degraded forest exceeds 600,000 ha. The dead forest stand is expanding, particularly to the south of **Norilsk**, and larch appears to be most vulnerable. In winter the Norilsk industrial region is clearly seen on space images as a dark spot. The area of snow pollution amounts to about 900 sq km (Gorshkov 1998; Gosudarstvenniy... 1998a). The whole territory affected by acid atmospheric precipitation associated with Norilsk emissions covers about 400,000 sq km, which is almost four times greater than on the Kola peninsula (Yevseev 1996). Contamination of soils by heavy metals, particularly copper, nickel and cobalt, is most severe within a radius of 3–5 km from the city (Poyasnitel'naya... 1996). Other authors estimated the size of the 'impact zone' as within 2–3 km radius (Yevseev and Krasovskaya 1998).

2.4.5 Overall ecological situation

The legacy of economic development and ecological change is summed up in Table 2.1. During recent decades, changes in the state of the landscapes have been so dramatic that the current overall situation can clearly be classified as an 'ecological crisis', while in the Norilsk region and in the vicinity of Monchegorsk and Nikel towns in the Kola peninsula the situation can already be considered as 'catastrophic'. Although all industrial sites occupy only 1% of the territory of the Murmansk oblast, their activity has negatively affected the environment within 3.2% of its area or 3,000 sq km (Yevseev 1998). The situation in parts of the Yamal peninsula can be still evaluated as an 'ecological crisis' although in the vicinity of major gas deposits rehabilitation of the tundra will require many decades.

The overall ecological status on the Yamal peninsula and around Norilsk is shown in Figure 2.3. In the Norilsk region the situation is clearly catastrophic in areas located to the south-west of the city. Pyasino Lake is very polluted and an additional threat comes from radioactive contamination of the former nuclear testing ground located to the north of the city. The complex of ecological problems within this region includes air and water pollution, contamination of soil and land destruction.

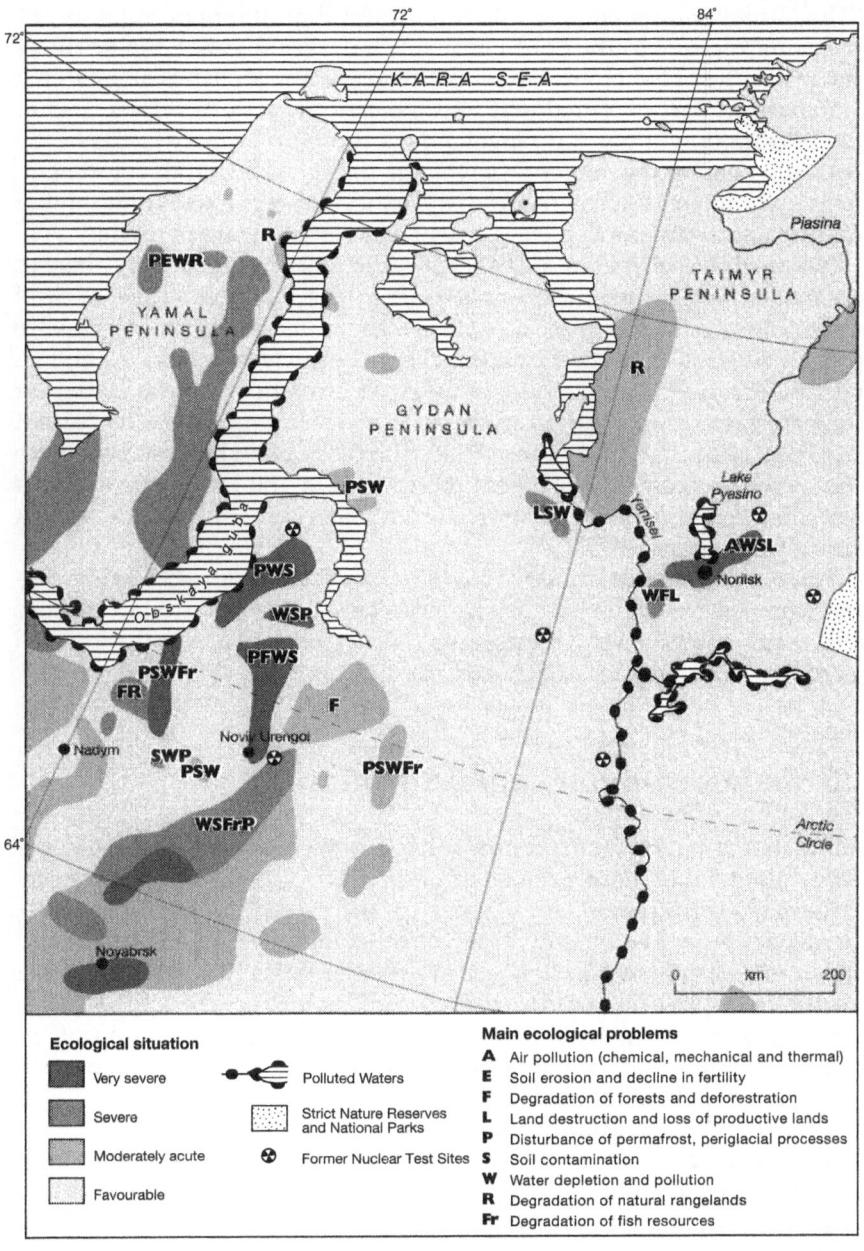

Figure 2.3   The present ecological situation on the Yamal peninsula and in the Norilsk region (adapted from Ekologicheskaya... 1999)

The Yamal peninsula is still less polluted than the main oil and gas producing regions to the south and south-east. However, the territory where the state of the environment is already severe stretches along much of the eastern part of Yamal and into the central western region (Figure 2.3). Degradation of natural rangelands is common in many parts of the peninsula despite the current early stage of its industrial development. Water pollution, soil erosion and disturbance of permafrost from mining activities and transport vehicles present additional severe problems.

One of the clear indices of the large-scale anthropogenic disturbances and transformation of nature is a shift in the location of the natural (zonal) boundary of landscapes. Generally, transitional landscapes are most vulnerable to human impacts. This is clearly reflected in the fact that the forested tundra zone has been affected most, the boundary of northern taiga having shifted to the south when compared to its current climatic or natural border in the 1930s. The dislocation had reached up to 200 km in the north of European Russia, up to 250 km in the north of West Siberia and up to 130 km in Central Siberia including the Norilsk region (Kryuchkov 1987).

The retreat of this boundary to the south actually started several centuries ago due to fuel wood cutting and reindeer pastoralism but its rates were much slower than at present, when technogenic pollution and mechanical disturbances can mainly be held responsible for the process.

2.5 Socio-economic and health issues

Although the Arctic North is probably one of the most inhospitable areas of the world, it has been populated and used by humans for many centuries. The indigenous population of the Arctic North is sparsely distributed over the territory. One of the specific features is the nuclei-type of settlements within the Arctic region. Immigrant population tends to settle in the few main industrial towns. Despite the overall low population density (1–5 people per sq km) a peculiar phenomenon of the 'overpopulation' of this harsh environment is noted by some experts (Golubchikov 1996; Prokhorov 1996).

From the socio-economic standpoint, the Arctic is one of the most depressed regions in Russia (Vil'chek et al. 1996b). Arctic development has generally been followed by some major economic and social implications, which have affected its population and particularly the indigenous nations. The social implications of environmental problems include an increase in ethnic tensions, particularly between indigenous nations and the European immigrants, growing alcoholism and drug addiction, a decline in the standard of living and migration from the region.

The dramatic changes that have occurred in the past ten years have affected the most vulnerable regions of the Russian Federation such as the Arctic North (Sallnow 1998). Every year, the problem of transporting enough fuel and food from the 'continent', which for 'Northerners' implies European Russia, becomes a survival problem. The navigation period is very short and lasts only for the summer period. While experts keep arguing whether it is economically feasible to subsidize northern development or whether it would be cheaper to resettle people from the Arctic, the everyday lives of these people present insurmountable problems. The rapid rate of inflation between 1992 and 1994 reduced their lifetime savings to zero, even allowing for the enhanced wages that could be earned under Soviet incentive schemes in the difficult environment of the North. Now the local population, despite being a skilled work force, feel that they are hostages because they cannot afford to move away from these hostile environments. It is estimated (Zaidfudim 1998) that over a period from 1991 to 1997 more than one million inhabitants migrated from the Russian North. The most recent data show that in 1998 over 120,000 people emigrated from the Russian North including 15,800 from Murmansk oblast, 3100 from the Yamal-Nenets autonomous okrug and 13,000 from Krasnoyarsk kray (Chislennost'... 1999).

However, the native people have been particularly badly hit. Most of the national minorities of the Arctic North belong to 'old nations' who feature very low adaptability to external changes and disturbances. Their main occupations for centuries have remained unchanged. Fishing, hunting and reindeer grazing directly depend on the quality of the environment they live in. However, intensive industrial development has in many ways dramatically undermined their traditional economies and modes of living. Furthermore, current economic transition and land tenure reform has made them even more vulnerable (Fondahl 1995; Osherenko 1995). While in the past Arctic development was heavily subsidized, changed economic conditions have pushed the indigenous peoples to the limits of survival. Bureaucratic governmental institutions fail to provide adequate support for the northern minorities (Poelzer 1995).

The direct impacts of air pollution on health are quite well known but it is difficult to precisely evaluate them. However, some authors estimate that 48% of the structure of all ecological damage is related to human health (Yevseev and Krasovskaya 1996). For the immigrant population the grave ecological situation, particularly air pollution and poor quality of drinking water, further complicates the impact of existing unfavourable natural conditions on human health. It was revealed that among children living around Monchegorsk and Olenegorsk on the Kola peninsula, frequency of anaemia had risen to up to two times more than the national average, and gastritis, asthmatic bronchitis and bronchial asthma up to four times. The overall illness frequencies in all age groups increased by 64%, while three-quarters of all illnesses are respiratory (Prokhorov 1996). In Murmansk oblast and Yamal-Nenets autonomous okrug, death rates from oncological illnesses increased by 1.3–1.6 times between 1990 and 1996 (Zelyoniy Mir 14, 1999).

In Norilsk, the population suffer most frequently from bronchitis, pneumonia, bronchial asthma and related allergies. Lung cancer is quite common. Frequencies of respiratory illnesses in children is 1.5–2 times higher than the national average. During the last 40 years cancer frequencies in males were recorded at 3.5–4.5 times greater than the Russian average. Heavy metals and radionuclides have a strong mutagenic effect – for example, in Norilsk the frequency of genetic defects in new-born babies amounted to 11.2 per 1000 persons compared to the Russian average of 6–8 (Gosudarstvenniy... 1998a). Concentration of mercury in contemporary people's hair was three times higher than in the people of the 15th century, of lead nine times and copper two times (Prokhorov 1996).

Compared to the immigrant population, the indigenous nationalities are less influenced by natural factors but for them various ecological, economic and social factors bear greater importance. Changes in the traditional diets of the native people, such as a reduction in the proportion of hunting and sea products aggravated by the rising consumption of alcohol, have ensued in increased frequencies of nutritional illnesses. The indigenous people are more vulnerable to infectious diseases, notably respiratory illnesses, which can easily become chronic. Tuberculosis is 2.5 times more prevalent among the indigenous than the migrant population. This is probably associated with the weakening of their immune system which is not adapted to life in a highly polluted environment. Accumulation of caesium-137 in rangers' bodies is 100 times greater than in the bodies of people living in the temperate latitudes. The transition to the market economy has further limited their access to health care, caused a decreased standard of living and increased unemployment (Bradshaw 1995 and Fondahl 1995).

At present there appears to be a clear trend in the reduction of population numbers, particularly within indigenous Northern minorities. During recent years birth rates in these groups have declined by 69% and death rates increased by 35.5% (Gosudarstvenniy... 1998a). Life expectancy is a reliable integral index of human health. During the last three decades the average life expectancy of the Nentsi people declined from 61 to 45 years (Golubchikov 1996). All the above facts give evidence that there is an overall health deterioration which is a clear sign of the *environmental crisis* in the Arctic North.

2.6 Prospects for the future

A purely technocratic approach to economic development of the Arctic during the Soviet period has already produced a number of catastrophic consequences for the environment of these marginal zones and indigenous peoples. The resistance threshold of Arctic ecosystems was exceeded decades ago. Even without human intervention, these fragile areas are at risk from natural causes. However, the intensity of interference here often exceeds the magnitude of impact in more environmentally stable landscapes. Additionally, as Vil'chek *et al.* have correctly stated, 'Even an ideal regime of environmental use and development of protected-area networks cannot resolve all the problems of the native minority peoples of the North' (Vil'chek *et al.* 1996b, 264). It seems that in the foreseeable future this situation is likely to mount further under the new challenges brought about by the transition to the market economy.

Non-ferrous metallurgy will remain one of the main sources of hard currency for the Russian Federation for years to come; therefore, unless a radical change in the attitude to the environmental issues takes place there are not many grounds to expect any major improvement in the state of the environment on the Kola peninsula and Norilsk region. Diverting existing scarce investments to mending past mistakes and improving the environmental situation seems in the foreseeable future to be an unrealistic scenario.

The continued large-scale exploration and production of new gas and oil deposits in West Siberia will inevitably result in an exponential increase in the total pollution load and in greenhouse gases in particular. It is expected that when Bovanenkovo and Harasaveiskoye gas condensate deposits in the Yamal peninsula reach their full production capacity, the total amount of emitted hydrocarbons, nitrogen oxides and carbon dioxide will increase by three, four and five times, even if modern technologies are introduced and protective measures are implemented (Myach 1996).

In addition, in view of forecasted climate changes, the increase in temperatures predicted for higher latitudes in the next century may dramatically affect the Arctic North through its impact on permafrost. It is expected that the depth of the active layer will increase, thus affecting the permafrost and threatening the existing infrastructure in the polar latitudes – pipelines, roads, settlements. The overall resistance of ecosystems will be undermined, which will result initially in changes in the vegetation cover, and will lead to migration of fauna, which is more flexible. Thus the contemporary environmental catastrophe could be further aggravated.

A holistic approach is needed in the future to achieve the objectives of sustainable development in this marginal environment. It should be based on the comprehensive study of its potential vulnerability and should fully take into account the needs of the indigenous and immigrant population.

CHAPTER 3

Lake Baikal: a unique ecosystem under threat in the Siberian taiga

'We have only one Baikal and it can be still saved.'

(N. Timashova, *Izvestiya* 26 December 1995, 1)

3.1 Introduction

This chapter will examine the current ecological problems of Lake Baikal, 'the pearl of Russia' and the state of the environment in its drainage basin. The aim is to demonstrate the global importance of Lake Baikal's unique ecosystem and its particular vulnerability to human impacts. The focus of this chapter is on the natural prerequisites and causes of various environmental problems within the Baikal basin; the current situation is related to the pollution of this precious lake and also to recent changes associated with economic recession during the transition to the market economy.

Many features of this lake cannot be described without the use of superlative terms. With a volume of 23,000 cubic km it is the world's greatest lake containing 20% of the global and 95% of the Russian freshwater resources. It is more than seven times greater than the volume of total global freshwater withdrawal. The total surface area of Baikal is 31,500 sq km which is approximately the size of Belgium; its length is 636 km and the average width is 48 km. The lake's coastline stretches for over 2,000 km and is indented with some major bays and inlets.

Maximum depth in the middle part of Baikal reaches 1620 m making it the deepest continental water body on the Earth. There are 27 islands in Baikal, and the largest of them, Ol'khon, has an area of 730 sq km. The area of the drainage basin accounts for an impressive 570,000 sq km, which is only slightly smaller than the area of the Iberian peninsula i.e. the combined area of Portugal and Spain. Its southern part stretches into Mongolia (Figure 3.2). Within Russia the lake's drainage basin is shared by Buryat Republic or Buryatia, which covers about 73% of its territory, Chita oblast 21% and Irkutsk oblast 6%. Baikal is worshipped by the Buryats who call it 'the Sacred Sea'. The lake is deeply embedded in their Buddhist religious beliefs, spiritual and material life, customs and traditions.

Biologically, Baikal is perhaps the most important lake on Earth, because it contains almost all types of animal life found in the freshwater bodies of our planet. Almost three quarters of the 2600 species of organisms which inhabit the lake are endemic. The unsurpassed richness of its fauna and flora, plus exceptional water purity due to the work of its organic life and the lake's ancient age, attract global attention to this beautiful and unrivalled place (Plate 5). Unfortunately, pollution of this unique lake has continued unabated since the construction of the Baikalsk pulp and paper mill in the 1960s. Despite a limited reduction in pollutants more recently, the overall situation has not changed for the better.

The pollution of Lake Baikal is probably the 'oldest' world-known environmental problem of the USSR. During the Soviet and post-Soviet periods, many Russian and Western authors examined the environmental problems associated with the ongoing pollution of this glorious lake, which is sometimes called 'the blue eye of Siberia' or 'the blue heart of Siberia'. Various issues related to the nature and problems of Baikal have been widely covered in scientific literature (Feshbach and Friendly 1992; Mnatsakanian 1992; Peterson 1993; Pryde 1991; Stewart 1992b,

Plate 5   A serene view of the 'Sacred Sea'

1992; Vorob'yev 1988; Yanshin and Melua 1991). A comprehensive atlas of Lake Baikal was published in 1993 (Baikal 1993). In the 1990s extensive studies were summarized in a series of annual environmental problem reviews (Problemy... 1994, 1995, 1996, 1997). Issues related to the state of the environment in the drainage basin of the lake were also covered annually in the Russian State Report on the Status of the Environment (Gosudarstvenniy... 1992–1998).

It should also be emphasized that Lake Baikal had especial significance for the creation of an emerging political opposition in the former Soviet Union. In the 1960s, long before censorship on ecological information was lifted, this lake served as a catalyst for the first awakening of the Russian environmental movement.

3.2 Physical geographical setting

The 'health' of Baikal to a large degree depends on the state of the physical environment within its drainage basin. Therefore, the application of a 'catchment' or 'drainage basin approach' is needed to assess the overall ecological situation. In order to explore the causes of vulnerability of the unique Baikal's aquatic ecosystem, it is essential to examine not only the physical geographical conditions of the lake but also of its drainage basin. Figure 3.1 illustrates the existing complex interrelationships between anthropogenic factors and the physical environment of the Baikal basin. The state or 'health' of the unique endemic organic life in the lake depends on its water quality. The latter is directly or indirectly determined by the state of the environment in the whole drainage basin which in turn affects the quality of water in all surface and ground waters ultimately reaching the lake.

3.2.1 Geographical location and relief

Lake Baikal is located in the heart of Eurasia, in the south of East Siberia. It is situated in the coniferous forest landscape zone called the 'taiga' (Figure 1.3). From all sides, Lake Baikal is surrounded by ranges of mountains. The lake's northern and north-western shores are surrounded by the Primorskiy range with maximum heights of about 1700 m and by the Baikal'skiy range with a maximum height of 2588 m (Cherskogo M.). The Barguzin range with average heights of over 2000 m stretches along the eastern coast of the lake. It has classical alpine shapes and various evidence from the last glaciation. In the south, Baikal is surrounded by the Khamar-Daban range which is composed of gneiss, granite and marble rocks with heights exceeding 1500–2000 m.

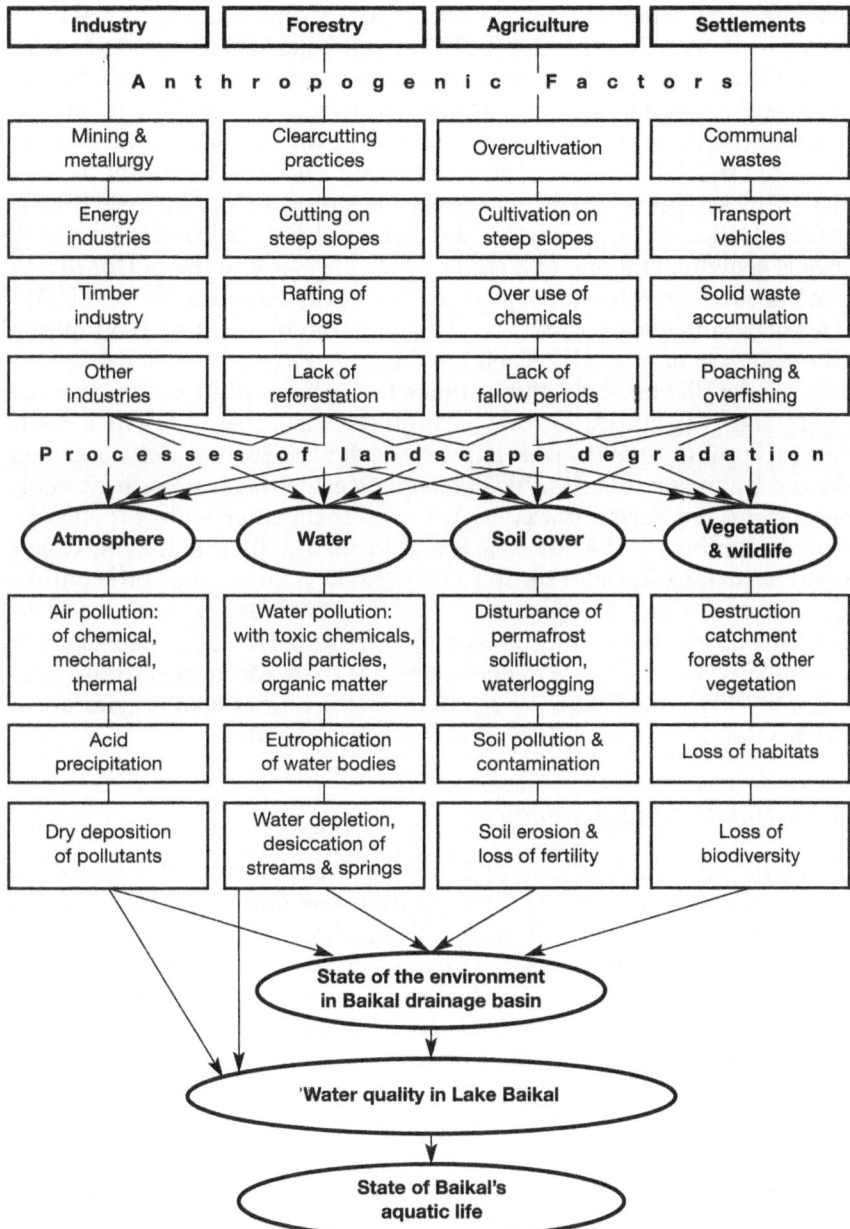

Figure 3.1 The scheme of interrelationships between anthropogenic factors and physical environment within the Baikal basin

The lake has a tectonic origin and is located in a rift valley, which explains its unusually great depth. The age of Baikal is estimated at 25–30 million years; the world's second oldest lake, Lake Tanganyika, is just two to three million years old. Despite its 'respected' age there is no evidence that the Baikal is ageing and will disappear from the surface of the Earth. This fact has enabled scientists to put forward a hypothesis that Baikal is an emerging ocean. This theory is supported by the recent geophysical studies and the fact that the distance between its east and west shores is growing at a rate of 2 cm per year, similarly to the rate of divergence between the continents of Africa and South America (Baikal 1993).

The region is featured by high seismic activity – up to 2000 minor earthquakes occur annually. Their average magnitude is estimated at 3–4 points on the Richter scale, but at times they can be quite destructive. In January 1862 the largest known earthquake was recorded, which measured 10.5 points. This disaster happened near the Selenga delta and was followed by major catastrophic consequences: extensive territories collapsed and were buried under Baikal's water together with settlements, people and animals. The bay which was formed in their place was called 'Proval' which in Russian means a 'downfall'. During the 20th century the largest earthquake was recorded in 1959 with a magnitude of 9 points. As a result, the middle section of the lake located at a distance of about 20 km from the shore deepened by 15–20 m. More recently, on 9–11 February 1999 an earthquake estimated at 6 points was registered in southern Baikal with an epicentre 74 km from Irkutsk.

3.2.2 Climate and hydrology

Climate in the drainage basin of Baikal has many features typical of the taiga zone, but it also shows some peculiarities due to the impact of its immense volume of water. Overall, the taiga has a warmer summer compared with the tundra and forested tundra zones, but winters are often even colder. The climatic regime has a clearly expressed seasonality. The climate of the East Siberian taiga is of an extreme continental type featuring a great range of temperatures between seasons. Very few places on the Earth can compete with this region in the severity of their climatic conditions. Location in the heartland of the Eurasian continent predetermines the intensive cooling of the huge land mass. In winter Siberia lies within the sphere of influence of the high-pressure Asiatic anticyclone with cold and dry atmospheric air.

The mean January temperatures range from –17 to –45 degrees C, being at least 6–14 degrees C lower than at similar latitudes elsewhere in the world. The absolute minimum recorded temperatures are –71 and –72 degrees C in Verkhoyansk and Oimyakon accordingly. In the Republic of Sakha (Yakutia) the range of absolute temperatures reaches 100 degrees C (Gvozdetskiy and Mikhailov 1987). For many days in winter severe frosts are accompanied by clear skies and windless weather. Temperature inversions are quite common particularly in the northern and eastern parts of this zone.

The depth of snow cover throughout the taiga is extremely variable ranging from only 30 cm in East Siberia to up to 1 m in West Siberia. In summer cyclonic activity is predominant. The main part of precipitation occurs in late summer – often two to three times more than during the cold winter period. The average values of July temperatures vary from +13 degrees C in the north to +19 in the south of the zone. The amount of precipitation is greater than in the tundra, ranging from 300 to 600 mm, and its value exceeds potential evaporation making this zone relatively moist.

The location of Lake Baikal in the centre of the Asiatic continent predetermines its high degree of continentality. However, some moderating influence is manifested due to the impact of its enormous volume of water and a specific pattern of atmospheric circulation within the Baikal depression surrounded by mountains. Hence, Lake Baikal has some specific climatic features i.e. a relatively mild winter and a cool summer and is probably the warmest place in East Siberia. The impact of its water body is expressed through a decrease in the amplitudes of seasonal and diurnal fluctuations of temperatures and air humidity and a monthly delay in the beginning of each season in the coastal regions compared to close continental areas. In summer Baikal is cooler than the surrounding land mass, which prevents the formation of ascending air. Therefore, the amount of precipitation over the lake surface in the north is as low as 200 mm. The maximum amount of moisture is precipitated on the leeward slopes of the Khamar-Daban range, which receive up to 1000–1200 mm.

One unusual feature of Baikal's climate is a prolonged duration of solar radiance. Over the northern part its value amounts to 1900–2000 hours per year, which exceeds the values for the Baltic Sea resorts and corresponds to the duration of radiance over some of the Black Sea resorts. The average air temperatures over the lake surface vary from –21 degrees C in winter to +15 degrees C in summer; in the coastal areas these values are –25 and +17 degrees C accordingly. The wind regime is seasonal: in winter winds blow from the continent to the centre of the lake, in summer it is the other way round. Storm winds are common in late summer and early autumn.

Because the lake actually represents a depression surrounded by mountain ranges, Baikal has one specific feature of climate, which is a high frequency of occurrence of atmospheric stagnation or still air. This phenomenon is responsible for the low ability of the air masses to effectively dissipate pollutants. It has been estimated (Problemy... 1996) that due to this effect, a similar magnitude of pollutant emissions would result in two to three times higher levels of atmospheric pollution in this region compared to the European part of Russia.

The **hydrological** network of Baikal incorporates its 336 tributaries, the largest being the Selenga, Barguzin and Verkhnyaya Angara. The former river provides approximately half of all water reaching the lake from the drainage basin. The only river flowing out of the lake is the Angara, a tributary of the mighty Yenisei river heading its waters into the Arctic Ocean. The hydrological regime of Lake Baikal is stable with fluctuations within a range of 94 cm. Its seasonal level is highest from late August to early October followed by a fall throughout the winter period up to April. Of its recharge, 83% comes from the river inflow, 13% from precipitation, while the remaining 4% come from other sources (Baikal 1993). The water exchange in its northern part can last up to 225 years, in the middle part 132 years and the southern part 66 years. Annually 8214 cubic km of Baikal's water is renewed.

Every year the surface of Lake Baikal freezes completely, covering up with ice from north to south. In late October shallow bays become frozen first, while the deepest parts get covered with ice only by mid-January. The depth of the ice reaches 1 m, which is less than on minor Siberian rivers. In late April ice gets broken initially in the southern part of the lake and this process continues for about a month. The water in Baikal is relatively cold: in August the open surface temperatures do not exceed +9 − +12 degrees C. They can reach +15 to +20 degrees in shallow bays and near the coasts.

In winter, water temperatures under the ice shield gradually increase to +3.5–3.6 degrees at depths of 250–300 m. In the bottom layers it becomes slightly warmer due to the impact of the Earth's internal energy and chemical processes. Low water temperature throughout the year limits the potential assimilative capacity and slows down decomposition processes in Baikal's water. It has been estimated that over one year, only 30–40% of all organic matter entering Baikal with river inflow become decomposed (*Zelyoniy Mir* 18–19, 1999). Mineral compounds practically cannot be decayed or need a very long period of time to do so. This feature explains why pollutants keep accumulating and why the impact zones around sources of pollution constantly increase in size.

Baikal's water is characterized by very high transparency up to the depth of 40 m, a low overall mineralization of less than 150 mg/l and an alkaline reaction. Its oxygen contents is never less than 9.5–10 mg/l and even at the greatest depths amounts to 70–75% of the saturation level (Baikal 1993). The period of the total water exchange in Baikal is 383 years. Every year approximately 60 cubic km of an exceptional quality water is reproduced. Physical properties and ion composition of water are stable both spatially and temporally, while biogenic and gaseous contents have clear seasonal fluctuations, which reflect the pattern of functioning of Baikal's main life forms.

Although permafrost is widely spread in the taiga and reaches the depth of 1500 m, the role of cryogenic and thermokarst processes in land formation in the southern part of Siberia is less important than in the tundra. Water erosion becomes the dominant geomorphologic factor in land and soil formation. The permafrost layer within the drainage basin of Lake Baikal is continuous in the north and discontinuous or sporadic in the south of the basin. In the coastal zone its depth does not exceed 10 m, but it increases towards the periphery of the basin.

·Thermokarst and other cryogenic processes are quite common throughout the region while mudflow activity is typical for the south-eastern and north-western coasts of the lake. When the Trans-Siberian railway mainline was under construction, 104 active mudflow basins were recorded, only in the southern part of the coast. After the railway exploitation started, nine catastrophic mudflows were registered between 1915 and 1971 (Problemy... 1996).

3.2.3 Vegetation and soil cover

The vegetation cover in the Baikal basin is mainly represented by the mountainous **taiga** (also spelt as tayga). The taiga or *boreal coniferous forest* occupies over 60% of the area of the Russian Federation extending from its western border and across Siberia for over 9000 km to the Pacific Ocean. This is the Earth's largest single coniferous forest formation, providing the greatest reserves of softwood timber. Sometimes the taiga is called 'the lungs of the planet', because much of the oxygen inhaled by Europeans and North Americans is generated in Siberia.

In the European and Asiatic parts of Russia the species composition of taiga differ. Fir and pine dominate in the European taiga while spruce and larch along with pine are most common in the Siberian taiga. Generally, the taiga forests are poorer in species composition compared to mixed and deciduous forests, but biodiversity is richer than in the tundra. Two types of taiga can be distinguished: light coniferous and dark coniferous. The former type is widely spread in central Siberia and the Russian Far East, while the latter is found in the European part and in the southern areas of Asiatic Russia. Narrow needle-like leaves of conifers are both frost and drought resistant and have evolved to reduce desiccation during the cold winter period. The location of the Baikal region in the southern part of the taiga zone and its mountainous topography explains a relatively greater variety of vegetation.

Altitudinal zonality is clearly expressed throughout the basin in the succession of landscape zones from the foothill areas up to the tops of mountains, generally depending on their location and height. For example, to the north-east of the lake where the Barguzin range is particularly high, the taiga vegetation on the lower downhill slopes up to the heights of 1400 m is replaced by forested tundra and further up-slope by the tundra vegetation. The latter, in turn, gives way to scattered Arctic tundra type of vegetation and glaciers closer to the highest mountain tops.

In the south, tundra vegetation is found only on the slopes of the highest mountains of the Khamar-Daban range, while taiga vegetation predominates on the mountains along the lower western and south-eastern coasts. Coniferous forests here spread up to the higher altitudes of

1900 m. The range of altitudinal zonality is more striking in the south where it is represented more fully. Here, the sequence of landscapes starts with steppe (grassland) vegetation of the inter-mountain valleys and is gradually replaced by forested steppes which sometimes spread as high as 1000–1200 m, and further up-slope by the taiga and tundra vegetation.

Trees in the light coniferous taiga, which in the Baikal basin is mainly composed of larch (*Larix dahurica* and *L. gmelini*), are sparsely spread providing an impression of a light forest (Plate 6). This type of taiga is common in the northern and north-eastern part of the region. In many places larch species are interspersed with such small-leaved deciduous tree species as birch, alder, willow and poplar.

Composition of dark coniferous taiga which is spread along the central eastern and south-eastern coasts of Baikal includes fir species (*Abies sibirica, Picea obovata* and *P. ajaensis*); the unique Siberian stone pine (*Pinus sibirica*) which is locally called the 'Siberian cedar' is common in the central and southern parts of the drainage basin. This pine has a multitude of uses but is particularly noted for delicious nuts which are gathered commercially. Larch and spruce along with pine species are more common in this type of the taiga.

In total, approximately 61% of the Baikal drainage basin is covered by forests, one third of which grow on slopes which exceed 15 degrees. The role of forests in the Baikal region should not be underestimated. Their functions within the Baikal drainage basin include water conservation and regulation, protection of soils from erosion, provision of timber and other resources, recreation, etc. Forest biomass accumulates pollutants and is a source of oxygen moderation and regulation of the overall gaseous balance which is disrupted by various anthropogenic sources.

Coniferous forests are more resistant to human impact than tundra ecosystems (Figure 1.4). Forests have a multi-tier structure including an underwood and a ground vegetation layer, which is normally composed of berry-bearing shrubs and lichens. Nevertheless, the vegetation of the taiga is vulnerable because of its low rate of metabolism and biological productivity due to cold temperatures. Additionally, the mountainous topography of the Baikal basin presents a potential threat to its ecosystems if excessive anthropogenic pressure is applied.

The **soil cover** in the north of the basin is represented primarily by mountain taiga soils or permafrost taiga soils. Soils are frequently oversaturated by moisture accumulating above the permafrost. Soil formation is slow, because of predominantly freezing temperatures and the tough nature of the waxy and resinous needles of the conifers. Calcium and other nutrient bases are leached by the various acids and form a podzolized horizon. A more compact illuviated sub-layer has enriched iron contents, thus creating an indurated iron 'pan', which inhibits plant root development and impedes soil drainage. In permafrost soils genetic soil horizons can be absent. Upheavals generated by seasonal alternation of freezing and thawing are quite common.

Plate 6   The typical light-coniferous taiga landscape

In the southern parts of the region peaty podzolic and dark grey forest soils alternate with chestnut type soils under forested steppe and steppe vegetation. The specific problems of soils in the Baikal region are due to the highly dissected terrain and predominance of high mountains and steep slopes. The soil cover is uneven and 'patchy' and erosivity risks are very high, making soils extremely vulnerable to any anthropogenic disturbance.

The uniqueness of Baikal is primarily attributed to the abundance of **organic life** and the degree of endemism in its aquatic ecosystem. Approximately three-quarters of the 2600 species of fauna and flora inhabiting the lake are endemic. There are 1085 species of algae, mostly diatoms and blue green and 450 species of plants (Baikal 1993). Animal life is particularly rich: over 1500 species and varieties are found in the lake and half of them are endemic.

It should be noted that the unrivalled purity of Lake Baikal's water is due to its filtration by the endemic *Gamerate* sponges and the *Epishura baikalensis* shrimp. This tiny creature also provides an oxygen supply for Baikal's water. *Epishura* cannot live anywhere else but inside the lake and it dies even if kept in Baikal's water in a laboratory retort. It is extremely sensitive to change in physico-chemical conditions and even more so to any amount of pollutants in water. The abundance of animal life in the bottom layers is exceptional but it greatly increases closer to the lake's surface.

There are also 53 species and subspecies of fish, including 27 endemic and 15 commercial ones. A unique 'golomyanka' fish (*Comephorus*) is scaleless and has no swimming bladder. It is viviparous and represents the most abundant fish species in Baikal. Although an average fish weight is only 15–25 g, their total biomass is about 160,000 tonnes, which is greater than the biomass of all other fish taken together (ibid.). 'Golomyanka' provides the main 'daily diet' for the Baikal seal. The ecosystem of the lake is non-uniform throughout the lake's aquatorium and highly dynamic. Fish stocks change due to cyclic alteration of natural conditions and anthropogenic factors. Contemporary fish productivity amounts to about 2.5 kg per ha (Gosudarstvenniy... 1996). 'Omul' is the main commercial species along with salmon, sturgeon, grayling, pike and less precious fish. In 1995 the 'omul' catch accounted for 3100 tonnes.

Baikal is the only place where the unique freshwater seal called 'nerpa' resides (Plate 7). It is the only mammal in the lake and the upper element in the lake's food chain. Morphologically and biologically nerpa is similar to seal species which inhabit the Arctic and the Russian Far Eastern seas. Adult animals reach 1.6–1.7 m in length and 130 kg in weight and are a particularly precious game for the local Buryat hunters. The total numbers of nerpa seals in Baikal are at present estimated at 82,000–114,800 (Gosudarstvenniy... 1996; Problemy... 1996, 1997). The latest available estimate was lower at 60–70,000 (*Zelyoniy Mir* 18–19, 1999).

Plate 7   The unique freshwater nerpa seals bathing in Baikal

3.2.4 *Overall vulnerability of landscapes*

On the whole, Lake Baikal and its drainage basin are characterized by a high vulnerability to human impact and a relatively low level of self-rehabilitative capacity. The aquatic ecosystems of this unique lake are particularly sensitive, because their evolution lasted for millions of years and they became adapted to a slow-changing environment with pristine conditions. Therefore they are reacting very intensely to even minor changes in natural conditions. In addition, endemic species cannot be reintroduced if they become extinct. To summarize, all direct and indirect factors which are responsible for a high overall vulnerability of the drainage basin and, ultimately the lake itself are as follows:

(a) a unique nature of the lake and subsequent fragility of the ecological balance in the lake's ecosystem itself;
(b) very high level of endemism in species evolved under conditions of pristine water quality, where disruption in one section of the food chain may cause destruction of other organisms;
(c) high degree of dependence of the lake's environment on the human activities ongoing within the whole catchment basin;
(d) lack of out-flowing rivers which could divert polluted water and already high levels of pollution in the Angara river;
(e) low assimilative capacity of rivers and the lake due to predominance of cold temperatures;

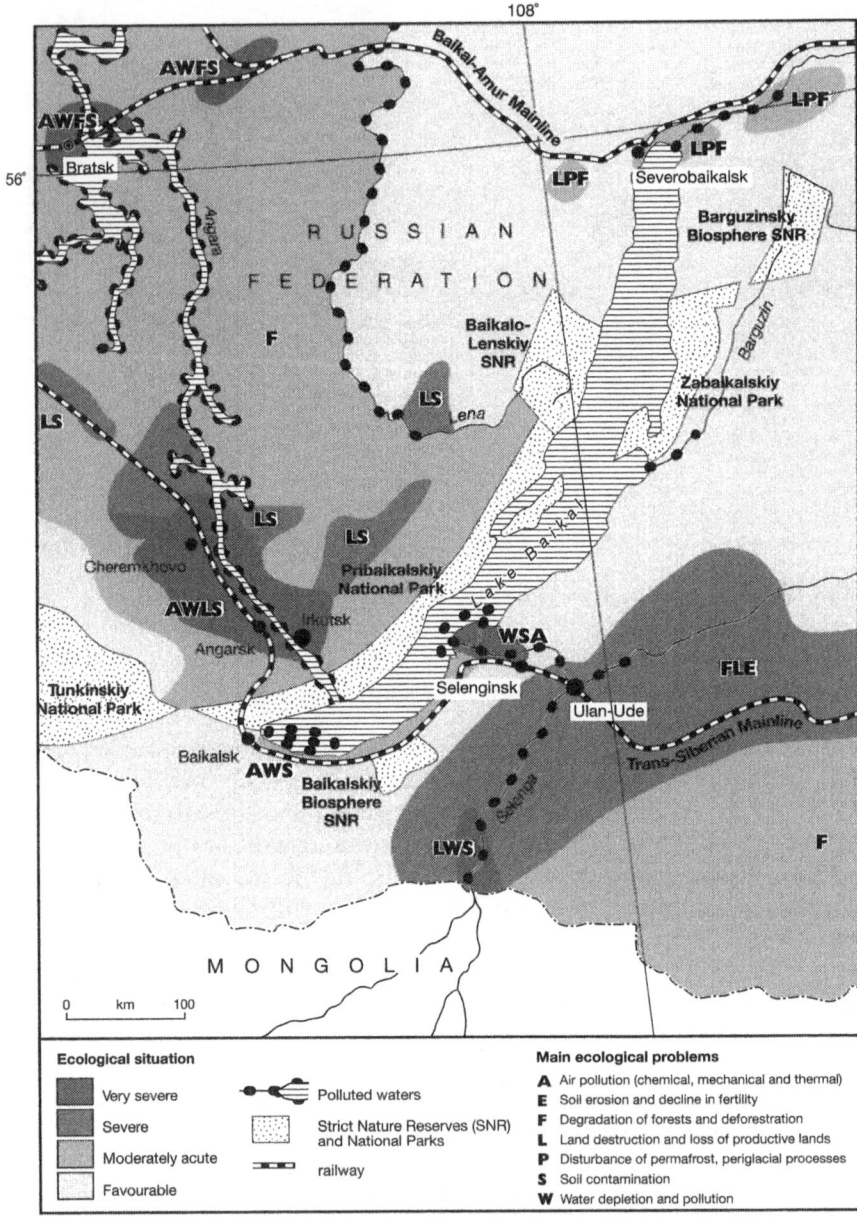

Figure 3.2 The present ecological situation in the Lake Baikal region (adapted from Ekologicheskaya... 1999)

(f) high erosion risks in soils on steep mountainous slopes;
(g) low rates of biological metabolism and productivity due to cold climate;
(h) high overall vulnerability of mountainous landscapes, particularly those located in the watershed areas;
(i) climatic predisposition to predominance of still air, particularly closer to the centre of the Baikal's valley;
(j) widespread permafrost;
(k) high seismic and mudflow risks.

A very important role in the protection of the Baikal's ecosystem is played by an extensive system of *protected areas* surrounding the lake (Figure 3.2). At present, it includes five state strict nature reserves: Baikalo-Lenskiy, Baikalskiy and Barguzinskiy biosphere reserves, Jerginskiy and, partially, Sokhondinskiy; three national parks: Pribaikalskiy, Tunkinskiy and Zabaikalskiy; and also 26 game reserves and 272 natural monuments. The most famous of these is Barguzin which was founded in 1916 to protect Barguzin sable. The total area of these protected territories amount to 4.1 million ha or about 13% of the drainage basin. This system represents half of the various natural land-scapes and ecosystems in the Baikal region and approximately 60–80% of the regional species biodiversity (Problemy... 1997).

3.3 Historical perspective and economic development

In 1954 the Soviet government adopted a decision to start the production of high-quality cellulose aimed for 'super-super' viscose cord used for tyres in supersonic aviation. Very high-quality water was needed for this purpose. After examining four options – lakes Baikal, Onezhskoye, Ladozhskoye and Teletskoye – a special governmental commission finally decided on the first option. Thus in 1959 the construction of the Baikalsk integrated pulp and paper mill (BPPM) started.

Unfortunately, a number of mistakes were already made during the planning and design stages. As Academician B. Laskorin admitted: 'We have made a range of mistakes in the construction of the BPPM. The main one – in scientific forecasting' (*Trud* 10 December 1992, 2). The target was to produce high-quality cellulose cord while in the West these were already being made from synthetic fibres.

Another set of mistakes was related to the location of the complex – first, there was no need for water of such a high quality and second, the quality of local timber was not adequate for the production of 'super cel-lulose'. The third mistake was in the technological design, as there should have been no illusion made about the quality of wastewater treat-ment. As N. Talyzin, the former head of Gosplan, the main Soviet planning body, said: 'Now it is clear even to a fool, that this cellulose plant should never have been constructed' (*Trud* 10 December 1992, 2). Moreover, aesthetic and ecological considerations were not sufficiently taken into account.

Initially BPPM aimed at an annual production of 100,000 tonnes of cellulose. In 1961 it was decided to increase the plant's capacity to 200,000 tonnes. Other amendments, such as increasing the expenditure on treatment facilities from 6.2 to 14.9 million roubles, were introduced before the construction was completed in 1966 (Problemy... 1995). These measures reflected the immediate concern of the general public about the safety of this unique lake which is considered sacred by the Buryat people inhabiting its eastern and southern coasts.

It should be stressed that there had been no precedent for such open opposition to a governmental decision in the whole of the former USSR before the Baikal case. It was the first time that the targets of industrial and military production had clashed with the need to protect a unique object of nature. Under increasing environmentalist pressures the plant was ultimately equipped with advanced water treatment facilities.

This unprecedented environmental campaign forced the Central Committee of the Communist Party of the Soviet Union and the Council of Ministers to adopt a decision in 1987 to cease cellulose production and retool BPPM by 1993, which was then put back to 1995. However, it is still operating and in 1996 the amount of discharged wastes was only slightly less than in 1986 at 55 million cubic m. It has been estimated (Gosudarstvenniy... 1997) that at present only 0.18% of all polluted wastewater is being treated to the standard level!

The Selenginsk integrated pulp and paper mill located downstream of the Selenga river started operation in the 1950s as a cellulose factory and in 1978 it adopted an antiquated purification system. In 1990 a closed water cycle operation was proposed to divert pollution from entering the lake. However, pollution still reached Baikal from this complex even although a 'sanitary protection zone' was established along the southern section of the lake (Pryde 1991).

One of the major projects of the former Soviet Union which had extensive environmental consequences was the construction of the Baikal-Amur Mainline (BAM). This was planned in the 1970s and it stretches for over 3500 km parallel to the Trans-Siberian railway, running very close to the northern coast of Lake Baikal, crossing the Trans-Baikal mountains, the Republic of Sakha and reaching the Russian Far East. The total volume of earthworks exceeded 220,000 cubic km which can be compared to the size of Mount Everest. Unlike many other Soviet projects, it initially took into account the need for ecological considerations and made provisions for the existing nature reserves located along its way. However, insufficient consideration for permafrost-related problems and for the potential threat to Baikal's ecology created a number of 'hot spots' along the route of this 'project of the century'.

3.4 Ecological consequences and recent changes

During the last 40 years Baikal has been threatened by the continued and complex impact of various anthropogenic activities. It has not only experienced direct pollution by industrial and communal water users, but has also been indirectly affected by various economic activities in different parts of its drainage basin (Figure 3.2). Therefore, it is important to apply the 'catchment approach' and examine the whole range of direct and indirect anthropogenic factors and sources of pollution operating throughout the region.

3.4.1 Water pollution

In 1996 175.3 million cubic metres of polluted wastewater was discharged into Baikal (Table 3.1). This amounted to a 64% increase compared to the official data for 1987 (Narodnoye... 1988). Approximately 23% of all discharged wastes underwent no treatment at all as the capacity of treatment facilities in the Baikal catchment basin is sufficient for processing only 38% of all generated wastewater (Problemy... 1997). However, an official source (Gosudarstvenniy... 1998a) suggested that the overall potential capacity of treatment facilities in the basin amounted to 240 million cubic m which exceeded the cited value of polluted wastes. At the same time it stated that, despite this, only 0.10% of all generated wastes are treated to a sufficient degree. This controversy can be probably explained by the distribution of processing installations, most of which are concentrated in Baikalsk while the Selenga, as will be shown later, is the primary source of polluted wastes. The total amount of pollutants discharged into the lake in 1996 was 55,200 tonnes. On average about 15 various pollutants reached Baikal – these include dioxins (organic chlorine compounds), sulphur compounds, phenols, etc.

It can be estimated from the data in Table 3.1 that in 1996 the share of polluted wastes from Buryat Republic which comprises the main part of the Selenga's basin was 58.6%, while 31.2% of pollution came from Irkutsk oblast where the Baikalsk PPM is located. The remaining 10.2% came from Chita oblast which incorporates the upper reaches of two tributaries of the Selenga. This proportion reflects a change in the structure of contributions from various pollution sources compared to the Soviet period.

The main sources of pollution within the **Selenga** catchment are Ulan-Ude, the capital of Buryat Republic, the Selenginsk pulp and cardboard complex (SPCC) and the Goosinoozersk industrial centre, along with numerous industrial enterprises and small towns along the Selenga, most of which lack treatment facilities and discharge oil products, sulphates, chlorides and phenols directly into the river. In 1987 their share in the overall pollution of Baikal was 21%, 12% and 3% accordingly, compared to 32% which was related to pollution from Baikalsk PPM

Table 3.1 Discharge of waste waters within Baikal basin, 1995–1996

Pollution index	Total in Baikal basin		Buryat Republic		Irkutsk oblast		Chita oblast	
	1995	1996	1995	1996	1995	1996	1995	1996
Total discharged								
waste waters, million cubic m	598.4	597.7	497.8	514.5	71.9	54.7	28.7	28.5
as % of Baikal basin	100.0	100.0	83.2	86.0	12.0	9.2	4.8	4.8
Total polluted in								
discharged, million cubic m	203.7	175.3	113.9	102.8	71.9	54.7	17.9	17.8
as % of all discharged	34.0	29.3	22.9	20.0	100.0	100.0	62.4	62.5
Including (a) untreated,								
million cubic m	40.2	40.6	25.0	25.8	0.5	0.1	14.7	14.7
as % of polluted	19.7	23.2	21.9	25.1	0.7	0.2	82.1	82.6
(b) inadequately treated,								
million cubic m	163.5	134.7	88.9	77.0	71.4	54.6	3.2	3.1

Source of data: Problemy... (1997)

(Narodnoye... 1988). In the late 1980s, Ulan-Ude annually dumped approximately 500 tonnes of nitrates into the river system. This volume accounted for 70% of the total nitrate load in Baikal (Mnatsakanian 1992). At the present time, the capital of Buryatia remains the major organic polluter of the lake.

At the same time it is clear that the overall contribution to Baikal's pollution from the Selenga has also grown compared to the Soviet period, despite the installation of new treatment facilities in Ulan-Ude and Selenginsk in the early 1990s. Thus the present state of the environment in the lake is determined by the chemical discharge of the Selenga. In 1997 the Selenga's flow carried into the lake approximately 400,000 tonnes of sulphates, 50,000 tonnes of chlorides and 320,000 tonnes of organic materials. The total amount of suspended matter has almost trebled compared to 1996 and reached about 700,000 tonnes (Rosgidromet 1998). The total amount of some pollutants discharged into the lake has also increased, for instance volatile phenols by 2.4 times and surfactants by 1.6 times.

It was recently recorded that the water in Selenga is highly polluted with oil products for the whole length of its watercourse from the boundary with Mongolia to the delta (Problemy... 1996). It has been revealed that there is a worsening trend in the environmental status of the Selenga's shoal waters. During the last 10 years frequencies of oxygen content lower than 10 mg/l increased from 5% to 30% and there is evidence of ammonium nitrate in the water. It has been estimated that currently the zone of active impact on bottom sediments and shallow shelf waters covers an area from 200 to 1500 sq km (Rosgidromet 1998).

The **Baikalsk** integrated pulp and paper mill is currently the second largest polluter of Baikal. However, some experts believe that it is still its main pollution source (*Zelyoniy Mir* 18–19, 1999). It appears that despite a reduction in its share in the overall polluted waste discharge, the mill is clearly the major contributor of toxic wastes, particularly harmful for Baikal's ecosystem. In 1992 this industrial complex had an effect on at least 200 sq km of the lake's surface (Baikal 1993). By 1997 the impact area increased to 220 sq km (Gosudarstvenniy 1998a). During its more than 30 years of operation a great amount of organic chlorine compounds, oil products, phenols and mercury accumulated in the impact area.

When arguing about the complex's impact, BPPM's director often states that it has the country's best treatment facilities. Indeed, by 1986 BPPM had an annual capacity to treat 98 million cubic m of wastewater and had installed advanced systems of closed and consequent water supply for over 220 million cubic m. Nevertheless, even from officially published data, it appears that the amount of insufficiently treated wastewater discharged into Baikal in 1986 still accounted for 58 million cubic m (Narodnoye... 1987). In 1991 over 230 million cubic m of wastewater, including 168 million cubic m of polluted wastes, were discharged into the lake (Gosudarstvenniy... 1992).

During the post-Soviet period the total amount of pollutants discharged by PPM has been gradually decreasing with the overall cellulose production decline and in 1996 it accounted for 55 million cubic m (Gosudarstvenniy... 1997; Problemy... 1995, 1996, 1997). However, this reduction was not always continuous, as is illustrated in Table 3.2. It shows that between 1994 and 1995, both cellulose production and the amount of many pollutants discharged into Baikal increased. Moreover, while the value of growth in cellulose production reached 125%, increase in the amount of discharged oil products and sulphates amounted to 156% and 168% accordingly. In addition, despite a decrease in the total cellulose production a year later in 1996, the discharge of sulphates still exceeded the 1994 level. There is an opinion that the average amount of wastewater generated by this combine in the 1990s can be compared to the volume of communal wastes from a city with a population of 500,000 people (*Izvestiya* 26 December 1995). Another threatening fact is that approximately 17,500 cubic m of polluted ground waters reach Lake Baikal every day from the industrial estate which covers 6.5 km of Baikal's shoreline (*Izvestiya* 12 January 1999).

Some of the problems associated with the construction of the **Baikal-Amur Mainline** (BAM) included soil compaction, erosion and solifluction that, in turn have caused irreversible damage to terrestrial ecosystems (Bridges and Bridges 1996). In addition, the increase in the surface runoff near the northern coast of Baikal has brought about extensive eutrophication and the silting up of the spawning grounds of precious Baikal fish species. Oil products and other chemicals used or transported along the railway were washed off directly into the lake affecting fish development and poisoning nerpa seals and birds (Pryde 1991).

Table 3.2 Cellulose production and pollutants discharged by Baikalsk PPM, 1994–1996 (000 tonnes)

Index	1994	1995	1995 as % of 1994	1996	1996 as % of 1994
Total cellulose production	147,798.0	184,963.0	125.1	130,969.0	88.6
Suspended particles	214.0	210.0	98.1	177.0	82.7
Easily oxidised organic compounds	531.0	581.0	109.4	444.0	83.6
Oil products	2,844.0	3,570.0	125.5	2,232.0	78.5
Phenols	1.3	1.2	92.1	1.0	76.4
Sulphates	6,668.0	11,177.0	167.6	8,791.0	131.8
Chlorides	6,222.0	7,208.0	115.8	4,634.0	74.5

Sources of data: Problemy... (1996, 1997)

Another kind of impact was linked with the creation of new **settlements** associated with the new railway. From around 50 of them built as temporary settlements to provide housing for the BAM workers, two, Severobaikalsk and Nizhneangarsk, were established close to the upper section of the lake and have caused considerable damage to Baikal ever since their construction. No water treatment facilities have been introduced at the construction stage and communal wastes were discharged directly into the lake. Mote noted that '... the Baikal'sk Pulp and Paper mill was equipped with a three-stage water pollution abatement system and still had problems ... But the by-products being generated in Severobaikal'sk and other BAM settlements in the Lake Baikal watershed are released without even the most primitive pollution control...' (Mote 1992, 47). Many boiler houses in Severobaikal'sk have no or insufficient filtration systems and air pollution has become a major health hazard and the cause of acidic deposition in the lake. However in recent years a few measures have been introduced in Severobaikal'sk, including the installation of a new sewerage works to treat effluents and the construction of scrubbers on the existing and new boiler houses.

The transportation of timber across the lake had caused serious environmental problems. Before a special regulation in 1975, logs were transported individually which resulted in 50% of wood being lost (Bridges and Bridges 1996). Degradation of the sunken organic matter by anaerobic bacteria led to estimated 250,000 tonnes of the lake's oxygen being lost annually. Since 1975 all logs have had to be transported in rafts or by ship. In 1993 a law was passed which banned timber rafting and oil transportation across Baikal. Nevertheless, in reality this practice continues up to the present time.

Timber rafting is the main log transportation method used throughout Siberia and on all tributaries within the Baikal basin. The effects of rafting include sinking of approximately 10% of all rafted timber; destruction of fish spawning grounds in the lower reaches of Baikal's tributaries. Between 1958 and 1968 an estimated 1.5 million cubic m of timber sank leading to the loss of 3500 km^2 of fish spawning area. This was followed by putrefaction of water with phenols from lignin degradation and underdevelopment of omul and its roe.

Another negative factor for the ecosystem's well-being during the last decade was related to a **change in the hydrological regime** of the lake, i.e. an increase in the lake's level associated with damming by the Irkutsk hydroelectric power plant. Construction of this power plant was completed in 1959 and had a tremendous impact on the hydrological regime of the Angara and Baikal. It had caused an increase in the lake's level and a large-scale intensification of coastal erosion processes not only along the western but also along the eastern coasts of the lake, as well as waterlogging and flooding of agricultural lands. During the last decade which was relatively 'wet' the problems of coastal destruction and flooding have increased in magnitude (Problemy... 1995, 1996).

3.4.2 Air pollution

One of the major threats to Lake Baikal comes in the form of air pollution through direct deposition of pollutants over its aquatic surface, in tributaries and acid precipitation within the whole catchment. It is estimated, that the overall area under the direct impact of anthropogenic air pollution within the Baikal basin covers nearly 6000 sq km. An important source of atmospheric pollution is an intra-regional air transfer particularly from the Bratsk and Irkutsk-Angarsk industrial centres (Figure 3.2). The latter affects approximately 3000 sq km. Another major source is Ulan-Ude which has an impact on one thousand square kilometres. In 1987 only 49% of all harmful pollutants from Selenginsk were intercepted (Narodnoye... 1988). All the above factories are also partly responsible for pollution of the lake through wet deposition of combustion gases containing fluorine and sulphates. The annual deposition from all sources with snow within the aquatic area of Baikal is estimated at 11,000 tonnes of sulphur and 2400 tonnes of nitrogen (Rosgidromet 1998). The maximum loads of deposition are found in the southern part of the lake as far north as the Selenga mouth.

In almost all industrial cities around Baikal pollutant levels exceed maximum permissible concentrations. In 1996 Irkutsk, Angarsk, Cheremkhovo and Ulan-Ude were included in the list of the most polluted Russian towns (Gosudarstvenniy... 1997). Almost every year during the post-Soviet period Bratsk was included in the list of Russian cities with the greatest level of air pollution (Gosudarstvenniy... 1992–1997). Although the overall amount of pollutants emitted from stationary sources in Angarsk and Bratsk declined between 1992 and 1998, as can be seen from Table 2.2, the value of Angarsk emissions in 1998 was greater than from such a large industrial city as Moscow. In 1997, recorded maximum concentrations of benzopyrene in Irkutsk reached almost 18 MPC and were even higher in Ulan-Ude at 22 MPC (Rosgidromet 1998). This was clearly due to an increased share of air pollution from transport sources evident throughout Russia and the former Soviet Union.

In 1997 Baikalsk had increased levels of methyl mercaptan, maximum concentrations of which in Baikalsk reached 110 MPC. This value can be compared with 42 MPC in 1995 (Problemy... 1996). A hydro-chemical survey of the snow cover around BPPM made in 1996–1997 has revealed that the zone of severe impact affecting the state of adjacent forests has increased from 150 sq km in 1996 to 220 sq km in 1997 while precipitation of atmospheric pollutants amounted to 30 tonnes per sq km (Gosudarstvenniy... 1998a). The total area within the influence of Baikalsk pulp and paper complex is currently 400 sq km.

3.4.3 Other changes within the Baikal basin

Soil contamination is a serious problem in industrial centres. Very severe pollution with heavy metals is recorded within a radius of five kilometres from Irkutsk e.g. in 1995 lead concentrations exceeded 10 MPC (Gosudarstvenniy... 1996). A similar situation is found around the city of Bratsk. In the same year mean concentrations of fluoride around Irkutsk and Bratsk aluminum plants reached 13–19 MPC, while maximum values ranged from 58 to 158 MPC. An immense underground accumulation of oil due to infiltration is located in Angarsk city.

Excessive woodcutting in the rivers' highly dissected watersheds and Baikal's catchment basin as a whole generate increased surface runoff and soil erosion, siltation and pollution of water bodies. With the transition to the market economy the number of illegal operations in the region have escalated. Although the overall amount of official harvesting has reduced during the post-Soviet period, the lack of control in timber industry and prevalence of clear cutting practices have accelerated soil erosion in many parts of the drainage basin. Out of 4700 control checks made in 1996 by regional forestry inspections 970 involved violation of forestry regulations or illegal wood cutting (Problemy... 1997). It was noted in the State Report on the Status of the Environment in 1992 that clearcutting practices in Russia amount to 90% of all cutting practices (Gosudarstvenniy... 1993). Forest fires due to human negligence present an additional threat to the catchment forests.

Around BPPM, air pollution has affected the overall area of 2000 sq km which includes the impact zone of pollutants deposition. As a result of this, 250,000 ha of forests in the Baikal basin have dry canopies and in an area of 35,000 ha the fir trees have completely dried up (*Zelyoniy Mir* 18–19, 1999). Disappearance of small rivers is also linked with excessive wood cutting in the upper reaches of Baikal's tributaries. Grigoriy Galaziy, a world renowned expert on Lake Baikal's problems, cited in Yanshin and Melua (1991), noted that while for millions of years of the lake's existence all 336 tributaries of the lake remained intact, almost 150 rivers ran dry or became close to extinction due to unwise human actions over the last 25–30 years.

Additionally, significant areas of agricultural lands within the Baikal's water-protection zone are subjected to processes of wind and water erosion (Problemy... 1995). It has also been noted that between 1993 and 1995 the share of moderately and severely eroded lands had increased by 10% (Problemy... 1996). The fertility of lands is declining due to insufficient use of mineral and organic fertilizers in the naturally nitrogen-deficient soils, along with poor management practices typical for the whole agrarian sector during the current transitional period. The changes in the land tenure system and the creation of private farms did not result in the improvement of land use practices because of the poor physical status of lands, a short-term profit-orientated approach and an overall lack of financial investments. In addition, agriculture is a major source of pesticides including DDT which tend to accumulate in the highest trophic levels of Baikal's aquatic system.

Increased anthropogenic pressure also affects the wildlife of the Baikal taiga forests – 60 species of animals and 150 of birds. The major threat relates to all commercially hunted animals. During the post-Soviet period the total numbers of reindeer have decreased by 16%, sable by 21%, elk by 33%, wild boar by 62% and bear by 44% (Problemy... 1997). The Soviet monopoly on fur-producing animals gave way to licensed hunting and private sales, which in many cases led to increased all-year-round pressure on some animal species e.g. musk deer for their gland. A major threat to the system of protected areas around the lake has come in the form of pollution within the coastal zone, illegal wood cutting, grazing of domestic animals, poaching, and uncontrolled tourism. Due to the reduction in state financing, weakened control of the protected regime presents an additional challenge.

An additional stress on the lake's drainage basin came with the dissolution of the former USSR and the transition to the market economy in the early 1990s when many former government regulations and environmental acts ceased to be implemented (Problemy... 1995). Also as a result of the reduction in funding of environmental conservation activities strict nature reserves and national parks adjacent to the lake have become the victims of poaching precious animals like sable, lynx and nerpa and over fishing of omul. It was estimated, that the reduction in the animal population of sable reached 30% of the 1975 levels (Bridges and Bridges 1996).

3.4.4 Impact on the lake's water quality and aquatic life

As a result of all above stated causes the overall **water quality** in many parts of the lake has worsened. For example, in 1994 increased concentrations of oil products were recorded throughout Baikal's open waters (Problemy... 1995). In 1996 concentration of oil products in some parts of the bottom layers reached 6.2 MPC, phenols 7 MPC and copper 9 MPC (Problemy... 1997). Despite an installation of advanced water treatment systems in the 1980s, in 1997 it was still considered that the state of water quality around the Baikalsk complex has not improved (Rosgidromet 1998). However, in the central deep-water part of Baikal water quality still remains high particularly when compared with other regions of Russia or West Europe.

A water quality monitoring survey carried out within a 35 km area from the Baikalsk PPM has revealed a worsening situation from 1995 to

1996 despite the fact that cellulose production during this period fell by 29% and the total volume of wastewater declined by 24% (Problemy... 1997). There was a reduced oxygen content and an increase in the area polluted by non-sulphate sulphur in 1996 compared to 1995. Hydrochemical analysis has shown accumulation of hard-hydrolyzing carbohydrates and humus-lignin compounds in the bottom layers which have a clear impact on the aquatic systems.

The influence of water pollution on the **aquatic life** in Baikal is greatest in the vicinity of the Selenga delta and the Baikalsk PPM. A trend of declining water quality in the shoal waters of the Selenga littoral during the last decade was also reflected in the reduction of the population numbers and biomass of zooplankton and in the predominance of the *Oligochaetae* class which can survive in polluted water in the bottom communities.

The overall hydro-biological situation has worsened and a survey of the impact area's size for zooplankton has shown a 1.5 increase compared to 1995. Wastewaters clearly have a stimulating (eutrophicating) impact on phytoplankton and a depressing impact on zooplankton. The average numbers of epishura in the polluted zone was 2–4 times lower in 1996 than in 1995. More recent studies revealed that epishura dies even in 'adequately treated' wastes after dilution by 100 times with Baikal water (*Zelyoniy Mir* 18–19, 1999). As a result of their deaths, the annual 'productivity' of Baikal as a producer of pure water decreased by 4.5 cubic km. Eutrophication was also reflected in the wide spread of the water thyme (*Elodea*) algae in the coastal waters of Baikal which started in the late 1970s (Problemy... 1994).

There have also been changes in benthos biodiversity. During the last decade only 10 species of zoobenthos remained out of 27 in the vicinity of BPPM and the total biomass has been reduced by three times (Problemy... 1997). The data from the Baikalsk Limnological Institute showed that the area of bottom pollution zone increased in 1998 by 25% compared with the previous year. According to Academician A. Yanshin, up to 90% of all species which used to inhabit the lake's bottom within 20 km from the complex have already disappeared, while the remaining few are rarely seen and have a low biomass (*Zelyoniy Mir* 18–19, 1999). It was also noticed that the anthropogenic impact on the Baikal ecosystem was reflected in the depressed status of the omul fish and in the decline of the pelagic omul reserves (Problemy... 1994–1997). It was recorded that the total numbers of mature reproductive omul spawning in Baikal's tributaries declined between 1969 and 1980 from 4.7 to 3.7 million; their average weight and fertility decreased along with the amount of emitted roe.

In the mid-1970s a trend in the reduction in weight and reproductive functions of the unique nerpa seal was first recorded (Problemy... 1994). In 1987–1988 an outbreak of viral distemper caused the population of nerpa seal to drop by 5–10,000 (Problemy... 1995). More recently, in late May and early June 1999 about 150 dead seals were found along the southern coast of Baikal (*The Moscow Times* 24 August 1999). Scientists disagree about what caused these deaths – pollution, virus or hunters. However, the deaths could at the least be linked to stress and the depressed state of immune system which make animals vulnerable to pollution effects (Grachev *et al*. 1989).

Recently it was also suggested in the mass media that the content of dioxins in the nerpa seal's fat is dangerous for human health (Problemy... 1995) and exceeds MPC for drinking water by 20,000 times (*Izvestiya* 26 December 1995)! This can be explained by the fact that up to 90% of their diet comes from 'golomyanka' fish in which concentrations of DDT and dyphenyl polychlorines reach 75 to 125 mg per kg of weight. In the nerpas' fat DDT concentration reached 54–62 mg per kg of weight (Problemy... 1994). It should be noted that the use of these toxic chemicals which become rapidly accumulated throughout the food chain is still widely spread in the agricultural practices throughout the lake's basin. Seals, in their turn, are commercially hunted by the top element of the food chain – humans. For example, in 1996, 3234 seals were killed by hunters (Problemy... 1997). A recent study also revealed that some water-fowl bird species, for example mallard and marsh sandpiper, are heavily contaminated with polycyclic aromatic hydrocarbons, phenols and organochlorines which come from their diet of aquatic non-vertebrates (Lebedev *et al*. 1998).

The transition to the market economy has brought about new environ-mental threats. Poaching has become an increasing threat to the ecosystem of Lake Baikal. Owing to its large-scale migratory pattern of movement omul is an easy target for poachers. What is even more worry-ing is the fate of such precious fish species such as the Baikal sturgeon which is included in the Red Data book. During 1988–1990 population of this fish was estimated at 14,000. A special fish farm was constructed in the Lower Selenga for restoration of the sturgeon population. However, while in the above period on average approximately 100 specimens were at reproductive age, all producer specimens caught between 1991 and 1995 were found immature for reproduction (Problemy... 1997).

Although no systematic studies of the anthropogenic impact on fish resources were carried out, individual surveys confirm the existence of a direct link between the two. It appears that during the last decade there is a clear tendency of a decline in the omul roe survival rates in the spawning grounds affected by polluted wastewater. The current critical situation within the Selenga river basin may result in the loss of genetic pool of the Selenga omul species and a decline in the fish-producing value of the river. A similar situation is found on the Angara river spawn-ing grounds (Problemy... 1996).

3.4.5 Overall ecological situation

The change in the ecological situation over the Soviet and post-Soviet periods is summed up in Table 3.3. From the above discussion it is clear that despite an economic recession and a decline in industrial production, the ecological situation in the Baikal region remains in the crisis stage. The situation is not uniform either within the lake or within its drainage basin. While the quality of the lake's water in deeper central parts is still considered very high, there are two areas of polluted water – one surrounding the Selenga's delta and another around Baikalsk.

The overall ecological situation in the Lake Baikal region is shown in Figure 3.2. There are four areas of very severe status: the greatest area by size is located in the industrial region along the Angara river extending from Cheremkhovo to Irkutsk. A large area in the south-east of the region around Ulan-Ude has a severe ecological situation as well as the northern regions adjacent to the Cheremkhovo–Irkutsk industrial region. The latter is further surrounded by a vast area of moderately acute or critical situation which is also typical of the northern and southern coasts of Baikal and the remaining part of the Selenga catchment basin. Additionally, areas around Bratsk have an environment in the severe state. The scope of ecological problems include deforestation, land destruction and loss of productive lands throughout the region, disturbance of permafrost in its northern part and air and water pollution around industrial centres.

It should be realized that, due to the unique character of Lake Baikal and the existence of a large number of endemic species, its irreversibility threshold is significantly lower than in other aquatic systems. Relict species evolved in the conditions of millions of years of isolation and pristine water quality. Standards for water quality appropriate for ordinary water bodies are not adaptable to Baikal because if an endemic species became extinct (a) it cannot be reintroduced, and (b) its disappearance will start an adverse chain reaction in the whole ecosystem.

An additional *potential* for a major catastrophe in the southern part of Baikal is related to its high seismic risk, because it has been estimated that during the operation period of BPPM approximately 3.5 million tonnes of lignin chloride concentrate were accumulated in the storage ponds located only 200 m from the lake's shore (*Izvestiya* 26 December 1995). If any major earthquake occurs in the vicinity of these poisoned reservoirs all waste concentrates will find their way directly into the lake. A recent comparison suggests that the degree of Baikal's pollution with dyphenyl polychlorines and dioxins is similar to a very polluted Baltic Sea (*Zelyoniy Mir* 24, 1998).

Table 3.3 The legacy of economic development, ecological change and environmental protection in Baikal region

Period	Legacy of economic development and protection measures	Ecological situation	Stage in ecological degeneration
Before 1917	Construction of Trans-Siberian railway mainline and settlements along the southern shore; industrial activities limited to Irkutsk oblast; beginning of industrial coal production and forestry activities in Baikal basin; extensive sable hunting; fishing; creation of Barguzin game reserve in 1916 to protect sable	Limited local impact of pollution and forestry activities in drainage basin; natural regrowth in wood cutting areas remains satisfactory; sable endangered overall unspoilt water quality in Baikal	Satisfactory state
1918–1945	Increase in industrialization, in part, due to evacuation of industrial enterprises from the European part of the FSU; expansion of mining activities in Irkutsk-Cheremkhovo region, partially based on forced labour; rural and urban development throughout basin; increasing agricultural pressures; growth of timber production	Locally significant impact e.g. from coal mining; water pollution near Ulan-Ude and air pollution near Irkutsk; limited erosion in agricultural lands; local destruction of catchment forests; beginning of Baikal's local water pollution	Strained situation
1946–1959	Intensification of industrial development: construction of Irkutsk HEP (1956), Ulan-Ude industrial complex; cultivation of 'virgin lands' in Buryatia; increased timber production throughout basin; increase in fisheries; decision adopted on construction of BPPM (1954) and SPCC (1959)	Pollution in Selenga delta by untreated industrial and domestic wastes; increasing air pollution; water and wind erosion in farmlands and clear-cut forest sections; signs of desiccation of upstream rivers; reduction in omul catch	Critical situation
1960–1969	Further industrial expansion: construction of Bratsk HEP (1962); start of operation at BPPM (1966); expansion of cultivated lands area by 2 times between 1950–1970; 3 times increase in sheep stock; authorization of timber-rafting across Baikal (1963); start of public campaign for	Beginning of biological and chemical pollution from BPPM, SPCC & Ulan-Ude complex; substantial air pollution in Irkutsk region; erosion on farmland on 200,000 ha; forest damage and erosion from	Critical situation

Table 3.3 Continued

	Baikal's protection (1959), first recommendations on protected regime (1965); of integrated use programme (1966); establishment of Baikalskiy SNR (1969); resolution on protection and sound use (1969)	clearcutting practices; increase in wastes from Ulan-Ude; start of eutrophication in Baikal's shallow waters; decline in pelagic omul reproductive functions	Ecological crisis
1970–1986	Intensification of economic development: 30-year programme to increase industrial production, urbanization, transport, forestry and agriculture; construction of BAM (1974–1984): Severobaikalsk – new threat to northern Baikal; decisions to restore fish stocks, introduce closed water circuit at SPCC; 'Temporary protection regulations' (1973); decision to cease polluted discharges by 1985 and abate air pollution; large-scale multidisciplinary research; establishment of Baikalo-Lenskiy SNR (1986)	Expansion of air and water pollution in basin; start of lake pollution around Severobaikalsk (126 sq km) increase in lake's bottom pollution: at BPPM (3 sq km) and around Selenga delta; desiccation of dozens of small rivers in basin; land degradation along BAM; acceleration of erosion: 600,000 ha of farmland; increase in agricultural pollution by 2.5–3 times; changes in Baikal's aquatic system; accumulation of pesticides in nerpa fat	Ecological crisis
1987 up to present time	Uncontrolled growth of population in Severobaikalsk and Ulan-Ude; decline in industrial production after 1991; weakening of state controls with transition to free market economy; increased poaching; introduction of advanced treatment system at both plants; decision to retool BPPM by 1993; resolution on protection and rational use for 1987–1995; status of 'World Heritage Site' (1996); adoption of 'Law on Protection of Baikal' (1999)	Increase in share of pollution from Selenga; rise in airborne pollution; decline in water quality around BPPM due to lack of treatment despite decrease in pollutants in 1988–89 and after 1991; desiccation of many small rivers; shifting sands on 100,000 ha; erosion on 650,000 ha of farm lands; various adverse changes in aquatic life, e.g. decline in pelagic omul reserves	Ecological crisis

3.5 Socio-economic and health issues

The general downfall trend in the environmental situation in the Baikal region inevitably tells on the quality of drinking water, sanitary conditions and human health. Over 30% of the population in the Baikal region lives in towns and settlements in which concentration of pollutants in the atmosphere permanently exceeds maximum permissible levels. The worst health situation is in Ulan-Ude where illness frequencies in children are almost two times greater than the average level in the Buryat Republic, while frequencies of newborn abnormalities are 2.6 times. In adults, frequencies of respiratory diseases are 2.7 times and blood diseases 5.2 times higher. The level of bone and muscle pathology in Bratsk is 9 times greater than the Russian average (Gosudarstvenniy... 1996). This is due to an extremely high volume of industrial emissions of fluoride in this permanently polluted city.

Only 75 out of 658 settlements in Buryatiya are at least partially provided with centralized water supply systems. Quality of drinking water is often inadequate due to the widespread pollution of surface and ground waters. Infectious diseases are up to 1.9 times more frequent in the Selenga region than on average in the Russian Federation. Death rates from oncological diseases in the Baikalsk region are 1.7–2.7 times greater than the background level (Problemy... 1997).

Similarly to other regions of the Russian Federation, there has been clear evidence of adverse demographic trends from the beginning of the post-Soviet period. In the Buryat Republic, Irkutsk and Chita oblasts, birth rates halved between 1985 and the late 1990s. Negative natural growth was first recorded in 1993 but the values of decline were slightly less than the Russian average (Rossiyskiy... 1998). This tendency has reflected both the decline in the standards of living due to the socio-economic crisis in the country and in the sanitary and environmental conditions in the region.

3.6 Prospects for the future

As Table 3.3 illustrates, over the last three decades numerous protective measures were being elaborated but very few actually implemented. Still, over the last decade or so, there have been a number of positive developments. Probably the main one was that no new giant industrial projects have been attempted in the Lake Baikal basin. Other developments include

(a) the elaboration of 'The territorial scheme of integrated utilization of natural resources of Lake Baikal' (TERSKOP) in 1988–1989;
(b) official banning of logging operations within the coastal zone and timber rafting across the lake;

(c) the introduction of strict regulations for clearcutting practices, only allowing them on slopes up to 15 degrees and allowing selective harvesting only on slopes up to 25 degrees;

(d) an introduction of an advanced four-stage waste treatment system at the Baikalsk PPM;

(e) the establishing of two nature reserves and three national parks;

(f) the opening of the International Baikal Centre for Ecological Studies: 20–30 expeditions are carried out annually;

(g) the inclusion in December 1996 of Lake Baikal in the UNESCO list of World Heritage Sites.

The main problem remains the operation of the Baikalsk pulp and paper mill. In 1987 the government adopted a decision to retool the BPPM by 1 January 1993. The dissolution of the FSU and transition to the market economy have halted these plans. An important factor is that the plant has been transformed into a share-holding society, thus effectively moving away from the direct control of the state. The main problem with the closure now is that half of all labour active population in the city of Baikalsk or 4500 people are employed either at the mill or in related activities and are dependent on it for their well-being.

Between February and July 1995 eight experts from the United Nations Industrial Development Organization (UNIDO) worked on recommendations for the retooling of the Baikalsk pulp and paper mill. Their estimates for the total cost of the plant's closure at the rates of 120,000 tonnes cellulose output were estimated at about US$600million (*Rossiyskaya Gazeta* 17 October 1995). This cost included export losses, as at present 56% of all produced cellulose is exported, dismantling and recultivation costs and expenses for support for the unemployed.

The conclusion of the UNIDO experts was that the major social cost would involve massive unemployment. Furthermore, the current share of BPPM in the municipal and regional budget is estimated at 80–95%. The plant is involved in the provision of heat, electric energy and waste treatment for the city. Thus after the plant's closure, for two years domestic wastes would be discharged into Baikal without any treatment. Taking the above into consideration, UNIDO came forth with a recommendation for modernizing the complex with an objective to replace chlorine bleaching of cellulose by ozone bleaching. The total cost of this measure was estimated at US$150 million. An introduction of a closed water circuit system at the plant would cost much more. However, both projects seem an unrealistic target under the existing conditions of economic recession.

A major positive development includes elaboration and adoption of the 'Integrated Federal Programme on the protection of Lake Baikal and ecologically sound utilization of its basin's resources' designed for 1995–2000. The structure of financial sources for its implementation is given in Table 3.4. One-third of all costs was to be covered from the

Table 3.4 The structure of financial sources of the Federal programme on the protection of Baikal, 1995–1996

Financial source	Expenditures (million roubles)*			% of total
	Total	1995	1996	
Total	1,641,048	355,051	329,576	100
Including:				
Federal budget	531,951	121,890	89,520	32.4
Regional budgets total	46,397			2.8
Republic of Buryatia	28,659	8,687	8,687	1.7
Irkutsk oblast	6,228	1,863	1,863	0.4
Chita oblast	11,450	3,455	3,455	0.7
Ecological funds	12,160	2,251	2,241	0.8
Enterprise budgets	809,500	138,745	138,755	49.3
Credit sources	241,040	70,260	80,260	14.7

Source of data: Problemy... (1995, 1996)
* at 1994 prices

Federal budget, while half of the needed expenditures had to be allocated from budgets of enterprises operating in the region. However, the actual expenses were much lower: for example, in 1995 less than one quarter of estimated federal investments were actually spent on the Baikal programme needs (Gosudarstvenniy... 1996).

All the above-mentioned positive developments still failed to stop pollution and environmental damage in the Lake Baikal basin. As Valentin Rasputin, a famous Russian writer stated: 'Communism was not kind to Baikal. Today the attitude is also indifferent' (*Rossiyskaya Gazeta* 10 September 1996). In December 1998 the Irkutsk Regional Committee on the Environment sued BPPM with the aim of having it shut down. It had estimated that in 1997 alone the ecological damage from BPPM amounted to 3.9 billion roubles; this was significantly greater than the cost of the combine's produce in that year (*Izvestiya* 12 January 1999).

Arguably, the most important development was the preparation of the Federal Law 'On the Protection of Lake Baikal' which was approved by both chambers of the Parliament on 24 June 1997, but it was declined by the President on 21 July 1997 and sent back for amendments. Clearly, the main reason for this was the lack of financial means for its implementation. Only very recently, on 1 May 1999, it was finally adopted and enforced. Current hopes for the future of Baikal can only be linked with the implementation of the Law on the Protection of Baikal. The main points raised by this document are the following:

- The law determines not only the legal, but also the economic and organizational basis for the protection of Baikal.
- Three ecological zones are established within the basin, each having its own regime of natural resource use and environmental protection: the 'Central zone', the 'Buffer zone' and the 'Zone of atmospheric impact'.
- In the 'Central zone', which includes all the aquatic system of the lake, its water conservation zone and also protected territories adjacent to Baikal, it is forbidden to: locate enterprises of the atomic industry; construct storage reservoirs of radioactive and toxic wastes; carry out explosive works in the aquatic system of Baikal; locate new enterprises pertaining to cellulose, petrochemical, metallurgical, organic synthesis or other toxic industries; locate animal farm complexes; construct main railway lines; carry out large-scale wood cutting operations; rafting of logs.
- Similar regulations are envisaged for the 'Buffer zone' which includes all the remaining part of the drainage basin and for the 'Zone of atmospheric impact' which incorporates the Russian territory 200 km to the west and northwest of the drainage basin.
- A special water regime for the lake is established which cannot be violated e.g. by changes in existing plans of hydraulic energy construction or new projects.
- Commercial fishing and hunting is allowed only on the basis of data from environmental monitoring.
- In the whole drainage basin of Baikal clearcutting practices are forbidden as well as any cutting of the Siberian pine and rare tree species, wood cutting on permafrost and slopes over 10 degrees, etc.
- Special stricter norms for pollutant emissions, discharges and other forms of impact are introduced.
- A 'Commission on the Baikal' is issued a status of a permanent state organ attached to the government of the Russian Federation.
- International co-operation involves protection of the Mongolian part of the drainage basin under the interstate agreement with this country.

It appears that if all the above regulations are actually enforced, there are real chances that Baikal can be still saved and the continuing environmental crisis can be overcome. However, as the first issue of the 'Baikal'skiy Region' bulletin stated: 'The economic profit always took over the interests of the unique lake; dozens of government resolutions of the USSR and Russia remained unrealized' (Baikal'skiy... 1998, 6). As long as this practice inherited from the past persists, the unique wealth in global terms of the endemic life of Lake Baikal will remain threatened.

CHAPTER 4

The Moscow region: complex problems of Europe's major conurbation

In prospect, in the 21st century the majority of mankind will be living in megalopolises, because a city, even the biggest one is the result of self-organisation in society.

(N. Moiseev, *Zelyoniy Mir* 18, 1997, 2)

4.1 Introduction

The objective of this chapter is to demonstrate a unique complex of ecological problems associated with major conurbations by taking the example of the Moscow region, which in the context of this chapter includes the city of Moscow (in Russian called 'Moskva') and the surrounding territory known as Moscow (Moskovskaya) oblast [province]. The author also attempts to evaluate some specific features of this city and examine a feedback mechanism resulting from very close interdependence between humans and their environment existing within an urban conurbation.

Moscow is one of the world's largest megalopolises and Europe's biggest capital by population. The total area of Moscow city accounts for 994 sq km including 878.7 sq km within the boundary of the orbital motorway called MKAD (Russian abbreviation for Moscow orbital automobile highway). The total built-up area is 531 sq km. The population of Moscow as at 1 January 1999 was 8,630,000 people (Chislennost'... 1999). The combined territory of the city and oblast amount to a massive 47,000 sq km which is comparable to Luxembourg and the Netherlands taken together. This area accommodates over 15 million people. It is very difficult to compare the area of Moscow with other major world cities, because in most cases the precise city boundaries either correspond to a relatively small area, e.g. Paris, or to a much larger area of territory e.g. Tokyo. London, like Moscow, is an example of a clearly defined city with an area 1.6 times greater than Moscow.

Moscow is the centre of various political, economic, scientific and cultural activities of Russia. But at the same time it is one of the few world's capitals with intensive industrial production based on a concentration of enterprises pertaining to heavy industry, mechanical engineering and

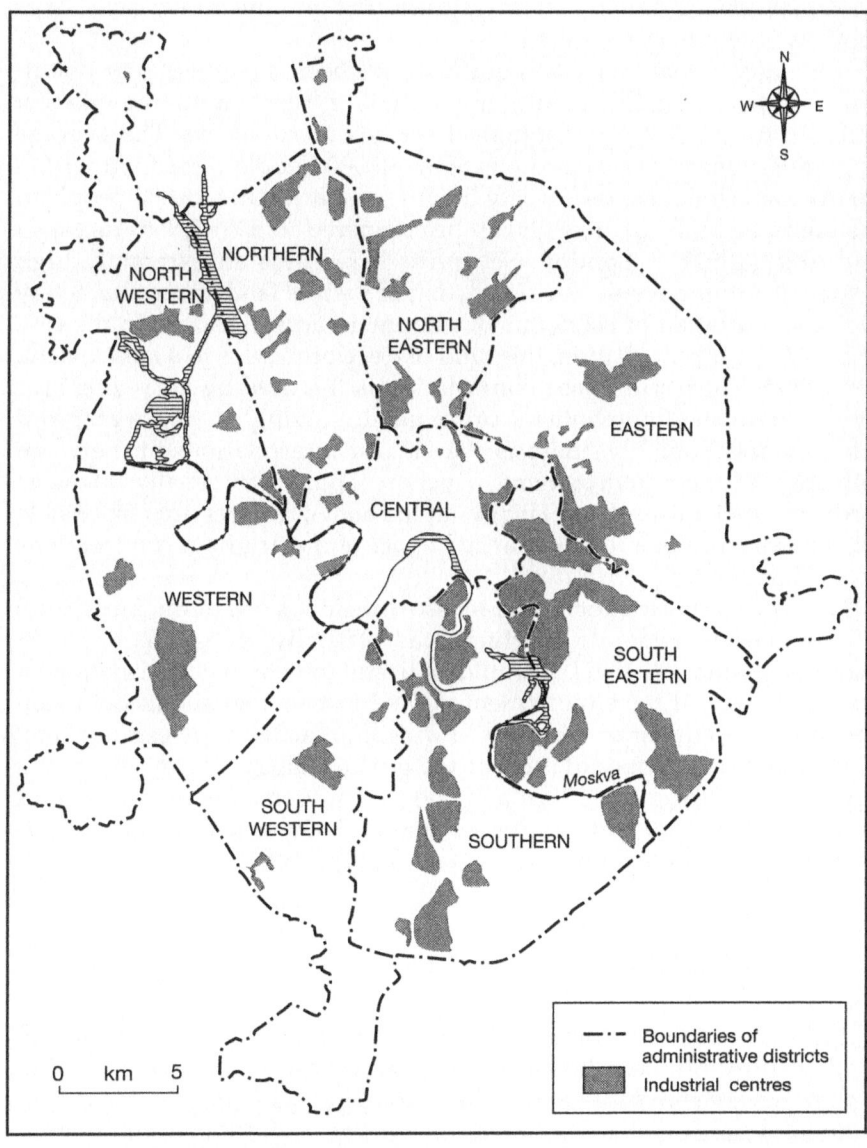

Figure 4.1 Distribution of industrial centres and administrative division of Moscow city (adapted from Moskovskiy... 1998; Moskvovedeniye... 1996b)

metallurgy, transport construction, machinery, chemical and petrochemical, electrical engineering and textile industries. Overall, it accommodates more than 14,000 industrial enterprises, including a major oil refinery and petrochemical production complex and an automobile production combine. It also includes about 17,000 construction and 2500 transport enterprises, 382 motor transport depots, 14 major and 48 district thermal power stations and many other facilities. The administrative division of Moscow is shown in Figure 4.1.

Generally, a major city has quite a few specific features that distinguish it from a rural area. Firstly, it is a place with the highest concentration of a single biological species, *homo sapiens*. The average population density in Moscow is 8500–10,000 people per sq km within MKAD, which delineates the city boundaries. It reaches 16,000 people in the high-rise built-up areas. It can be compared to 5351 (1995) in Tokyo and 4228 (1991) in London. Secondly, it features an extremely high degree of anthropogenic transformation of natural landscapes and a very high concentration of economic development and transport. In the case of the Moscow conurbation, this kind of development is to a large extent, industrial. Thirdly, a major conurbation is featured by very intensive dynamics of the 'metabolism' processes involving energy, water and other cycles. Fourthly, there is the closest interrelationship between humans and their environment, which explains an unusually intensive feedback mechanism there. Finally, all the above features are responsible for the existence of a long-distance impact of major urban conurbations on the surrounding territories.

The ecological situation in the Moscow region is one of the most acute in the Russian Federation which was officially recognized in 1991 (Gosudarstvenniy... 1992). It would be useful to note that information on the actual state of the environment in the Moscow region was not available until the advent of 'glasnost' during Gorbachev's 'perestroika' era. Thus, the gradual degradation of the city's ecology, which was being aggravated during the 1960s and the 1970s, went, for the most part, unnoticed by its inhabitants until the mid-1980s, when in 1986 censorship on environmental information was finally lifted.

4.2 Physical geographical setting

The Moscow region is **located** in the central part of the East European plain within a 'mixed forest' geographical zone. Forest composition within this zone involves a mixture of coniferous and deciduous species with an increasing share of broad-leaf varieties, and it is found only in the European part of the former Soviet Union. In the north it is bordered by the European taiga, in the south by the forested steppes, and in the west by the 'broad-leaved forest' zone.

The **relief** of the Moscow oblast is predominantly plain but the topography is rather dissected. In the north and west the Smolensk-Moscow upland is the highest but flattened landform. The average heights within Moscow do not exceed 175–185 m, but outside the city its absolute maximum heights reach 285 m. In the south is the Moskvoretsko-Okskaya moraine erosional plain with absolute heights of 200–250 m, and in the east a waterlogged Meshcherskaya lowland with heights not exceeding 160 m. The latter is a flat sandy depression with a few raised moraines and patches of peat bogs.

The **climate** of the Moscow region is moderately continental, with relatively cold and snowy winters and warm summers. On average freezing temperatures last for 103 days during the year. Mean January temperature is −10 degrees C but fluctuations from the norm can be quite significant. Thus, long thaws have become quite common in December. The multi-year average July temperature is +17 degrees C and in summer relatively cold periods and prolonged rains can be common. The annual amount of precipitation varies from 450 to 650 mm in different parts of the oblast. In Moscow the multi-year annual mean is the greatest at 644 mm and it has further increased during the last five-year period to 760 mm, which represents a 20% growth (Moskovskiy... 1998). The main rivers are Oka, Moskva and Klyaz'ma. The Moskva river is the left tributary of the Oka river which ultimately drains into the mighty Volga.

The typical **soils** in the Moscow oblast are 'peaty podzolic'. They are formed in the conditions of excessive moisture regime that becomes closer to the optimal balance only in the southern margin of Moscow oblast. Their mechanical composition varies from loamy, sandy loams and sands on different landforms. Boggy and gleyic podzols are formed on waterlogged terrain and in the south of the region grey forest soils have developed. The latter combine the features of peaty podzolic soils and chernozems or black earths.

The **vegetation** cover is typical of the mixed forest geographical zone and has a greater biodiversity than in the taiga landscape zone. Many centuries ago, the territory of the Moscow region was completely covered with dense forests including those composed of fir, fir and broad-leaved species, oak groves, and pine stands. Forests alternated with bogs and lakes. Only a few original coniferous and deciduous forests remain now. Many formerly forested areas are ploughed or replaced by secondary formations of small-leaved deciduous species, mainly consisting of birch and aspen. The remaining oak groves have a poorer species composition and an increasing component of maple, ash and lime, which were not part of the pristine forests.

At the present time, the forested territory accounts for 50% of the total area of the Moscow region, which compares favourably with the neighbouring oblasts. Conservation of a vast forest massive called the 'green belt' around Moscow was an important policy throughout the Soviet period. The most predominant species are fir, pine and birch, which in the south are also supplemented by oak and linden. Over 1000 plant species including 366 species of trees are found directly within the city limits. Only 43 of the latter are local species (L'vov 1998).

The mixed forest zone is generally more stable than the taiga, but it is still quite vulnerable to such intensive human impact as urban use. In the case of the urban territory the landscape is completely modified and the issue of **vulnerability** is related to the human environment. There are several factors that contribute to the aggravation of the ecological situation in Russia's capital. The most important of these is climate. As noted by Shahgedanova *et al.* (1999), meteorological conditions have a strong effect on the levels of air pollution. For example, high concentrations of carbon monoxide are caused by the surface-based temperature inversions and low wind speeds associated with anticyclones. The latter are typical of the atmospheric circulation pattern over the Moscow region.

A long relatively cold winter results in the accumulation of pollutants in the snow cover and their excessive peak effect on water bodies during the spring flooding. Indirectly, it increases heating requirements, further contributing to the overall load of air pollution. Icing of streets presents another major problem as salt compounds used to defreeze the ice cover pollute the water and soils in the city. In addition, the period of active photosynthesis is limited to five to six months a year, which is insufficient for vegetation to recover from enormous pollution pressures. There are also human causes, such as the high-rise built-up areas in the city centre which reduce the wind speed and prevent effective dispersal of pollutants.

4.3 History of urban development

The first record of Moscow dates back to 1147 when it was founded by Prince Yuriy Dolgorukiy. Moskva was the official centre of the Russian lands from 1328 to 1713 when Peter the Great made St Petersburg the capital of Russia. It regained its status in 1918 when it became the capital of the former Soviet Union. After the dissolution of the FSU in December 1991 it became the capital of the Russian Federation. Intensive urban development of Moscow arose in the late 19th century. By that time over 10,000 small and large industrial enterprises with more than 120,000 employees had settled in this city. The textile industry ranked high in the list of various branches with the construction, timber processing, machinery construction and food industries following.

The concentration of industry in the pre-Soviet period peaked by 1910 when almost half of all the labour force were employed in six per cent of enterprises. During the Soviet period rapid urban growth focused on further industrial development, with priority given to textile manufacturing, metallurgy, machinery and automobile construction and, later, the chemical and oil processing industries. According to Simagin (1997), the administrative territory of Moscow increased by 18 times over the last 100 years, while in the majority of West European cities it changed only a little. The specific features of urbanization during the Soviet period included (a) a continuous expansion of the urban area; (b) fast growth of suburban centres at the expense of migrants from outside the region; and (c) large-scale construction of summer houses ('dachas') in the suburban area around Moscow. As a result of these processes, Moscow is at present surrounded by a wide circle of so-called 'satellite cities' at regional level; most of these are industrial centres. In addition, Moscow oblast is the major provider of the capital's food requirements, and accommodates an intensive agricultural economy.

In the course of the economic reforms of the 1990s Moscow, like many other Russian cities and towns, was affected by the problems of production decline and economic recession. These were associated with rocketing inflation and consumer prices during 1992–1993, a rapid decline in the population's standard of living, an increase in the housing costs and many other problems. Between 1990 and 1998 the total volume of industrial output decreased by more than half while the number of employees in this sector reduced by 40% (Moscow… 1999).

In August 1998 Moscow was hit by the financial crisis from which it is now gradually recovering. Moscow remains the leading Russian city in attracting foreign investments. In 1996 65% of the total foreign investments into Russia's economy, amounting to US$4.3 billion, were directed at its capital. This has enabled the city government, effectively headed by its Mayor Yuriy Luzhkov, to introduce a multitude of measures for Moscow's further development. As stated in the Moscow City Ecological Profile, 'The history of Moscow's development indicates that, on the one hand, a large city acts as a severe polluter of the environment and, on the other hand, it is a centre of scientific and technological progress that aims at the solution of its ecological problems' (Moscow… 1999, 176).

It should be noted that during the post-Soviet period new trends in urbanization patterns have emerged. According to Simagin, the transition to the 'classical' western suburbanization processes became common (Simagin 1997). Among these are resettlement of wealthy city dwellers, called the 'New Russians', to suburban areas, massive cottage construction and accelerated increase in car numbers. These trends reflected a rapid stratification in the Russian society typical of the country as a whole but most obvious in Moscow. An important point is that many of these 'New Russians' prefer to keep their accommodation in the capital and buy a second 'mansion' located only 20 to 30 km from the city boundaries, thus creating a new intensive pressure on the city's 'green belt'. This kind of urban encroachment adversely affects the ecological situation in Moscow oblast because the most 'prestigious' regions are often those located within the water and soil conservation territories.

4.4 Ecological consequences and recent changes

The range of problems that currently affect the city itself and Moscow oblast includes air, water and soil pollution, waste accumulation, degradation of vegetation cover, etc. They have different levels of magnitude in the administrative districts of Moscow and throughout the region. Within Moscow oblast the current ecological status is also very severe, particularly in the towns of Podolsk, Elektrostal, Voskresensk and Kolomna.

4.4.1 Air pollution

The Moscow region contributes 25.2% to the total load of air pollution in the Central Economic Region of the Russian Federation. Annually from 1.5 to 2.5 million tonnes of various pollutants reach atmosphere of the Moscow region. Despite some reduction in the industrial activities in Moscow, an overall pollution continues to grow and generally exceeds the maximum permissible concentrations (MPC) (Table 4.1). During the last seven to eight years persistent air pollution was recorded with mean concentrations of nitrogen oxide at 2–2.5 MPC, ammonium 1.25–3.25

Table 4.1 Changes in concentrations of pollutants in the atmosphere over Moscow city, 1990–1997

Pollutant	Mean annual concentrations (number of MPC*)							
	1990	1991	1992	1993	1994	1995	1996	1997
Carbon monoxide (CO)	0.7	1.0	0.7	1.0	1.3	1.0	1.0	1.0
Nitrogen dioxide (NO2)	2.1	2.3	2.0	2.3	2.5	2.5	2.8	2.2
Sulphur dioxide (SO2)	0.2	0.4	<0.2	<0.2	<0.2	0.2	0.6	<1.0
Phenol	1.3	1.0	2.0	1.7	0.7	1.0	1.7	<1.5
Formaldehyde	4.3	2.3	2.0	3.0	2.0	3.3	3.3	2.3
Ammonium (NH3)	2.3	2.5	3.0	3.3	1.3	0.8	1.3	1.0
Benzol	3.4	2.3	2.2	2.5	3.1	1.9	2.3	1.0

Source of data: Moskovskiy... (1998)
*MPC – maximum permissible concentration

MPC, benzene 2.2–3.4 MPC and formaldehyde 2–4 MPC (L'vov 1998). A substantial increase in the concentrations of carbon monoxide and nitrogen oxides since 1990 were also noted by M. Shahgedanova *et al.* They revealed that in 1996 CO concentrations in central Moscow were twice as high as in 1990 (Shahgedanova *et al.* 1999). They stated that the major sources of nitrogen oxides were power generation and the increasing vehicular emissions. According to these authors, power generation in 1994 accounted for 63% of the total nitrogen oxide emissions with transport contributing another 30%.

The distribution of industrial centres inside the city limits is relatively uneven (Figure 4.1) and the main proportion of the heavy industry enterprises including a major oil refinery is concentrated in the Eastern, South-Eastern and Southern administrative districts. Industrial landscape dominates the view in many parts of the city even in the most ecologically favourable districts such as the Western, as seen in Plate 8.

The main industrial polluters in Moscow are the Moscow NPZ (Russian abbreviation for an oil processing plant) which emitted 36,100 tonnes of pollutants in 1997, TETS-23 (Russian abbreviation for central thermal power plant) with 17,000 tonnes and 'ZIL' (Russian abbreviation for 'plant named after Lenin') automobile plant with 3200 tonnes of pollutants. Approximately 60% of all pollution from stationary sources came from thermal power plants (Gosudarstvenniy... 1998c). A positive change was associated with a general move from coal to gas as a fuel for power generation. However, these enterprises still have alarmingly high levels of pollution with nitrogen oxides, sulphur dioxide and vanadium pentoxide.

Plate 8   The industrial landscape dominates this view in the city of Moscow

Bridges and Bridges (1996) reported that in 1993 Moscow had high levels of concentrations of ammonia, carbon monoxide, phenol, nitrogen dioxide and organic cyanides. A study of air pollution with carcinogenic N-nitrosamines has shown that their average concentrations varied from one MPC over ecologically clean areas, e.g. parks and plantations up to 30 MPC near emission sources of resin and tyre industrial enterprises (Khesina *et al.* 1996). It should be emphasized that despite this fact, an assessment of the integral state shows that the atmosphere in the centre is most polluted (see Table 4.2 on page 110).

It is interesting to note that the overall load of pollutant emissions from stationary sources in the whole of Moscow oblast in 1997 was equivalent to only 40% of the total of 764,000 tonnes emitted in 1991 (Gosudarstvenniy ... 1992). Airborne emissions from stationary sources in the Moscow city have also significantly declined accounting in 1997 for only 41% of the total value recorded in 1987. However, over the same period the share of pollution from transport had dramatically increased particularly in the Russian capital (Figure 4.2). In 1993, 72% of all emis-

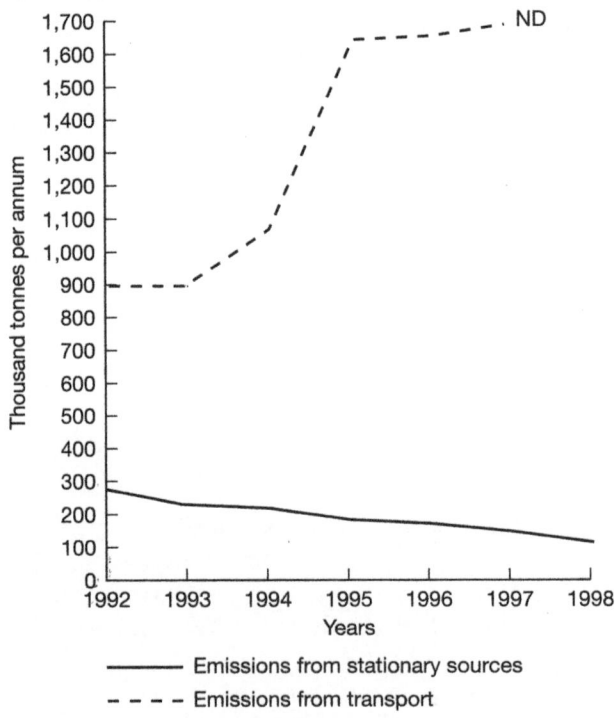

Figure 4.2   Atmospheric pollutant emissions from stationary sources and transport, 1992–1997/1998 (compiled by author from data in: Moscow... 1999; Rossiya... 1999)

sions came from transport (Gosudarstvenniy... 1996). Systematically high air pollution was recorded over Moscow with concentrations of several pollutants, i.e. ammonium, carbon oxide, phenol, nitrogen dioxide and hydrogen cyanide, exceeding 10 MPC.

During the Soviet period transport was responsible for over half of the total pollution load. However, in the post-Soviet times its share in the overall pollution load has dramatically increasing due to a number of reasons: (a) rapid growth in the total number of vehicles, particularly private cars with increase in 'consumerism'; (b) poor quality of fuel, mainly leaded petrol; (c) an increasing number of second-hand cars cheaply imported from Central and Western Europe; (d) an inadequate system of traffic control, particularly in the city centre. The latter reason was, in part, historically related to the ring-radial urban structure within Moscow.

Before 1990 there were about 600,000 cars in the city including 400,000 in private ownership. By 1993 the total number has increased to 1.3 million and in 1997 the automobile park has exceeded 2 million cars. In addition, there were also approximately 200,000 visitors' cars. An important factor is that there is a tremendous influx of commuters from the Moscow oblast during the working week. It was estimated that in 1993 transport was responsible for 72% of the total volume of pollutants while in 1997 the share of transport had already reached 88% (Gosudarstvenniy ... 1994, 1998c).

Low octane fuel and leaded petrol and diesel represent a major threat to air quality. Used foreign cars have been imported in vast quantities during the last decade. Inadequate traffic control generates extensive jams in the city centre and all along the Garden Ring as shown in Plate 9. During peak hours concentration of various pollutants in the lower atmosphere would normally reach 10–12 MPC. Another 750,000 cars, including 84% privately owned, are registered in the Moscow oblast. In 1997 the total load of emissions from transport here reached one million tonnes (ibid.).

Some pollutants like benzapyrene are emitted both by industrial enterprises and transport vehicles. The detailed study mentioned above (Khesina *et al.* 1996) assessed Moscow air pollution with respect to 15 main carcinogenic polycyclic aromatic hydrocarbons. It has revealed that, for example, average concentrations of benzapyrene in the atmosphere varied from one MPC over forests and parks and 3 MPC in the ecologically clean city districts, to 10 MPC over the city centre and 20–30 MPC near large industrial enterprises which deal with high-temperature processing of organic fuel, and also over heavy-traffic crossroads.

4.4.2 Water pollution

Water pollution is another major problem that does not seem to be affected by the overall decline in industrial production during the post-Soviet period. In 1997 over 2341 million cubic m of polluted wastewaters were discharged into the Moscow water bodies. Unlike the situation with

Plate 9 The role of traffic in urban air pollution has dramatically increased over the last decade: the Russian Foreign Ministry on the Moscow Garden Ring

airborne pollutant emissions from stationary sources discussed above, the total amount of discharged waterwastes had not decreased over recent years but remained at approximately the same level (Okhrana... 1998). Although in comparison with previous years no major decline in the quality of water supply sources was recorded, the quality of the city's surface waters had worsened.

On average, water pollution with suspended particles amounts to 1.5–2 MPC, concentrations of oil products exceed 2 MPC and biogenous substances over 5 MPC (Gosudarstvenniy... 1998c). It was also recorded that in the Moskva river pollution level sharply increased after Lyublino aeration station. In its lower reaches the annual concentrations of oil products and ammonium reached three to five maximum permissible levels while that of nitrites exceeded it by five to ten times (Gosudarstvenniy... 1996).

Water quality in the small rivers in and around Moscow is in critical condition because of the post-Soviet large-scale expansion of cottage construction, urban encroachment and lack of control over the land use within many water-conservation and coastal zones. Decline in the environmental state of small rivers had already started in the 1970s when anthropogenic pressure on water resources dramatically increased (Gosudarstvenniy... 1992). This process was made evident by the reduced discharge of the rivers, degradation of river beds and siltation of water. However, the stated recent changes present a potential major threat for the future quality of water supply for the Russian capital and the oblast as a whole. It was estimated that approximately 430,000 tonnes of suspended matter reached water courses in the Moscow oblast in 1994.

4.4.3 Problems associated with water supply

In 1997 the overall volume of water withdrawn from surface and ground water sources amounted to 1865 million cubic m in Moscow and 4520 in the Moscow region (Okhrana... 1998). This represented a decrease by almost 24% in the capital and by 13% in the region as compared to 1993. However, excessive water consumption by Moscow remains a serious problem. Daily water supply of Moscow via water pipelines amounts to 6.5 million cubic m (L'vov 1998). The total water consumption in the Russian capital in 1997 included one third used for industrial purposes and the rest for communal needs.

The current economic recession and decline in industrial production has resulted in the decreased industrial water consumption by 20% during the stated period. At the same time, communal water use has slightly *increased* reaching 1992 million cubic m (Okhrana... 1998) despite a simultaneous decrease in the total population numbers by over 100,000 people (Table 4.3, page 113). An important point is that per capita water consumption in Moscow is higher than in most Russian cities. According to Bridges and Bridges (1996), in Moscow per capita daily water use in 1994 was 350 litres

Table 4.2 Environmental pressures and health in Moscow city by district

District	Air pollution from stationary sources*		Ecological pressures by four-point scale					Rank in illness frequencies*	
	tonnes/year	%	Total air pollution	Water pollution	Soil contamination	Radiation background	Total	Total	Children
Central	4,756	2.2	3.6	3.4	2.5	4.0	13.5	1	1
Northern	31,654	14.5	2.2	2.0	2.5	1.6	8.3	6	6
North-Western	7,475	3.4	2.3	1.2	3.0	1.3	7.8	5	5
North-Eastern**	5,582	2.6	2.7	3.6	2.3	1.7	10.3	3	2
Eastern**	34,975	16.0	2.6	3.0	4.0	2.3	11.9	5	4
South-Eastern	54,996	25.1	2.9	2.0	3.5	2.9	11.3	4	7
Southern**	36,137	16.5	2.1	2.4	2.5	2.6	9.6	2	3
South-Western**	9,735	4.5	2.8	2.6	1.8	2.0	9.2	7	9
Western	32,260	14.7	1.8	1.2	2.3	2.3	7.6	8	8

Sources of data: L'vov (1998); Moskvovedeniye... (1996a)

* data for 1993

** districts with highest population density

compared to 24 litres in Kalmykia. Some confusion in the estimates may be due to the fact that in Moscow many communal and industrial enterprises use drinking rather than technical water in their technological processes. This also explains why drinking water is generally of low quality and is not enriched with iodine or fluoride as in many other cities of Europe.

During the last decade the volume of water use reached a value of 760 litres per person per day and was officially limited to 520 in 1996 (Moscow... 1999). The share of drinking water consumption is also very high and reached 400 litres compared to approximately 200 litres in Paris, 180 in London and 120 in Vienna. It was estimated that all available sources of drinking water within the Moscow boundaries have already been depleted (Gosudarstvenniy... 1998c).

Excessive water consumption affects the hydrological balance of the hydrographic system within the city and the region as a whole. It had also affected the quality of ground waters and resulted in their depletion in a number of districts. According to L'vov, the decline in the ground water levels over the last two decades in Moscow amounted to values from 50–70 m to 90–100 m (L'vov 1998). At the same time, daily losses of water leaking into the ground from a system of water pipes, which are typically worn-out or in a state of disrepair, in 1997 reached 440,000 cubic m or 15% of the total water consumption (Gosudarstvenniy... 1998a). As a result of these adverse processes, some sections in all districts within Moscow are being inundated, which means that the closest aquifer is located within less than 3 m from the surface. This is currently a major problem, as an overall inundated or flooded area is estimated at approximately 40% of the city's territory (L'vov 1998).

Pollution of ground waters which was first recorded in the 1950s is linked with the penetration of polluted surface waters. Water pollution by heavy metals, dioxins, oil products, nitrates and, in the last decade, with bacteriological agents has greatly exceeded the natural assimilative capacity of rivers throughout the region and particularly in major industrial centres. The interaction between polluted surface waters and ground water sources used for drinking water supply has already presented a major problem. For example at some intake water points in the Moscow region oil products content exceeds 50 MPC. Because of the presence of multiple potential pollution sources that have existed for decades, cases of oil pollution in ground waters within the city limits are quite common (L'vov 1998).

4.4.4 Changes in the vegetation cover

During the recent period there has been a tendency towards impoverishment of the local flora and an increase in the proportion of alien species. There appears to be a trend in the per capita vegetated space within urban areas. While the standard urban norm in Russia is equal to 24 sq m of 'green lands' per person, the current value in Moscow is about 18 sq m.

The worst situation in this respect is in the city centre, where only 9 sq m of green plantations per person is available (*Zelyoniy Mir* 30, 1998).

Most typically, areas previously reserved for tree plantations are used for construction, while many existing vegetated areas are cut down. For example, approximately 2000 ha of land reserved for afforestation were taken away for construction during 1995–1996. Intensive automobilization in the Russian capital in recent years has been reflected in the rapid growth of new garages replacing trees in the cramped yards of central Moscow and elsewhere. Almost half of the territory of the Fili-Kuntsevo park zone is under houses, garages and other constructions.

At the same time, many plantations become degraded mainly due to pollution. During the last decade many plant diseases have been recorded that stem from soil degradation and contamination. For instance, in 1995 and 1997 massive tree deaths were due to the extensive use of potassium and sodium chlorides to thaw ice in the streets of Moscow. It was noticed that birch, poplar and maple are most vulnerable, while chestnut appears most resistant. Currently only 30% of the forest within the city limits retains the ability for restoration. Partial (10–15%) desiccation of both deciduous and coniferous trees is recorded along practically all central motorways and roads located close to the city's green massives. Degradation of birches is estimated at 40% of their total numbers. One of the main reasons for this situation is the use of salt and other chemical deglaciation reagents during winter periods. Trees planted within two metres from the road within the Garden Ring, Noviy Arbat and Tverskaya streets have almost completely perished. In plantations along the widest central boulevards like Tverskoi and Gogolevskiy the share of viable trees is only 45–50% (L'vov 1998).

During the current transition period a major cause of vegetation cover decline is associated with the often unregulated land seizure for suburban construction, industrial development, transport infrastructure and garages being built very close to the city limits outside MKAD. In this way, the main protective functions of the forest-park zone which surrounds Moscow are fast diminishing. Aggravating the situation, intensive urban construction around the existing satellite towns of Moscow threatens to result in the creation of a united front of urban areas outside the current city limits, thus effectively leaving the capital without any environmental protection. Over the last three to four years a major trend of a sharp reduction in the numbers of certain species, particularly of hoofed animals, and of an increasing depletion of fish resources has been revealed (Gosudarstvenniy... 1998c).

4.4.5 *Other ecological problems*

Contamination of soil with heavy metals, dioxins and radionuclides presents a growing problem in some districts of Moscow and regions of Moscow oblast. In 1997 50% of all investigated soils in the Moscow oblast were polluted with heavy metals. Inside the capital concentrations of copper exceed MPC by 4.5 times and zinc by three times (Gosudartsvenniy… 1998b). Snow heaps are often the main sources of soil contamination. Within the city's boundaries the most polluted regions are Krylatskoye in the Western administrative district and large areas in the Eastern administrative district (Table 4.2, page 110).

Another problem is related to the **excessive pressure on the lithosphere**. In the course of urban construction there have often been recorded unsound changes of the natural terrain, which have resulted in the loss of its stability and changes in the properties of the lithogenic basis. This results further in the increased frequency of landslides and floods, and technical accidents such as fires and explosions. Seismic activity has accelerated and earthquakes in the Moscow region have sometimes reached four to five points on the Richter scale. There have also been considerable changes in the structure of catchment basins and denudation conditions. Overall changes have included general flattening of the relief, confinement of rivers into underground beds, damming of rivers and construction of reservoirs, etc. The 'weight' of artificial creation within the urban boundaries, that is multi-storey housing, asphalt pavement, transport and infrastructure, have ultimately pressed down the surface of the city by an estimated 50 cm during the last 50 years.

Another serious problem is an increase in the total **waste load** during the period of transition to the market economy, when the increased availability of imported products and an apparent growth in consumerism have changed people's attitudes, such as towards the previous practice of reusing plastic bags. It was estimated in 1998 that approximately 300 million cans, including 210 aluminium cans, are consumed annually in Moscow. Another 30% of this amount is additionally used in Moscow oblast (*Zelyoniy Mir* 30, 1998, 3). None of these precious wastes are currently being recycled.

Table 4.3 Changes in population numbers in the Moscow region, 1975–1998

Region	Population numbers (000 people)									
	1975	1985	1991	1992	1993	1994	1995	1996	1997	1998
Moscow city	7,862	8,972	9,003	8,957	8,881	8,793	8,717	8,664	8,639	8,630
Moscow oblast	6,040	6,689	6,718	6,707	6,682	6,644	6,626	6,597	6,579	6,564
Total	13,902	15,661	15,721	15,664	15,563	15,437	15,343	15,261	15,218	15,194

Sources of data: Rossiyskiy… (1998); Rossiyskiy… (1994); Chislennost'… (1999)

To worsen the matter, accumulation of solid wastes at over 200 official and unauthorized dumps and tips around the city presents a major challenge. In 1994 these occupied an area of approximately 900 ha within city limits adding to waste kept at 108 municipal dumps located around the capital's boundaries. Every year over 20 million tonnes of industrial and 6 million tonnes of solid municipal wastes is being accumulated in Moscow oblast (Gosudarstvenniy... 1998c). Allocation of new land for tips and dumps within and outside the city limits creates an increasing pressure on the natural landscape.

Degradation of the urban and suburban landscape due to excessive recreational pressures presents a serious problem for the contemporary parks within the city limits, but it also increasingly endangers suburban forest massives, which used to provide an effective protective belt around the growing capital in the 1970s and 1980s. More and more land around Moscow is being annually withdrawn for the construction of thousands of mansion-type 'dachas' [countryhouses] for the rich 'New Russians'. This is often carried out within the boundaries of the most precious water conservation or other protected areas, at the expense of existing forest massives, etc. Development of the suburban economy, orientated towards the provision of the ever-increasing demands of this huge city, also requires more land allocation for agriculture, mining activities, etc.

In addition the impact of urban activities on **climate** has been recorded. The climatic implications include, firstly, an increase in air temperatures. Over the last thirty years the mean multi-year annual values have increased by one degree Celsius, while the average winter temperatures increased by 1.5 degrees. Secondly, there has been a recorded increase in the values of air humidity and the overall amount in precipitation. Thirdly, there has been a decrease in the total amount of solar radiation due to increase in air turbidity by about 20% and an associated increase in mist frequency and reduction of light reaching the surface (L'vov 1998).

4.4.6 Overall ecological situation

The overall ecological situation in Moscow and Moscow oblast worsened gradually during the 1960s and 1970s, when it clearly reached crisis point. In the 1980s this was complicated by the decline in environmental investment and the new challenges associated with the transition to the market economy. Although in a limited number of areas some improvements were recorded during the post-socialist period, e.g. the amount of pollutant emissions from stationary sources, these were negated by other sources of ecological degradation.

Inside the Russian capital, ecological pressures mounted at various rates in different districts as can be seen from Table 4.2 and Figure 4.3.

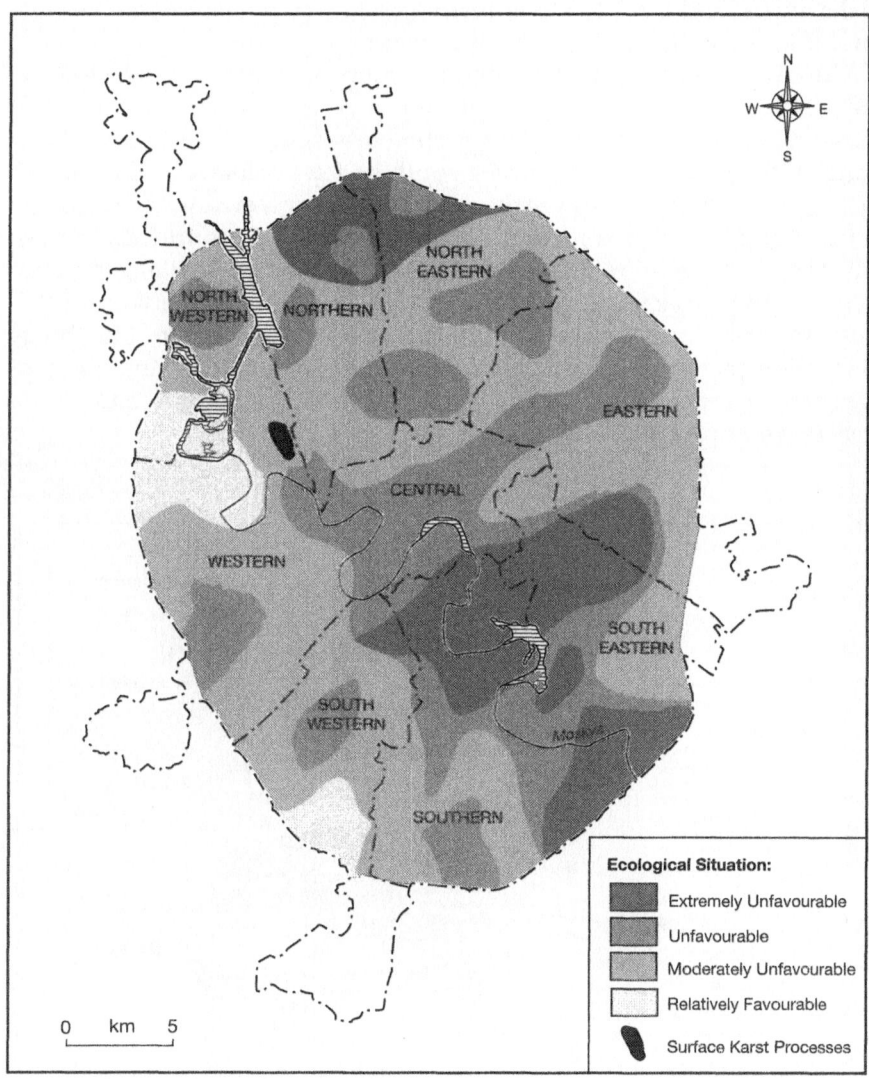

Figure 4.3 The present ecological situation in Moscow city (after Moskovskiy... 1998)

The worst area in terms of the overall ecological situation is the Central district due to its heavy load of traffic pollution. The North-Eastern, South-Eastern and Southern districts are confronted with both sets of problems, transport and industrial, which affect large areas closer to the Central district. It should be noted that the predominantly westerly wind direction explains why this territory has particularly acute ecological problems. Another area of severe pollution is located in the northern part of the capital and is related to industrial pollution in the city and the surrounding suburban centres, e.g. Mytishchi.

The present ecological situation in Moscow oblast is illustrated by the map in Figure 4.4, which shows that the whole area around Moscow is severely affected. It is particularly acute around Klin, in the north of the oblast, Podolsk in the south and particularly to the east of the capital, which is also partly due to the above-mentioned climatic feature of this geographical region. It is focused around the major industrial centres like Elektrostal (which is translated as 'electric and steel'), Orekhovo-Zuevo, Noginsk and Voskresensk. To the west of the Russian capital the situation is satisfactory. It appears from the above discussion that the ecological situation in many parts of Moscow and the central part of Moscow oblast can be classified with due cause as an *ecological crisis*.

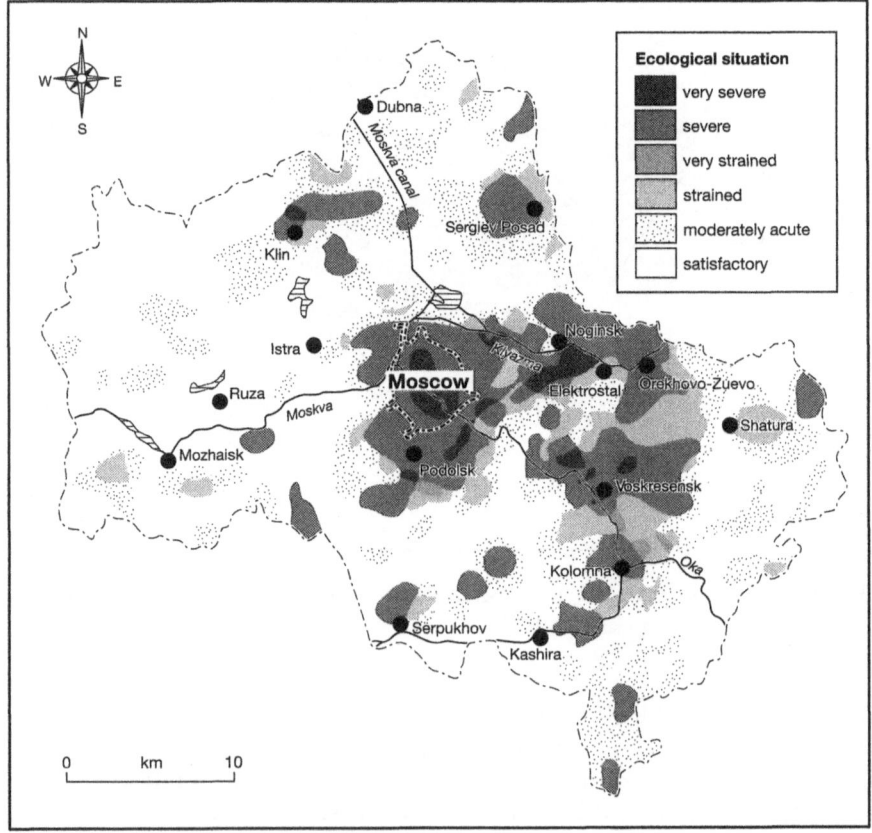

Figure 4.4   The present ecological situation in Moscow oblast (after Moskvovedeniye... 1996a)

4.5 Health and demographic issues

The various ecological problems have already had a severe effect on human health. As is clear from Table 4.2, illness frequencies in both adult and child age groups in Moscow correlates with the worst polluted districts. Central district ranks first in both categories while the North-Eastern and Southern districts share the second and third places. Next come the South-Eastern and North-Eastern districts. A study has revealed that in some regions of Moscow the contents of lead in children's hair in 50% of cases exceed the maximum permissible concentration. Other pollutants, e.g. cadmium which affects mental activity and bone structure, and nickel, were also found in excessive amounts. The overall health status in Moscow and Moscow oblast is threatened by massive allergies, immune deficits and lung illnesses which are recorded both in adults and children. For example, in Mytishchi, a small city to the north of Moscow, a study of children's health in a polluted district carried out in 1992 revealed that only 38.6–50% of children were practically healthy compared to the background value of 73.9% (Gosudarstvenniy... 1993).

The current health problems along with the declining socio-economic conditions during the transition to the free market economy have, in turn, affected the demographic trends in the Moscow region. These adverse tendencies have been also evident in the Russian Federation as a whole. As illustrated by Table 4.3 a gradual decline in the overall population numbers dates back to 1991 in the capital of Russia and 1992 in Moscow oblast. Figure 4.5 gives further detail on the structure of these

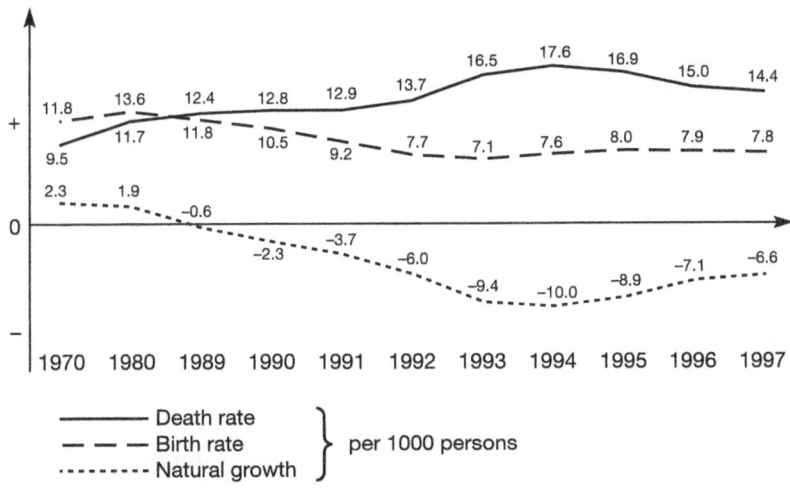

Figure 4.5   Demographic change in Moscow city, 1970–1997 (compiled by author from data in: Rossiyskiy... 1998; Moskvovedeniye... 1996b)

changes. Already in the 1970s death rates had started to increase while in the same period birth rates generally went down. As a result, the natural growth rates remained at negative values from the late 1980s. This figure also shows that after the lowest peak in 1994 it gradually increased.

4.6 Prospects for the future

The overall Russian trends of air pollution show that, starting from 1986, pollutant concentrations over many cities stabilized and began to decrease. However, the adverse ecological situation created during the last two decades of the Soviet period have remained in place for many industrial centres including Moscow. It can be expected that, in the future, the restoration of economic production industry will exert an increasing pressure on the ecological situation in Moscow and Moscow oblast. The average concentrations of nitrogen oxide and dioxide continuously increase as has been shown in this chapter. In addition, all-time high levels of pollution are often created by spontaneous and uncontrolled specific emissions from various industries.

Moreover, increased pressure from 'new' powerful pollutant agents associated with the transition to the market economy have begun to operate, transport in particular. Private and public vehicles will continue to threaten the population's health unless a shift to the use of unleaded petrol is achieved, which is a highly doubtful scenario in the short term. Heavy air pollution in the city centre cannot be overcome if the current traffic strategies persist.

In future, the current suburbanization trends will continue. Some experts envisage that the areas of existing suburban centres will increase at the expense of the 'adjoining' cottage settlements and, in many cases, amalgamation of the neighbouring settlements will take place (Simagin 1997). In view of this scenario, it can be expected that the ecological conditions in Moscow oblast and indirectly in the city itself will deteriorate further unless these processes are put under strict control. The latter is doubtful because of the common practice of bribing poorly paid regional officials in order to obtain permission for cottage construction in a forbidden area e.g. within a protected territory.

In addition, it appears that increasing challenges will be associated with flooding, tectonic activities, landslides, severe storms and other natural disasters. The area of influence of the city on the neighbouring natural landscapes will also increase in future. Taking into consideration the process of further ageing of the equipment installed in many plants and enterprises, the danger of further acceleration of the frequency of technical accidents should not be underestimated. For example, in Serpukhov of the Moscow region, an artificial lake which contains 300 cubic m of accumulated sewage is currently in a catastrophic state. Breakage of the dam which holds back these wastes could result in a major ecological disaster, because only four metres separate this lake from the Nara river, which flows into the Oka and, ultimately, the Volga.

Environmental awareness in Moscow is still relatively low. Opinion polls carried out in Moscow during the last two to three years have revealed that Muscovites are primarily concerned with two problems: criminalization of the society and worsening of the economic and sanitary conditions of living. At the same time only 9.2% of respondents consider their environmental awareness sufficient and 43.8% stated that the main source of this information comes from personal observations and opinions of relatives and acquaintances. The majority of respondents considered the current ecological situation as dangerous to health in different degrees, while inhabitants of the Central, Southern and North-Eastern districts considered it very dangerous (L'vov 1998). The main causes for the decline in the state of human health are air pollution and the poor quality of drinking water. This is confirmed by the fact that in all age groups the predominant illnesses are those related to the respiration, digestion and nervous systems. However, it is expected that the problems of drinking water quantity and quality will become more acute in future (*Zelyoniy Mir* 18, 1997).

However, there are also positive signs that can be observed from the activities of the current Moscow administration and leadership. A recently adopted programme for the improvement of the physical environment in the centre of Moscow by 2010 envisages the creation of tree plantations along streets and highways, inside the city's yards and other territories; restoration of gardens and parks around water bodies, ponds, etc. According to this programme, realization of which has already started, it is planned to create approximately 600 'green islands' in the capital's centre. One of the environmentally friendly projects which is close to completion, is related to the construction of a can-producing plant in Dmitrov, the Moscow region. It envisages that about 40% of the initial material for the annual production of 1.5 billion aluminium cans will be provided by their reuse (*Zelyoniy Mir* 30, 1998, 3).

A 'Programme of reorganization of the water supply and sewerage system' has recently been developed by the Moscow government. The programme aims at the reduction of the daily norm of water consumption in the capital to 235 litres per person from the current standards. It is also planned to improve the drinking water quality through the application of advanced ozone processing technology. Instead of using only two treatment stations for cleaning the sewage waters of Moscow, it is envisaged that every newly erected region will, in future, have its own local aeration station. Other measures over the last few years have included the elaboration of the Project of its Sustainable Development in 1998 and also various environmental protection actions e.g. the introduction of ecological requirements for car exhausts. It should be emphasized that most other urban centres in Russia are much less fortunate compared to the Russian capital.

CHAPTER 5

Chernobyl: the world's worst nuclear accident

Chernobyl has exploded not only belief in the peaceful atom – Chernobyl has exploded belief in communism, in science, in an equitable social idea, in a multinational state, in the might of a great empire.

(L. Annenskiy, *Izvestiya* 19 June 1998, 5)

5.1 Introduction

The objective of this chapter is to examine the specific nature of environmental impacts associated with radioactivity by the example of the Chernobyl catastrophe. Although the implications of this disaster were evident throughout the northern hemisphere, the focus here will be made on the contaminated territory inside the former Soviet Union, which currently includes parts of three present-day countries – the Ukraine, Belarus and the Russian Federation. The total area of contemporary contamination with caesium-137 is more than 137,000 sq km which is equivalent to 70% of the territory of the United Kingdom.

It should be emphasized that before the accident very limited information existed on the impacts of radioactive contamination on nature. Some studies of effects on humans had been carried out in Japan in the aftermath of the Hiroshima and Nagasaki bombings. However, in the Soviet Union, where a 'peaceful atom' was a political ideology, insufficient data existed on both topics. As noted by Medvedev (1994 and 1995 in *Zelyoniy Mir*, 25–26) the actual magnitude of the radioactive release was greatly underestimated particularly in terms of the initial dose effects. Moreover, the 'distant consequences' started to mount much earlier than expected. Since that time the contaminated territory has become a 'scientific laboratory' for national and international teams of experts. Therefore, the focus of this chapter is on some recent aspects of findings related to the ecological and human implications of the disaster. The issues of the natural vulnerability of landscapes and rehabilitation after the accident are also examined.

Since the event, Chernobyl has been the focus of public and scientific attention throughout the world. A substantial number of publications have been devoted to this notorious catastrophe; to mention just a few, those written by Medvedev (1990, 1992), Marples (1996), Savchenko (1995), Kovalevskaya (1995), Yaroshinskaya (1994, 1992) and many others. Recently, two major atlases were published in the Ukraine and Russia, the former related to the ecological situation in the exclusion zone (Atlas... 1996), and the latter (Atlas...1998) to the status of radio-active contamination within the whole affected area in the European part of Russia, Belarus and the Ukraine.

It would be useful to mention that although censorship on environmental information was lifted in Russia as early as in 1986 during the period of 'glasnost', much of the factual data relating to the Chernobyl catastrophe was still not available to the public until the early 1990s (Yaroshinskaya 1992). The scale of the problem was enormous and so were the impacts of this disaster, both for present and future generations; it seems that we are only now approaching the point at which they can be assessed in more detail.

5.2 Physical geographical setting

Chernobyl is **located** in the Ukraine, about 100 km to the north-west of Kiev, near the border with Belarus. The contaminated territories of Belarus, the Ukraine and the Russian Federation are located in the central part of the East European plain within the mixed forest landscape zone. The southern part of the contaminated area pertains to the forested steppe landscape zone. The main physical-geographical province within this zone, which is shared by the former two republics, is called 'Poles'ye'.

'Poles'ye', which in Russian means forested terrain, is a flat-bottomed depression with heights in its central part of 100–150 m above sea level. Moraine landforms here have been, for the most part, flattened by fluvio-glacial activity. The southern section of the affected area is located in the foothill area of the Pridneprovskaya upland on the right bank of the Dnieper river. Predominantly low-lying topography explains the prevalence of flat river valleys and poorly drained or boggy surfaces particularly in Poles'ye where low-land bogs are widely developed.

The **climate** of this territory is relatively mild, featuring warm humid summers with average July temperatures ranging from +18 degrees C in the north to +21 degrees C in the south. Winters are also mild with average January temperatures between –5 and –7 degrees C. Snow cover stays for a period from 100 to 120 days per year, but the snow cover is not very thick and frequent thaws are typical. The mean annual amount of precipitation normally exceeds 600 mm. The southern part of Poles'ye boasts significant thermal resources for agriculture as the sum of temperatures over +10 degrees C here reaches 2400–2500. A large proportion of the land is used for agricultural production.

Vegetation in the mixed forests of the Belarus and Ukrainian Poles'ye differs from that of the Moscow region in species composition and, particularly, by a greater share of the broad-leaf species, e.g. hornbeam (*Carpinus betulus*) which is not found as far north as Minsk. In the south field maple (*Acer campestre*) becomes part of the forest composition. Predpoles'ye woods are composed of mixed fir-grab-oak forests. Coniferous species are found in the north, namely the European fir (*Picea excelsa*) and common pine (*Picea silvestris*). A lesser share is of the small-leaved tree species like birch (*Betula verricosa*) and aspen (*Populus tremula*). The prevalence of sandy and sandy loam soils in Polesye in Belarus determines the predominant composition of pine found in 60% of all wooded areas. Second comes black alder at 13%; secondary birch groves at 15%; to a lesser degree oak groves alternate with fir and aspen forests. There is a greater share of pure oak groves and fir and pine forests and transitional fir forests with oak element either in the upper tier or as an undergrowth. Bogs are predominantly grassy and not sphagnum. Typical bogs on the right bank of Pripyat are composed of reed (*Phragmites communis*), bulrush (*Scirpus lacustris*), rush (*Typha latifolia, T. angustifolia*), and grasses (*Calamagrostis lanceolata, Gliceria aquatica* and *Acorus calamatis*). Forest bogs are found in woods composed of alder and birch. A typical feature of woods is that they are composed of several tiers, i.e. from upper, and several undergrowth layers. In addition there are shrub and semi-shrub layers and grass/moss cover. This contributes to a greater diversity and a more complex structure which provides a greater stability than in the taiga zone.

In the southern forested steppes, oak woods composed also of such broad-leaf species as ash (*Fraxinus excelsior*), maple and elm groves along the river valleys alternate with open spaces of grassland composed of steppe vegetation.

In the direct vicinity of the Chernobyl NPP, in the 'alienation zone', half of its area is covered by forests; about one-third of the zone is meadows and fallows (Atlas 1996). Of the forests, 85% grow in poor and relatively poor habitats. On average, pine plantations constitute about 80% of the forested area. Originally, this territory was covered by dense forests, but its development was associated with massive logging. By 1913 only 11–12% of forests remained in the central part. Afforestation measures were started in the 1920s, but were greatest during the 1950s–1960s when the total forested area was increased by four times. The role of trees in the stabilization of the radiobiological situation and rehabilitation of the landscapes after the accident cannot be underestimated.

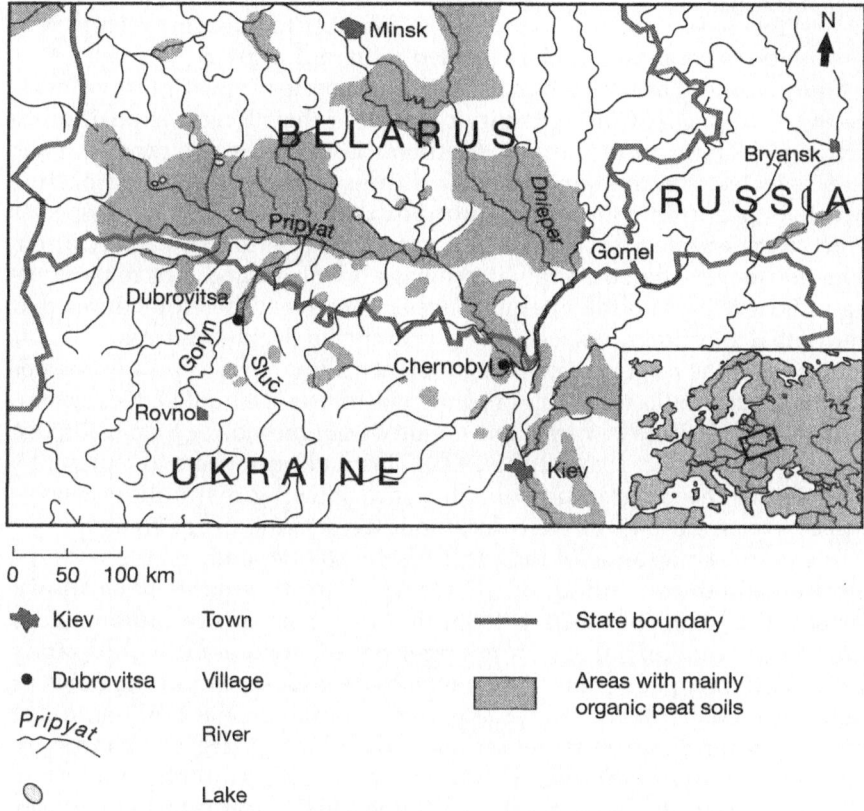

Figure 5.1 Distribution of organic peat soils in the Pripyat catchment (after Burrough *et al.* 1999)

The **soil cover** in the Pripyat catchment consists predominantly of peat organic soils shown on the map in Figure 5.1. These soils feature a high absorbing capacity in relation to radioactive contamination. Peaty and turf podzolic soils are also quite common throughout the territory. In the north the soils are oversaturated and alternate with bogs. In the south grey forest soils and podzolized chestnut soils are found within the forested steppe zone. Further to the south these give way to black earths or chernozem soils.

5.2.1 Overall vulnerability of landscapes

In general the types of landscapes found throughout the affected territory are more stable to human impact compared to coniferous forests, for example. Greater phytomass and biological production and a greater biodiversity along with a more complex structure explains their relatively high level of resistance to human impact. However, it appears that even these stable landscapes are extremely vulnerable to the complex and universal type of contamination associated with radioactivity.

Many studies have shown that coniferous species appear most vulnerable to the impacts of radioactivity because of their longer exposition to ionizing radiation and retention of radionuclides by the tree canopy. It has been revealed (Savchenko 1995) that during photosynthesis, radioactive carbon-14 and tritium are also absorbed by their foliage. Deciduous species which were affected during the first year after Chernobyl, namely birch, alder and aspen, shed the irradiated foliage on the ground thus redistributing radioactivity into the litter and upper soil layer. Therefore, doses that were lethal to coniferous species were bearable to deciduous trees.

A study of the responses in different plant communities to γ-ray exposures has shown that radio-resistance of coniferous forests is ten-fold lower in comparison with deciduous forest, and hundreds of times lower than oak forest or grass communities (Savchenko 1995). This finding appears to represent a reflection of the geographical zonality law, showing that radioresistance increases from the taiga through deciduous forest to the steppe zone.

The natural migration of radionuclides in soil depends on the soil type and the penetration ability of the former. Various studies have shown (Ivanov *et al.* 1997; Savchenko 1995) that rich soils absorb radionuclides more easily and keep them for a longer period compared to poor sandy soils. Therefore, peat organic soils which are widespread in the Poles'ye region are particularly vulnerable. Sandy soils, on the other side, are quite common in the direct vicinity of the plant. A degree of radionuclide's solubility is another factor and according to their capacity to dissolve, strontium-90 comes first with iodine-131, caesium-137 and caesium-144 to follow. At present the ecological situation is determined by the gradual disintegration of caesium and strontium.

Widely spread peat bogs which have a very high absorbing capacity made the landscape more vulnerable to the effects of irradiation. Sandy soils contribute to easy access of radioactivity to the lower layers of soil horizon and ground waters. Mushrooms have an increased absorbing capacity and their collection is traditionally part of the human diet. Overall this type of landscape is much more stable and resistant to human impacts than tundra or taiga. Nevertheless, reality has showed that even this type appeared quite vulnerable to such an integral and complex type of impact as radioactive contamination.

5.3 The causes of the accident

Detailed reconstruction of the events before, during and after the accident has been made by Medvedev (1990, 1992). The explosion of the fourth reactor of the Chernobyl nuclear power plant took place at 1 am on 26 April 1986 (Plate 10). The plant was located 15 km from the town of

Plate 10 Chernobyl: two weeks after the explosion

Chernobyl (currently transliterated as Chornobyl in accordance with its Ukrainian spelling). The type of reactor was RBMK, meaning water-moderated, and it worked on Uranium-235. The causes of the explosion as believed at that time and still most frequently cited were two-fold: personnel mistakes and the reactor's design and construction faults (Medvedev 1990). Since that time there have been many speculations as to the causes of the disaster. As Shcherbak noted, 'our today's under-standing of the causes of the catastrophe are much deeper ... More and more scientists agree that in Chernobyl a nuclear rather than thermal explosion took place' (Shcherbak 1991, 22).

The official cause of the accident was 'gross violations of the safety regulations committed by the engineering and technical staff of the CNPP, combined with faults in the construction of the reactor'. One recent hypothesis, however, has raised doubt about this, suggesting that a local earthquake which occurred on the night of 25/26 April with an epicentre located 10 km from the plant might have caused or contributed to initiation of the disaster. It was revealed that six months before the accident, the administration of the Chernobyl power station has approached the United Institute of the Earth's Physics (UIEP) with a request to carry out urgent studies of the causes of 'the abnormal geological deformations in the foundation of the Fourth block of the station' (*Zelyoniy Mir* 30, 1998, 3). These studies were planned for the summer of 1986. But just nine years after the disaster, analysis of seismograms recorded by a seismological expedition revealed that just 10–20 seconds before the explosion there was an earthquake with a strength of 2.4–2.6 points on the Richter scale. Therefore, the cause could be a seismic impact on the reactor, unprotected from vibrations.

As a result of this explosion a 1–2 km plume containing hot radioactive gases and aerosol particles was blown initially in a north-westerly direction reaching the Baltic Sea. During the next ten days, while the reactor was still burning, the wind direction changed to south-westerly, reaching the Balkans, north-westerly reaching the UK and south-westerly reaching the Mediterranean region (Atlas... 1996). As a result of complex and highly changeable meteorological conditions the overall impact of radioactive fallout was quite extensive. In the west, it was limited by the Atlantic Ocean, in the south by the Mediterranean Sea, in the north by the Arctic Ocean and in the east the radioactive cloud reached Siberia, China and Japan. Later the traces of the Chernobyl accident were also found in Greenland, Canada and even South America. It can therefore be concluded that the magnitude of the disaster was global.

The initial estimates of the total radioactivity vary between 44 million curies and 50 million (Medvedev 1990; Atlas... 1996). Some assessments have significantly increased the previous estimates given by Soviet and Western experts soon after the Chernobyl disaster. It is now estimated that at least 170–185 million curies were released into the environment during the accident (Medvedev 1994), 22 million by the initial explosion. The scope of implications was also extremely varied and multifaceted, because the Chernobyl catastrophe affected both the environment and human health. Moreover, it had a major impact on agriculture, economy, social life and politics.

5.4 *Ecological consequences and recent changes*

The radioactive fallout from the accident at Chernobyl consisted of a 'condensed' component (mainly radiocaesium and radioruthenium) and small fuel particles (uranium oxides) containing a full range of radioactive fission and activation products. The most important radionuclides from Chernobyl NPP were caesium-137 and strontium-90 because of their long physical half-lives, their amounts in the radioactive debris, their role in the biological pathways and adverse ecological effects. In general, the caesium was in both condensed and fuel form and the radiostrontium was mainly associated with fuel. The precipitation of radioactivity took the form of shedding the largest particles with dry fallout and the small particles and gaseous radioactive nuclides with wet fallout.

The initial impact was mainly associated with the action of isotopes of iodine 131, 132, 133 and 135, which are short-lived and have a half-life from 10 days to two weeks. They are easily absorbed by the human body, particularly of children, and affect the thyroid gland if no precautions are taken. Another major impact was related to contamination with caesium-137, the initial distribution of which in the reconstructed form is shown in Figure 5.2. The map shows two long 'veils' stretching into two directions from Chernobyl – one to the north-east and another to the south-west. In the Ukraine alone, 5 million ha of land were contaminated.

The dispersed nuclear fuel contained over 100 different radionuclides and transuranic elements with half-lives ranging from hours to hundreds of years, and even more; for example, the half-life of ameritium-241 is 433 years, while that of ameritium-243 is 7380 years (Chernobyl... 1989). Therefore the long-term effects from them can be expected for hundreds and thousands of years. An important factor of the impact was a very high degree of irregularity in dispersion and deposition of radionuclides via precipitation and dust particles transfer. Because of this, the spatial distribution of different intensity contamination of the environment is also extremely uneven and often highly contaminated areas were located in the direct vicinity of relatively 'clean' surfaces.

The impact of the Chernobyl accident on the environment or the radio-biological effect varied depending on the distance from the reactor, degree of radioactivity, type and form of the radionuclide, time span passed after the accident and the specific features/vulnerability of the

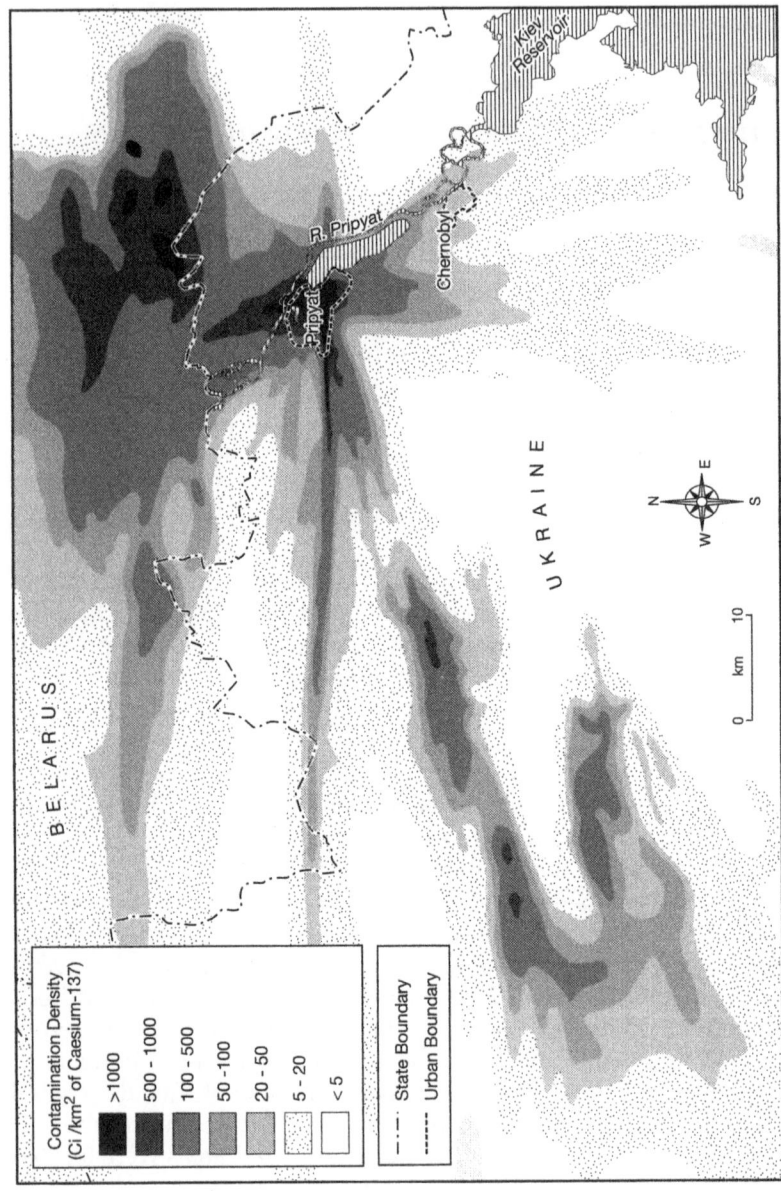

Figure 5.2 Radioactive contamination of the Ukraine and the adjacent regions of Belarus with caesium-137 on 10 May 1986 (adapted from Atlas... 1996)

affected landscape. Within the scope of this book, the focus of study is on the effects of the radioactivity on vegetation and soils in relation to their overall vulnerability. Another issue is the associated socio-economic and health implications.

5.4.1 Impact on vegetation

The most immediate impact of acute irradiation during the explosion and fuel fire was a total destruction of two massives of pine forest growing in the direct vicinity of the plant and stretching 7–8 km to the north along the axis of the initial plume (Atlas... 1996). What is called 'red forest' – a 5 km stretch of dying pine forest with an area of 38 sq km – appeared several days after the accident. The dose absorbed by tree crowns in the epicentre of the impact zone was estimated by different experts at 80–100 Gy (Savchenko 1995) to 100–110 Gy (Medvedev 1990). In 1987 the forest in one of these areas was cut down and replanted. Currently the other one is being naturally regenerated by deciduous softwood species although the rate of restoration is very slow (Atlas... 1996).

The radioactive threshold for the radiobiological effect on trees appeared to be 20 mRh/hr (Atlas... 1996). Within forests located in the near vicinity of the NPP, five subzones were distinguished in relation to the absorbed dose, status of tree crowns and phytomass growth rate. Total destruction of *Pinus silvestris* and *Picae excelsa* species was observed within a short time span in the acute irradiation subzone where the total dose absorbed in 1986 exceeded 60 Gy. This section formed the 'red forest'. The total dead forest area covered over 4000 ha. Complete defoliation of dead needles and stratification of bark were recorded in the following year. Only by 1988 did the undergrowth and grass cover appear. After 1989 dead tree trunks started to fall down, destruction of coniferous groves continued and four to five years after the accident regeneration by self-seeded softwood deciduous species started. Currently, a completely new plant community consisting of meadow associations which had replaced the dead forests and young softwood tree species is close to being formed.

The subzone of severe impact, which received between 11 and 60 Gy, was characterized by the partial death of pine forests, lack of regrowth and discolouring of needles. Trees which survived showed major deviations in growth and development. Invasion of pests and plant diseases followed, which further weakened the groves. A new plant community was formed during 1991–1995.

The remaining territories, which cover the largest share of the 30 km 'exclusion zone', received less than 10 Gy and were moderately or weakly affected. Nevertheless, even in the low impact subzone radiobiological effects were felt for one to two years. Depression of plant growth, a change in the formation of generative organs and occasional radiomorphoses like the one shown in Plate 11, were observed here and

Plate 11 One of the genetic effects of Chernobyl: a radiomorphose in the
plant's stem

rehabilitation took three years. One of the important changes in the post-accident period was also associated with forest fires that occurred over an area of 17,000 ha including the 'red forest' area. Another ecological change included the spread of pests and tree diseases throughout the 'exclusion zone'. The present intensity of contamination of the exclusion zone with β-radiation within this 'exclusion zone' in 1994–1995 is shown in Figure 5.3. It was revealed that outside the 'alienation zone' the overall impact on the environment was also most severe in forest ecosystems particularly within coniferous groves (Savchenko 1995).

5.4.3 Impact on soil cover

In Russia grey forest soils still contain 60% of radioactive material in the forest litter, over half of which is in the lower layer, while about 30% is in the soil. Studies carried out (Ivanov *et al.* 1997) inside the 30-km restriction zone of the Chernobyl Nuclear Power Plant (CNPP) and in other contaminated regions of Ukraine, Belarus and Russia have shown that much of the caesium-137 and strontium-90 deposited by the accident in 1986 has been retained in the upper thin layers of the soil and is likely to remain there for a long time. However, in wet organic soils there has been considerable downward movement. Field observations around Chernobyl in late summer 1986 indicated that all the radionuclides showed a maximum penetration into soil of about 1.5 cm. Migration then was probably unusually slow due to the soil being very dry from a lack of rain. Later (1988 and onwards), certain radionuclides, such as those of caesium-137 and strontium-90, became depleted in this topmost 1.5 cm layer as downward migration took place.

Contamination of soil was recorded: in sandy soils the upper 3 cm; in peat soils up to 30 cm; change in the soil fauna was noticed. Wide distribution of agricultural lands made the situation worse, because ploughing of land removed the radioactive soil to the depth of 20–40 cm (Savchenko 1995). On undisturbed land, the downward migration of both caesium-117 and strontium-90 is slow, and seven years after deposition, both the radionuclides are largely confined to the topmost 5-cm soil layer. In both the sandy and peaty soils, the strontium moved faster than the caesium, but the differences were least in the peaty boggy soils where movement of caesium-137 would not have been hindered so much by clay minerals.

The observations made by Ivanov *et al.* showed also that the migration of caesium is strongly dependent on the water status of the soil substrate. The experimental results showed that the type of soil and its water content had a significant influence on the radionuclide distribution pattern in the soil profile. In undisturbed well-drained sandy and sandy loamy soils, the radionuclides were retained in the upper soil layers. However, in peaty boggy soils and flooded meadows, there was a greater

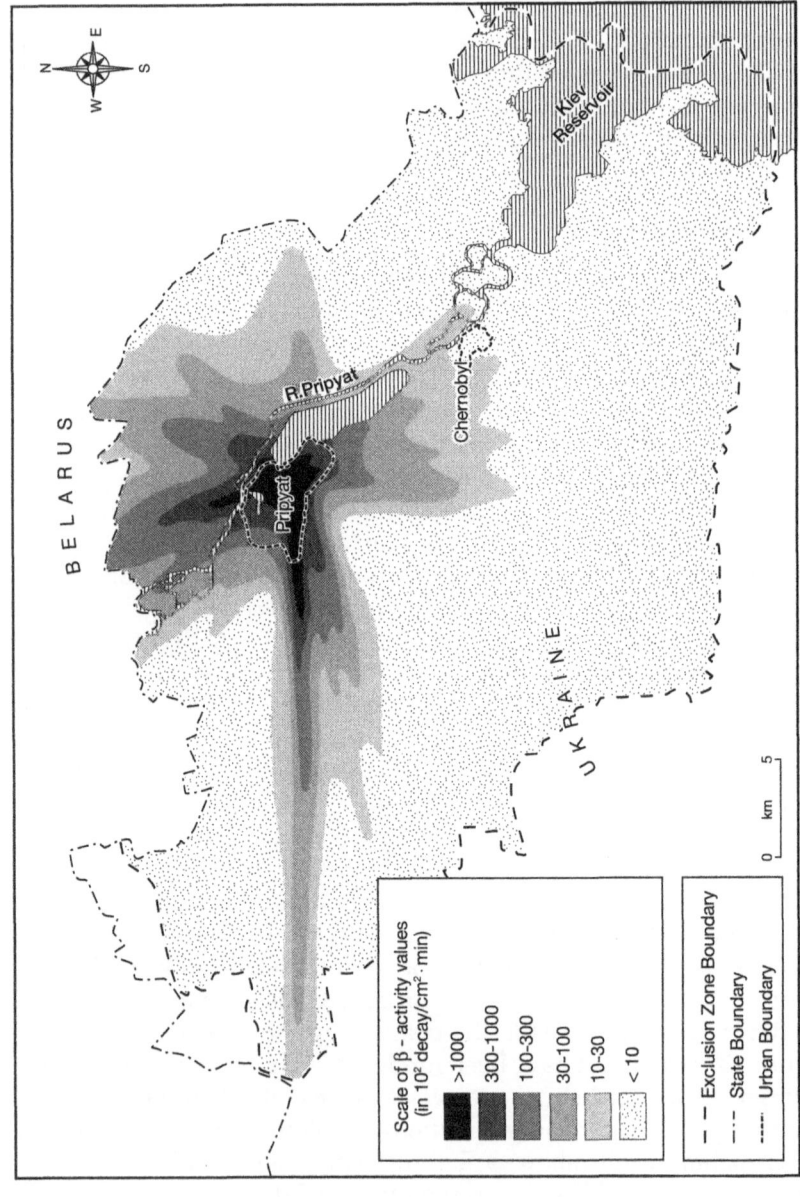

Figure 5.3 Intensity of β radiation within the 'exclusion zone', 1994–1995 (adapted from Atlas... 1996)

downward migration. In tilled soils, the radionuclides were distributed more or less homogeneously within the cultivated soil layer; the depth and homogeneity of the nuclide distribution depended on the soil texture and the way that the soil had been managed.

The vertical migration rates of strontium-90 were always higher than that of caesium-137. The downward movement of both isotopes varied according to the type of soil. The field observations, made by Ivanov *et al.* (1997) showed that, *several years after deposition, the radionuclides were largely in the top-most layer of soil (0–5 cm)*. It was also revealed that in lowland *peaty soils (periodically flooded) and peaty boggy soils of a high organic matter* and low clay content, the caesium mobility was 3–5 times higher. This fact supports the point made earlier that the natural vulnerability of organic soils is responsible for the extended long-term effect of radioactivity in the landscape.

5.4.3 Overall ecological situation

The overall impact on the environment was greatest in the direct vicinity from the reactor. The total contaminated area inside the FSU was almost 28,000 sq km, and there has also been a lesser impact on other European countries (Medvedev 1990). Figure 5.2 shows the density of contamination with caesium-137 of the Ukraine and the adjacent regions of Belarus on 10 May 1986. It shows that there were two major centres of contamination, one in the Ukraine inside the Pripyat region and another in Belarus close to the border with the Ukraine. Both centres recorded values of contamination exceeding 1000 curies per sq km. Contaminated areas in Russia amount to 1.6% of the Russian European territory and include about one million ha of forest areas (Gosudarstvenniy... 1993).

By area and by total absorbed dose, the country most affected by the accident was Belarus. It was estimated that approximately 16,500 sq km of its territory were contaminated with caesium-137, including over 2000 sq km with concentrations of more than 40 curies per sq km (Pryde 1991). The territory of the Ukraine was also greatly affected. The Russian Federation was least subjected to the radioactivity from Chernobyl, although areas of very severe contamination were found in many parts of the Bryansk oblast. More recent data on the distribution of radioactive areas is shown in Table 5.1. It shows that the total contaminated area covers almost 140,000 sq km which includes over 3000 sq km of heavily contaminated areas which have doses over 40 curies per sq km and almost 7000 sq km of those with 15 to 40 curies. It is apparent that the scale of the disaster as measured by the affected area and its ecological implications can be classed as a catastrophe.

Table 5.1 Areas contaminated with caesium-137 from Chernobyl accident by country (sq km)

Country	Density of contamination, ci/sq km			
	over 40	15–40	5–15	1–5
Belarus*	2,150	4,210	10,170	29,915
Including:				
Gomel oblast	1,625	2,760	6,740	16,870
Mogilev oblast	525	1,450	2,900	5,490
Russia**	310	1,900	5,326	49,509
Including:				
Bryansk oblast	310	1,900	2,700	6,680
Kaluga oblast	–	–	1,350	3,400
Ukraine*	571	882	3,177	37,205
Including:				
Zhitomir oblast	154	336	1,780	9,192
Kiev oblast	417	546	957	7,695
Total	3,031	6,992	18,673	116,629

Source of data: Atlas… (1998)
* as of January 1993
** as of August 1995

5.5 Health and socio-economic issues

Medvedev noted that the previous nuclear accident at Three Mile Island in Pennsylvania on 28 March 1979 'was only an "economic" catastrophe. No human lives were lost and the emission of radioactivity into the environment was limited to inert radioactive "noble" gases with a short half-life, thus not constituting any serious danger' (Medvedev 1992, 174). However in the case of Chernobyl, it appears that the most important consequences were related to health issues.

Detailed **health** studies have recorded hundreds of cases of thyroid cancer in children that can be associated with the Chernobyl catastrophe. There was an extremely serious exposure of the population to the effects of chronic external and internal radiation. According to the official report published in 1986 31 people were killed and 237 fell ill. The people who participated in the liquidation of the explosion and its consequences were called 'liquidators' and their numbers reached 400-600,000. According to unofficial data of the 'Chernobyl Union', by 1991 about 7000 liquidators had already died and 50,000 had fallen ill (Shcherbak 1991).

The 'late effects' include birth defects and other pathologies and many have already been noted. The reality of this disaster is that various health implications expected to appear in a matter of decades can be witnessed at the present time. In Chernobyl the average thyroid dose received by a child was 80 RAD from the radioactive iodine 131. The maximum dose reached 3000–4000 RAD. Already by 1992 a nine-fold increase in child thyroid cancer had been recorded in Belarus. In Kiev thyroid cancer in children doubled in frequency; in adults rectal cancer increased by 20%, cancer of the large intestine by 20%, cancer of the mouth cavity 80%, and skin cancer by 56%. Acute radiation disease and leukemia have affected people directly involved in fire extinguishing and immediate response measures. Other impacts include birth deformities, pregnancy complications, lung cancers, stomach and blood diseases.

Currently almost 800,000 persons are under the daily impact of radiation. Thirty-eight million people drink water from the Dnieper river. Ultimately the implications of water contamination can affect the people of the Black Sea, the Mediterranean and the world oceans. Recently radioactive sources of γ-radiation were discovered on the bottom of the Black Sea within its coastal zone, not only at a depth of 50 m but also at the depths of 1.5 km. These levels of radioactivity were dozens of times greater than the permissible level (*Izvestiya* 22 December 1998).

The Chernobyl accident was also a major social disaster, which undermined the security of hundreds of thousands of people who had to be resettled and who lost their jobs and homes. Table 5.2 shows the population still living in the contaminated zones. It reveals that in 1996 4.4 million people lived in polluted territories within three countries. This value included *over one million people* living in the zones of compulsory and voluntary resettlement. Hundreds of old people returned to their homes in the exclusion zone because they did not find a better life elsewhere. Hunger strikes among the liquidators of the Chernobyl disaster have become a 'common' feature of life over the last few years.

Table 5.2 Population settled in contaminated areas by country, 1995–1996 (000 people)

Country	Total population		Including in zones of compulsory resettlement		voluntary resettlement	
	1995	1996	1995	1996	1995	1996
Belarus	1841.0	1625.9	41.6	24.4	314.2	298.6
Russia*	415.5	413.2	8.5	8.5	81.7	81.1
Ukraine**	2405.0	2382.1	19.5	15.6**	653.7	649.9
Total	4661.5	4421.2	69.6	48.5	1049.6	1029.6

Source: Okruzhayushchaya... (1996)
* in Bryansk, Kaluga, Tula and Oryol oblasts only
** plus 492 people in the exclusion zone

Belarus received the main blow: 70% of all radionuclides were deposited on its territory. This made 485 settlements uninhabitable; 131,000 people were resettled from these areas. In the Khvoiniki region arable lands used to be the pride of Belarus. Grain productivity in some years reached 3000 kg per ha and the local farmers had started to grow peaches and apricots. All these lands are abandoned now. The overall distribution of contaminated agricultural lands in Belarus and the Ukraine is shown in Table 5.3. It illustrates that while the Ukraine has a greater overall contaminated area, the severely polluted land area is greater in Belarus. Agricultural production continues on the 'less severely' contaminated territory.

The magnitude of socio-economic and health implications and particularly their expected extremely long-term impact enables us to classify the overall situation in the worst affected areas contaminated by radio-caesium doses of over 40 curies per square km as an *environmental catastrophe*. Criteria for this category are given in Table 1.4.

The **economic** implications included expenses for resettlement of about 130,000 people and other post-accident measures to deal with the consequences of the catastrophe. It was estimated that in pre-devaluation Soviet prices, the total direct costs over the Soviet five-year period after the disaster amounted to 7.2 billion roubles. The indirect costs accounted for another 22 billion (Shcherbak 1991). By the time of dissolution of the FSU, an additional 5.2 billion roubles were needed to implement the resettlement programme in the Ukraine and about the same amount in Belarus. However, after both republics gained independence, all these costs had to be borne by themselves. For example, the overall expenditure in Belarus alone was estimated at 16 annual republican budgets (Marples 1995). The losses were considered comparable to those in the Second World War.

The Chernobyl accident also had far-reaching **political** consequences. The former Soviet Union was the first country in the world to create a power plant operated on nuclear energy: on 27 June 1954 the Obninsk

Table 5.3 Radioactive contamination of agricultural lands in Belarus and the Ukraine

Country	Area of contaminated agricultural lands, 000 ha			
	1–5	5–15	15–40	>40
Total in Ukraine*	3737.2	174.1	27.1	14.1
including arable lands	3341.8	87.3	19.7	10.8
% of all agriculture in category	89.4	50.1	72.7	76.6
Total in Belarus**	955.2	444.8	254.3	57.2

* as at 01/01/1991
** as at end 1998

nuclear station in the Kaluga oblast started its operation. This achievement demonstrated the might of Soviet science in the 'most progressive' type of society. Many hopes were associated with the 'tamed' nuclear energy in other countries of the globe. However, the realities of the Chernobyl disaster and its further implications have revealed to the people of the former Soviet Union the true hypocritical nature of the prevailing 'humanistic' proclamations of the Communist Party.

The growing disillusionment with the failure of the Soviet system to protect its own citizens was, arguably, one of the main driving forces for many former Soviet socialist republics behind their striving to gain sovereignty which ultimately culminated in the dissolution of the USSR in December 1991. Broader political implications for the world included abandonment of the construction of 53 reactors out of 149 planned or under construction in 1986 (Medvedev 1992).

5.6 Prospects for the future

In a recent book entitled *Global Environmental Crises* the authors stated that 'The Chernobyl event provided a dramatic lesson on the risks of nuclear power, but it would be wrong to overemphasize its importance when more subtle pollution continues daily, particularly in the Northern Hemisphere, from acid rain' (Aplin *et al.* 1996). However, this assessment overlooks the large-scale long-term nature of 'invisible' radioactive contamination that will continue to affect hundreds of generations of humans in the years to come.

Ecological changes reveal that there is a vicious circle of degeneration processes in the landscape. While the degree of radioactivity decreases in the upper layers of some soils, concentrations in vegetation increase, because isotopes reach the root system and are absorbed by plants. According to Ivanov *et al.* (1997) in peaty podzolic sand soils by 2010 about 99% of the activity will remain in the first 8 cm, while the maximum activity will be found at a depth of 1.5 cm. For the peaty podsolic gleyic soils, the maximum caesium activity will be concentrated at a depth of 3 cm, and the contamination will reach a depth lower than 12 cm.

On agricultural lands, contaminated crops are further consumed by humans or animals and accumulate in the food chain with an ultimate adverse effect on health. In the aftermath of the accident, contaminated food was distributed throughout the Soviet Union and 'dirty' products were mixed with 'clean' ones. It was revealed that in 1986, as stated by one classified document, 'the USSR *Health Ministry recommends the maximum spread of contaminated meat* across the country and to use it for production of sausages and meat products ...' (Yaroshinskaya 1992, 243). It should be realized that agricultural production on many contaminated areas continues at the present time. Furthermore, poaching of animals within protected zones e.g. in the Khvoiniki region of Belarus, contributes to the spread of contaminated meat throughout the republics and elsewhere (*Nezavisimaya Gazeta* 25 April 1998). In addition, contaminated timber from Belarus is sold cheaply in other countries of the FSU.

It has been estimated that the costs of dealing with the consequences of the catastrophe in the period from 1986 to 2016 will amount to the equivalent of 32 annual budgets (Marples 1995). Environmental surveys in the Former Soviet Union carried out over the last few years have also significantly increased the initial estimates of the total contaminated area and have included many new regions in the list of affected territories. In Belarus alone, 106,000 people need resettlement. Approximately 800 people are living or working within the 'alienation zone'. There are 679 settlements within the zone of strict control. It was estimated that Belarus *will require 320 years* to recuperate after the Chernobyl accident (*Nezavisimaya Gazeta* 25 April 1998).

A major problem is the security of the sarcophagus which contains approximately 185 tonnes of fuel. Already in the early 1990s Medvedev noted that the sarcophagus was not hermetic and had approximately 1000 sq m of various openings and cracks (Medvedev 1992). Other estimates suggest that the total area of holes and fractures reaches 1500 sq m (Shcherbak 1991). The latter author estimated that the sarcophagus contained 35 tonnes of radioactive dust with a total radioactivity of 1.7 million curies. In addition, approximately 800 'radioactive graves' within a 30 km zone which contain buried waste materials from the power plant present another major potential danger. Only a few of these 'graves' correspond to international security standards.

The problem of the operating Chernobyl reactor is still unresolved. The closure of this plant has been in the process of negotiation with Western nations for many years but the amount of compensation claimed by the Ukraine is unreasonably high, thus the 'trade' continues. It should be noted that fuel-deficient Ukraine heavily depends on nuclear energy. The contributory share of nuclear power plants in the overall electric energy generation in this country kept increasing after independence. In 1994 it reached 33.9% compared to 11.2% in the Russian Federation (Okruzhayushchaya... 1996). This will remain the main obstacle to achieving this goal.

New long-term threats are also mounting. Initially, the impact was due to the activity of plutonium-239 that amounted to 414 kg and plutonium-241 at 34 kg; half-life of the latter is 15 years. Disintegration of plutonium-241 produces ameritium-241, which has been already revealed in the Black Sea. It has a half-life of 434 years and can create caverns in the tissues of living organisms. It is expected that its share of impact of α–pollution will further increase and 50–70 years after the accident its activity will have increased by 6 times compared to the initial level. It will be twice the value of the total activity of the three plutonium isotopes, 238, 239 and 240 (Atlas... 1998). Very little is known about the impact of many consequent isotopes but what is clear is that the impact will last for at least another 2000 years.

The recent financial crisis has hit the population both directly and indirectly, and has created additional dangers in the sphere of nuclear energy production. Strikes of technical and engineering staff of nuclear stations because of chronically accumulating wage arrears have become common, thus threatening safe operation and maintenance of equipment. As was recently reported, the situation in the Russian NPPs has worsened which is reflected in the increased number of technical accidents; for example in 1998 102 disruptions in their operations were recorded, or 23 more than in 1997 (*Nezavisimaya Gazeta* 14 January 1999, 2). In 1999, one person died and three people received burns as a result of a fire at the Kalininskaya NPP.

It should be noted that there are also other sources of radioactive contamination in Russia. As noted in the State Report on the Status of the Environment in 1997, the contemporary threats are related to the following: (a) the global radioactive background associated with the former nuclear tests; (b) the existence of contaminated areas from Chernobyl and technical accidents at 'Mayak' in the Ural mountains in 1957 and 1967; (c) the operation of nuclear-based enterprises and power stations, nuclear energy ship installations and regional radioactive waste storage. 'Mayak' is potentially the most dangerous source, because during its operation a large amount of liquid and solid radioactive wastes have been accumulated with combined radioactivity of one billion curies (Gosudarstvenniy... 1998c). An additional danger comes from the recently increased activity of terrorists in different parts of this country.

CHAPTER 6

The 'Black Triangle' of Central Europe: air pollution comes under control?

... the improvement in environmental quality in Central and Eastern Europe is likely to be related more to the form of the necessary socio-economic development of these countries than to environmental policies and programmes.

(Europe's Environment 1998, 7)

6.1 Introduction

The objective of this chapter is to attempt to analyse the magnitude of environmental crises in the industrial heartland of Central Europe. It will examine the situation in the so-called 'Black' or 'Sulphur Triangle', which was arguably the worst affected region within the former 'socialist bloc' in terms of air pollution. The 'Black Triangle' includes the Silesian district of Poland, northern Bohemia in the Czech Republic and the south-eastern part of the former German Democratic Republic (GDR).

One introductory point is that the concepts of 'Central' and 'Eastern' Europe are currently undergoing change and are used with different meanings by various authors. As stated by Mellor (1975, 3), in the 19th and early 20th centuries Central Europe 'was a real image'. The division of Europe associated with the Cold War resulted in the abandonment of this term. The concept of 'Eastern Europe' that was broadly used during the post-war period incorporated eight socialist countries thus effectively excluding the enormous territory of the European USSR. With the creation of the New Independent States (NIS), and a dramatic change in the geopolitical situation, both concepts have been revised once again.

In this chapter, the term 'Eastern Europe' refers to the newly independent countries that previously formed the Soviet Union, including the European part of Russia, while 'Central Europe' is used in relation to the former socialist bloc partners of the USSR. It should be emphasized that in the latter region pristine nature is still estimated to cover about 30% of the area (Klarer *et al.* 1994). It contains the greatest reserve of biological diversity in Europe.

The magnitude and nature of the inherited environmental problems were extremely variable in each of the Central and Eastern European countries, but the overall situation could with good reason be called a crisis, particularly within the so-called Black Triangle. This heavily polluted region actually represents a polygon rather than a triangle, as noted by Manser (1993). Intensive industrialization after the Second World War, enforced by the Stalinist regime and carried on for four decades of socialism, has resulted in extensive ecological damage.

It would be also useful to point out that a rapid growth of environmental awareness had occurred in the countries of Central Europe towards the end of the Cold War period. By that time, when the public at large knew the real scope and magnitude of environmental problems, this environmentalism temporarily acquired the strength of a political opposition. However, during the post-socialist period the influence and strength of this movement has gradually weakened.

Dramatic political and socio-economic transformations that were evident in these countries throughout the 'era of transition' had a clear impact on environmental matters. While in the former GDR these were associated with the unification of Germany, in Poland and the former Czechoslovakia some positive changes were mainly linked with their aspiration to join the EU and the need to harmonize their environmental legislation with its policies. Therefore, while being the worst example of air pollution during the socialist period, the Black Triangle at the same time represents probably one of the few examples of at least a partial 'success story' throughout the whole of post-socialist Eurasia.

By the onset of the 'Velvet Revolution' and over the last decade, the problems of the Black Triangle have been widely covered in Western scientific literature (e.g. Carter and Turnock 1996, 1993; Levy 1994; Manser 1993). Along with these overall assessments, regional and country-by-country approaches have been used to study the multitude of environmental problems of this region (e.g. DeBardeleben and Hannigan 1994; Fisher 1992; Klarer and Moldan 1997; Vari and Tamaz 1993). Various political issues are discussed in these publications along with environmental problems.

6.2 Physical geographical setting

The region under study is **located** within the territories of three countries in Central Europe: the Silesian district of Poland, the north-eastern part of the former Czechoslovakia and Northern Bohemia of the former German Democratic Republic. Characteristic altitudes in the southern

Bohemia reach 250–390 m. The Sudeten Mountains, which rise to the altitude of 1602 m, form the boundary between Poland and the former Czechoslovakia. Generally, the relief in the south of the region features gently rolling Hercynian uplands with elevations at 150–170 m above sea level. They are separated by broad basins or deeply incised valleys.

To the north, the open areas of the North European plain with glacial landforms impart a clear contrast in landscape. Since the end of the Ice Age they have been greatly modified by melt waters and strong winds so that in the south there are large expanses of a thin layer of fine yellow dust, or loess. Along with sand and silt, loess covers an undulating lowland of the upper Elbe (Polabi) in the interior of Bohemia. The Bohemian massif is a broken and faulted plateau composed of fractured volcanic rocks.

The region has a transitional **climate** between the Atlantic western European oceanic type and the continental Russian plain type. Generally, the warm Atlantic Gulf Stream exerts a moderating influence on the climate of Central Europe. Compared to that of the Russian plain, the climate of Central Europe is substantially milder. At the same time, the yearly weather pattern has a more obvious seasonality compared to Atlantic coastal Europe. In addition, in the mountains, elevation plays an important modifying role on the climate. The Azores high-pressure system has a major influence in summer, when the pressure is highest. Prevailing winds are mostly westerly and the cyclonic atmospheric movement mainly follows from the west. This is also the main direction of transboundary pollution.

Winter becomes colder towards the east and lasts for two to three months, most of this time snowy. January temperatures average –1 to –5 degrees C with an absolute minimum temperature falling to as low as –29 degrees C. Temperature inversions with accumulation of fog are quite common in winter in the piedmont plains near the Sudeten mountains. Average July temperatures amount to +17 to +19 degrees C. Precipitation mainly occurs in late spring and summer. Average precipitation in the plains amounts to 500–600 mm increasing to about 1000 mm in the mountains. The **hydrological** network is quite dense and some major rivers originate in the mountains of Central Europe, e.g. the Elbe and Oder. During cold winters, near Dresden the Elbe freezes for about one month.

Over the North European plain **soils** are mostly podzolic grey-brown with more fertile brown forest soils closer to the mountain piedmont areas. Within mountain ranges the soil cover is more complex and diverse with high erosion risks at greater altitudes. While at lower elevations in the Bohemian uplands soils are mainly podzolic, poor in humus content and acidic, on the highest areas skeletal podzols are common.

The 'zonal' or original type of **vegetation** according to Romanova (1997) is the sub-boreal broad-leaved deciduous forest, now practically destroyed on the European plains. These forests were mainly composed of oak (*Quercus robur*) and beech (*Fagus silvatica*) with the inclusion of the occasional hornbeam, elm, ash, maple, lime, etc. They are still found within the Central European uplands at the altitudes up to 500–600 m. This type of landscape is arguably one of the most stable compared to other geographical zones. The broad-leaved deciduous forests have a complex many-tier structure that also includes a shrub and a rich grass layer. The overall amount of the phytomass at 300–500 tonnes per ha is the greatest compared to any other geographical zone of northern Eurasia. The value of plant biodiversity, estimated at 700–800 species per 100 sq km, is also very high and is greater only in the forested steppe and steppe zones (Table 1.2).

However, generally, these landscapes have been significantly modified by human influence due the long history of European settlement and economic development. The present forest in many parts has a quite different composition compared to what might be expected in its natural condition. Ploughing of large deforested areas throughout Europe has substantially changed the original landscapes. Furthermore, according to Mellor (1975, 32), 'there is a natural poverty of flora in Central and Eastern Europe conditioned by the wholesale annihilation of species during the Pleistocene glaciations. Time has been too short for a full recolonisation by the rich flora of those areas from which the ice last retreated'.

Transformation of the original landscape also occured with the planting of quick-growing coniferous trees even in areas of indigenous deciduous vegetation. This has resulted in the creation of extensive stands of coniferous forests and a greater participation of coniferous species in the mixed forests further east of the region. In the mountains, mixed forests have been replaced by stands of conifers with spruce and fir characteristic of higher elevations. For example, in Slovakia the area of forests covers about 20,000 sq km or 40.2% of the country's territory; this incorporates five national parks and 16 other protected areas. Within these forests, which are predominantly mountainous, the actual tree composition consists of 57.7% deciduous and 42.3% coniferous species (Mankovska 1997). The main deciduous trees include beech (29.8%) and oaks (14.3%). Coniferous trees are represented by the Norway spruce (*Picea excelsa*) at 27.2%, silver fir (*Abies alba*) at 5%, pines and larch.

Despite an overall high degree of transformation, the overall **vulnerability** of this type of landscape is less than in comparison with other geographical zones. This is due, firstly, to the favourable climatic conditions associated with the Atlantic influence and secondly, to the more stable natural vegetation communities described above. However, even in the relatively resistant type of landscape such as deciduous forests, various components have a different degree of vulnerability to intensive anthropogenic influences. For example, in the case of the impact of the Sudbury smelters described by Colls (1997), the damage pattern included the different reaction of differing vegetation tiers. In the forest formations, the trees had the greatest potential gas deposition fluxes because they experienced the highest windspeeds and had the lowest boundary-layer resistance. The undergrowth of bushes was not only lower but also had some protection from the trees. The ground layer composed of lichens, mosses and grasses managed to survive the effect even to within a few km from the source of pollution.

At the same time, temperature inversions during the winter period contribute to the accumulation of pollutants in the lower atmospheric layers and prevent their dispersal over larger areas. A wide distribution of conifers, which are more vulnerable to air pollution (see Chapter 2), is an important feature as well. In addition, mountainous topography presents a major erosion risk factor and aggravates atmospheric pollution by serving as a barrier to pollutant dispersal.

6.3 Historical perspective and economic development

The territory of the Black Triangle is relatively rich in certain energy resources, particularly in coal. It contains 110,000 billion tonnes of hard coal, 90% of which is concentrated in Poland, and almost 100,000 billion tonnes of brown coal or 'lignites', mainly concentrated in the former Czechoslovakia and GDR. Much of the high-quality anthracite coals extracted in Poland are exported (Plate 12), while the lower quality

Plate 12 Poland possesses Europe's largest reserves of coal

coals have sometimes been used locally for energy generation and industrial needs. In other countries, brown coal has served as the main energy basis for industrial production. The former GDR was the world's leading lignite producer; in that country brown coals were not only a source of energy but provided an important raw resource for the chemical industry.

Industrial development, which started in the 19th century in the former Czechoslovakia and GDR, was further accelerated by intensive industrialization during the Soviet period. Industrialization in the three countries of the region happened at a very fast rate, which is reflected in the data on change in their industrial production. Between the 1950s and the 1980s the total industrial production in Poland increased by 12–13 times, and in the former GDR and Czechoslovakia by 8–9 times (Nefedova and Treivish 1994). Another index of industrialization is employment in industry. Thus in 1980 in Poland it reached 54.9%, Czechoslovakia 65% and GDR 68.7%.

The Stalinist model of economic development, adopted in these countries under pressure from the former Soviet Union, included assigning the greatest priority to heavy and energy industries and was based on a very high consumption of energy and raw materials. This type of economy created a typical pattern found in all countries of the former socialist bloc, which involved the concentration of industrial and mining enterprises in a single geographical region (Nefedova 1994). Furthermore, energy production in Central European countries was in many cases based on highly polluting, sometimes even obsolete, thermal power stations. They often lacked proper treatment facilities and advanced re-use technologies with the exception of the GDR in the late socialist period.

In general, big state subsidies and the absence of competition have created a total lack of interest among enterprises and the population at large in saving energy, raw resources and water. Ineffective use of energy resources contributes to increased air pollution, primarily with carbon oxide. It was estimated that in the countries of the FSU and Central Europe consumption of energy per unit of production is approximately 1.5–2 times greater than in the USA and two to four times greater than in Japan (*Zelyoniy Mir* 25, 1996). Ideological dogma was another hindrance to environmentally friendly economic development. As noted by Carter and Turnock, 'the Marxist concept of regarding natural resources as "free goods" has only added to the problem by encouraging waste' (Carter and Turnock 1993, 1).

But probably the most aggravating circumstances and the most important specific reasons for the current situation were the sources of energy used for industrial development and electricity generation. Industries and power stations generally burned brown coal and low-quality lignite with a high sulphur content. This is a poorer source of fuel and, due to its lower calorific value, one needs to burn more lignite than hard coal to obtain the same amount of heat. The lack of locally produced oil and other energy sources, along with the acceleration of industrial development, has resulted in the expansion of lignite consumption during the years of the socialist regimes. For example, in the former Czechoslovakia 50% of all energy needs were covered from domestic sources, which were mainly limited to coal production (EIU Country Reports 1996d).

During the Soviet period production of lignite in the latter country increased and reached its peak value of 102.9 million metric tonnes in 1984. In 1991 brown coal and lignite made up over 80% of the total coal production (EIU Country Profile 1996a). In the former GDR, the total brown coal and lignite production during the covered period was even three times greater (OECD 1991). As a result of this, sulphur dioxide emissions per production unit in the countries of the Black Triangle were five times greater than in the USA and 20 times more than in the former West Germany. Against this background, it was estimated (Alcamo 1992) that a 70% reduction in sulphur dioxide in the region as a whole will cost annually approximately US$ 8.6 billion.

To summarize, there have been several reasons for the gravity of the environmental situation in this region, as discussed by the author earlier (Saiko 1998a). The most important of them are:

(a) a long history of industrial development in some countries of the region;
(b) intensive industrialization under the socialist regime;
(c) high priority attached to heavy and energy industries in economic development;
(d) lack of investment in industry against a background of a persistent requirement to increase output;
(f) a large share of ageing and obsolete machinery and equipment in industry;
(g) flawed regional development that has allowed concentration of mining, industrial enterprises and power stations within a single geographical region;
(h) predominant development of high resource- and energy-consuming monopoly industries;
(i) heavy regional reliance on thermal power stations which operate on brown coals;
(j) poor-quality lignite with a high sulphur content used for industrial purposes;
(k) lack of adequate treatment facilities and advanced re-use technologies in industry;
(l) lack of economic incentives for environmental improvement;
(m) lack of public environmental awareness and effective legislation.

6.4 Ecological consequences and recent changes

It should be mentioned that severe pollution was not a complete stranger to the socialist countries of the Black Triangle, as industrial development in some of them had started in the 19th century. As stated by Carter and Turnock (1993, 1), 'some pollution black spots already existed in this area prior to the Second World War but they were mainly confined to the limits of the Budapest-Lodz-Leipzig industrial triangle'. However, the magnitude and scope of ecological problems have dramatically increased over the socialist period. There is a whole variety of environmental implications that has stemmed from the nature of post-war economic development. This industrial heartland of the former socialist partners of the USSR has probably paid the highest air pollution cost for its rapid industrialization between the 1950s and the 1980s.

6.4.1 Air pollution

Airborne emissions of sulphur dioxide, nitrogen oxides, particulate matter and fly ash have historically been the main sources of air pollution in the Black Triangle. In general, European sulphur emissions increased steadily from 1880 up to a maximum of 60 million tonnes per year in 1980, when they started to decline sharply. The Central European countries featured a similar pattern although with a 10–15 years' delay. In 1971 in the GDR 4 million tonnes of dust and 4–6 million tonnes of sulphur compounds were emitted into the atmosphere by industry (Fullenbach 1981). By 1988 the former GDR provided 74% of the total volume of East European sulphur dioxide emissions (Levy 1994), and in general it would appear that some countries of the Black Triangle were even more seriously affected by this problem than the former Soviet Union.

As was emphasized in an IUCN report, Czechoslovakia produced more air pollutants per head of population than any other European country (IUCN 1990). In the GDR in 1989, energy installations produced 94% of sulphur dioxide, 73% of dust and 81% of nitrogen oxides (Manser 1993). According to Levy, 'a single power plant in East Germany emitted more sulphur than all sources in Sweden' (Levy 1994, 313). Within each country of the Black Triangle 'hot spots' accounted for the worst pollution. For example, the Katowice region of Poland which occupies only 2% of the country's area produced 22.4% of all emissions of sulphur dioxide.

The values of total emissions of sulphuric and nitrogen oxides in the selected countries of Europe in the 1980s based on the estimates of the Dobris Assessment are shown in Table 6.1. It illustrates that in 1985 the former GDR was the record holder in sulphur dioxide emissions with 5.4 million tonnes, coming close to the amount emitted from the enormous territory of the European part of the USSR. By 1990 it had already exceeded the Soviet Union's level despite an overall reduction in

Table 6.1 Total emissions of SOx and NOx in selected countries of Europe, 1980–1990 (000 tonnes)

Country	SOx emissions			NOx emissions		
	1980	1985	1990	1980	1985	1990
Czech Republic	2257	2277	1876	937	831	742
Germany	7650	7900	5800	3640	3700	3190
Former GDR	4650	5400	4800	540	600	590
Poland	4100	4300	3210	–	1500	1280
Russian Federation	7161	6191	4460	1734	1903	2675
Slovakia	700	606	539	–	–	245
Sweden	507	267	135	424	426	400
United Kingdom	4898	3724	3780	2365	2392	2779

Source of data: Europe's Environment (1995b)

emissions. If comparison is made with the United Kingdom, it is evident that while in 1980 emissions of the latter were greater than in any of the three Black Triangle countries, five years later both the former GDR and Poland emitted more sulphur oxides than the UK.

Table 6.1 also shows that the main share of emissions in the former Czechoslovakia, which was third in terms of the overall emission values, came from the territory of the present Czech Republic. For nitrogen oxide emissions the situation was rather different; most emissions were associated with the former West Germany, while Poland took the lead among the Black Triangle countries. In addition in the early 1990s more than 400,000 metric tonnes of soot ash, dust and 700,000 metric tonnes of polluting gases smothered Bohemia each year. Furthermore, an important kind of pollution in the region was associated with persistent smog hanging over the industrial cities and centres during the winter period. Air pollution with heavy metals, more specifically iron, zinc, lead, copper, chromium, nickel and cadmium, presented an additional problem in the Black Triangle.

6.4.2 Water pollution

Water pollution has been a serious problem as well. By the 1980s approximately 80% of the rivers in the Black Triangle countries were contaminated (IUCN 1990). Often water pollution was also related to areas of coal production. For example in Poland, the high salinity of river water is due to runoff pollution from mine sites (Nowicki 1997). Contamination of water by heavy metals is another major threat that has wider implications. In the former GDR the river Mulde directly received untreated waters, while the Elbe had mercury levels 250 times above EU standards (Carter and Turnock 1993).

Two of Poland's largest rivers, the Vistula and the Oder, are highly polluted and discharge into the Baltic Sea further contributing to its adverse water quality. In 1989 the Polish government classified 40.7% of their rivers unsuitable for industrial use by physical and chemical criteria; of the remaining managed rivers 27.6% were suitable only for industrial use, 28.3% for agricultural and recreational application and only 3.4% were pure enough for municipal usage. According to biological criteria only 0.5% were suitable for municipal water supply. More recently, for instance in the former GDR, half of all available water sources were still unusable for drinking and another 38% usable only after substantial treatment operations (Jones 1995).

6.4.3 Soil pollution

Acidification of soil associated with air pollution by sulphur and nitrogen oxides remains a serious problem in all countries of the region. Figure 6.1 illustrates the distribution of acid rains in Central Europe in the 1980s at the peak of air pollution. It appears that due to the prevailing westerly wind direction and transboundary nature of pollution, the focus of acid deposition shifted further east. The area with extremely high acidity in

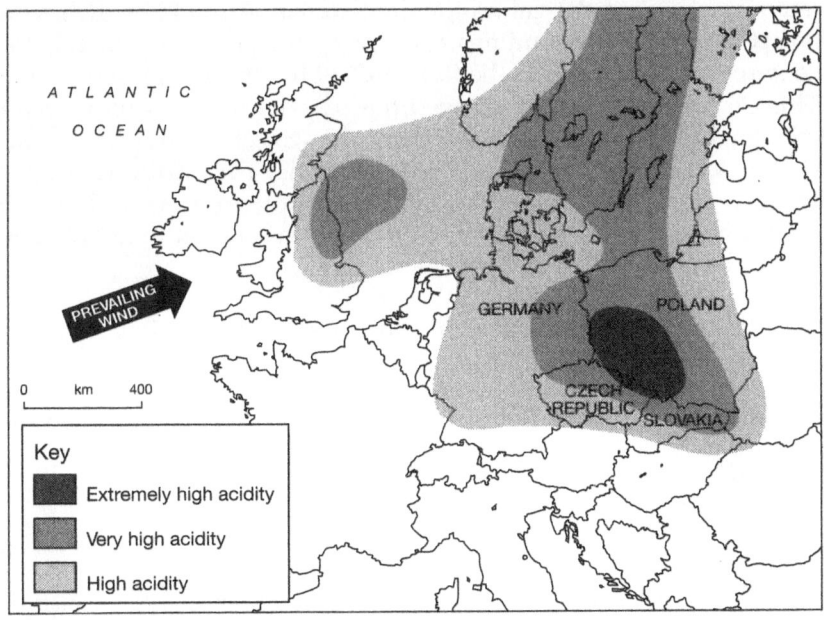

Figure 6.1   Acid rains in Central Europe, 1988 (after Thompson 1991): 1 – high acidity level; 2 – very high acidity level; 3 – extremely high acidity level

1988 covered much of the south-western part of Poland and a narrow band along the north-eastern border of the contemporary Czech Republic. At the same time, the areas of high and very high acidity effectively comprised almost the whole territories of the former Czechoslovakia, Germany and Poland. It should be noted, that along with wet deposition, dry deposition normally within the boundaries of the main polluting centres presents a serious threat to the local soils and vegetation.

More recently, it has been noted that while deposition of sulphur compounds have decreased, the input of nitrogen has remained high and has even increased in some areas (Matzner and Murach 1995). The latter can probably be associated with a dramatic increase in the number of private vehicles during the transitional period. It should be emphasized that there is a clear relation between air pollution, deposition of pollutants in soil, change in soil acidity and forest damage.

Furthermore, soil pollution with heavy metals is another major problem. A recent survey of concentrations of heavy metals in moss in the countries of the Black Triangle (Markert et al. 1996) has revealed an increase in heavy metal concentrations and higher average levels in the eastern parts of these countries. In addition, as shown in Table 6.2, the highest values of such pollutants as cadmium, chromium, copper, iron, lead and zinc were recorded in Slovakia. Only in nickel concentration values did the Czech Republic take the lead. In almost all categories except zinc pollution the values for Germany appeared to be the lowest. This fact supports other findings that during the post-unification period the environmental quality in the former GDR has been improving.

This discovery has even allowed the authors to speak of a 'second Black Triangle' at the point where the Czech Republic, Poland and Slovakia meet. While during the socialist period the two upper 'angles' were 'made up' by the GDR and Poland and the lower by Czechoslovakia, more recently (or probably in relation to heavy metal

Table 6.2 Mean concentrations of heavy metals in moss in countries of the 'Black Triangle', 1991–1992 (mg/g)

Heavy metal	Czech Republic	Germany	Poland	Slovakia
Cadmium	0.36	0.34	0.59	1.38
Chromium	2.29	2.11	2.54	5.17
Copper	8.70	9.49	10.4	20.4
Iron	809	720	1500	1870
Lead	19.1	14.6	30.0	60.5
Nickel	3.62	2.61	2.49	2.23
Zinc	52.5	55.2	66.4	173.0

Source of data: Markert et al. (1996)

contamination), there has been a change to another shape of triangle with one upper 'angle' being formed by Poland and the two lower ones by the Czech Republic and Slovakia. This latter 'triangle' has the highest level of heavy metal pollution.

6.4.4 Forest deterioration

Forest degradation is a much more widespread concern than deforestation. Defoliation and discoloration along with reduced vitality indicate a general worsening in forest conditions. Increased acidity of soils affects the root system of trees and has implications for their drought susceptibility. Additionally, there is a direct effect of gases on leaves, nutritional imbalances and indirectly on pest–plant interaction. According to Matzner and Murach (1995), a survey of leaf and needle losses of European forests in 1993 revealed that 23% of the total forested area had defoliation of more than 25% with a focus in Poland, Slovakia, Czech Republic and Germany. A survey carried out in the latter country in 1994 has also shown that there was recorded an increasing leaf loss in oak and beech species between 1984 and 1994. As noted by the above authors, oak stands have the highest visible damage with about 30–40% of the trees having severe defoliation.

In the Czech Republic, between 1985 and 1993 the share of healthy deciduous forest stands has decreased from 89.4% to 42.9% while the percentage of coniferous trees without damage dropped from 50.2% to 36.6% (Table 6.3). The proportion of trees with first symptoms and slight damage increased in both the coniferous and deciduous forests. At the same time, the share of severely and very severely damaged conifers has slightly declined. Table 6.3 also gives convincing evidence that coniferous species

Table 6.3 Forest damage by air pollution in the Czech Republic

Degree of damage	Damaged deciduous, % of total area			Damaged conifers, % of total area		
	1985	1990	1993	1985	1990	1993
No damage	89.4	54.6	42.9	50.2	38.6	36.6
First symptoms	10.6	36.8	46.7	28.6	35.0	37.0
Slight	–	8.0	9.9	13.1	19.0	20.0
Medium	–	0.6	0.5	6.1	5.6	5.1
Severe	–	–	–	1.4	1.3	1.0
Very severe	–	–	–	0.3	0.2	0.2
Dying	–	–	–	0.2	0.1	0.1
Dead	–	–	–	0.1	0.1	–

Source: After Moldau (1997)

appear to be more vulnerable to air pollution than deciduous. Another study of sulphur content in the foliage of deciduous and coniferous trees in Slovakia (Mankovska 1997) has shown that even within protected areas concentrations of sulphur in coniferous species were alarmingly high even when compared with the data from 1975. These results suggest that despite a substantial overall reduction of atmospheric emissions, only a limited improvement in the forest's 'health' has taken place.

The impact of air pollution on forests is illustrated by Figure 6.2. It shows that the shape of the area of the most severely damaged forests actually represents a polygon rather than a triangle. The centre of this polygon coincides, however, with the point where Poland, Germany and Czechoslovakia meet. It also demonstrates that extensively damaged areas spread well past this territory covering much of the former socialist

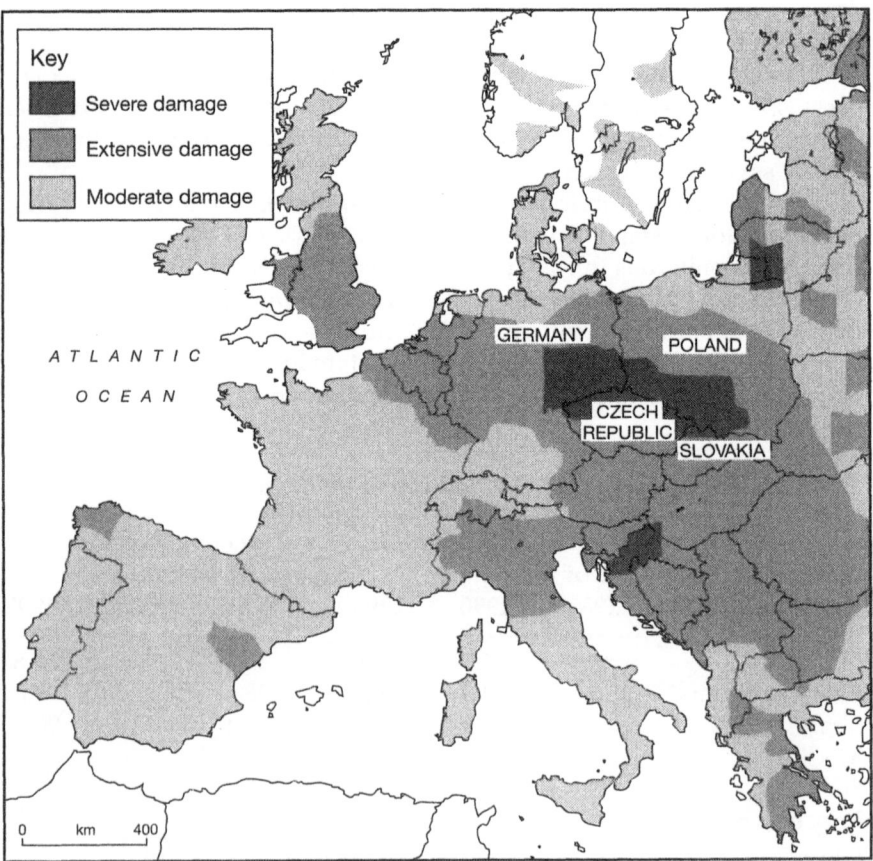

Figure 6.2 Impact of pollution on forests in Central Europe (after Wolfson 1994)

states to the south of Central Europe. An alarming recent finding has revealed that 'in spite of emission reductions, the monitoring results show a general increase in tree defoliation' (Europe's... 1998, 74).

6.4.5 Overall ecological situation

The overall ecological status in the region by the end of the socialist period could be with good reason called 'a crisis'. At the same time, in several 'ecological disaster areas' of the Black Triangle e.g. in the copper mining and smelting area around Glogow and Legnica in Lower Silesia and in the Krakow district in Poland, it was catastrophic. As noted by Manser 'In reality it was neither a triangle nor a rigidly defined area: other nearby districts, equally degraded, included Cottbus and the Halle-Leipzig district in Eastern Germany and the Ostrawa/Katowice/Krakow area in Moravia and southern Poland' (Manser 1993, 22). It should be pointed out that in comparison with the Black Triangle, the severity of air pollution problems in the Donetsk Basin, which is the 'worst' polluted industrial region of the Ukraine, or in the industrial centres of the Ural mountains in Russia appear to be substantially less.

However, the situation has probably best been described by the above author (Manser 1993, 18):

> In 1989, it was easy though misleading to paint a completely black picture of Eastern Europe. The region contained several large heavily polluted districts and a fair number of polluted cities and towns. Both within and without them, there were also smaller zones where the natural habitat had been or being totally destroyed. However, between these areas as well as sometimes within them, there were stretches of unspoilt forest, secondary growth and clean water. These included protected areas – national parks, nature reserves and the like with varying restrictions on human activity.

After the Velvet Revolution the crisis in the region has weakened and reduced in acuteness in some areas, particularly in the former GDR after unification and in Poland in the course of economic restructuring. Nevertheless, in the worst degraded areas of Poland and former Czechoslovakia a lot more still has to be done to make the radical improvement in the ecological situation. In the first place this relates to the state of forest vegetation and soil pollution.

It has been estimated that environmental rehabilitation in East Germany to bring it into line with Western standards would require 10 years and US$249–308 billion. For Poland, Czech and Slovak Republics it would take several decades and over US$20 billion (Zelyoniy Mir 25, 1995). Both values appear immensely high, considering the adverse economic realities in the period of transition. According to Halkos, in terms of percentage of GDP, the highest costs of imposition of West European standards (over 0.5%) were also to be borne by Germany, Poland and other Central and Southern European countries (Halkos 1996). It should be noted that while overall radical improvement in the worst affected areas clearly requires decades, some recent progress in pollution control associated with post-socialist restructuring and change is already evident.

6.5 Socio-economic and health issues

Although it remains a difficult task to distinguish between the roles that ecological deterioration and other factors play in affecting human health, there is a clear link between the two. As noted by Fodor (1994), as a result of complex synergistic effects the number of tumours in Central Europe doubled between 1976–1986. More than half a million people in the region are believed to suffer from pollution-induced ill health.

One of the main causes was smog composed of dust and soot that enveloped large areas, such as in the former GDR. According to Carter and Turnock (1993), due to this, in Leipzig four-fifths of the children aged six years developed chronic bronchitis or heart problems. Also according to this source, over the last four years, paediatric hospitals in the North Bohemian coal mining area of the Czech Republic have up to 12 times the number of sick children compared to the Czech average; infant mortality rates are more than 40% higher than the national norm. A reliable overall health indicator is life expectancy. Recent studies have indicated that life expectancy in urban areas in Poland and the Czech Republic is significantly lower than the average for these countries (e.g. Herzman 1995). Other assessments showed that, generally, life expectancy in the worst polluted regions was five to seven years less than national levels.

Polish economists have estimated that the annual damage to health and economy from pollution is equal to 10–15% of the country's GDP (*Zelyoniy Mir* 25, 1995). In the former Czechoslovakia a similar expenditure is estimated at 5–7% of GDP. Therefore, rehabilitation of the environment should be considered not as an obstacle but as one of the main prerequisites for economic revival.

6.6 Prospects for the future

Recent data indicate that there has been a decrease in the emissions of sulphur dioxide in Western, Central and Eastern Europe (Figure 6.3). Even in the Black Triangle there has been a noticeable improvement. According to the results of the second assessment of the state of the environment at the pan-European level which was made by the European Environment Agency, emissions of sulphur dioxide in Europe *halved* between 1980 and 1995. In Central Europe the decrease accounted for

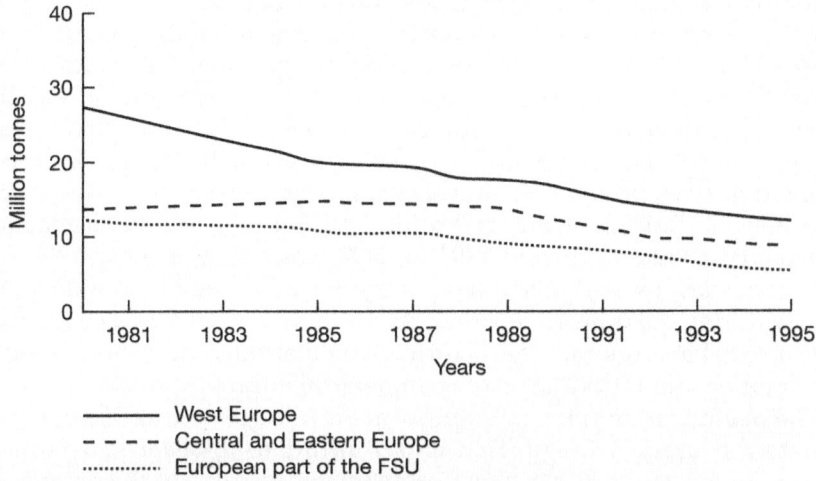

Figure 6.3 Reduction in sulphur dioxide pollution in West, Central and Eastern Europe, and the European part of the FSU (after Europe's Environment... 1998)

40% of the 1980 level. The value of sulphur deposition has also decreased. For example according to Colls (1997), in the Erzgebirge mountains the annual deposition declined from 125 kg per ha in 1989 to around 60 more recently. The latter value can be compared to the UK maximum of about 30 kg.

Concentration of sulphur in the precipitation has decreased by 40% while in air it was even greater at 70% (Brueggemann and Spindler 1999). The wet deposition has declined by 30% between 1991 and 1995. A study of the effects of wet deposition on ecosystems carried out in the former GDR between 1992 and 1995 showed a 20% to 70% decrease in concentrations of sulphate and calcium in the rural and mountain areas (Brueggemann and Rolle 1998). In addition, no increase in acidity was observed.

At the same time, total nitrogen emissions including nitrogen oxides and ammonia, which remained roughly constant during the 1980–1990 period, fell only by about 15% between 1990 and 1995, the largest falls occurring in Central and Eastern Europe. However, critical loads, or the levels of deposition above which long-term harmful effects can be expected, are still being exceeded in 10% of Europe's total land area, mainly in Northern and Central Europe (Europe's ... 1998).

Although many previous hopes of a rapid and dramatic environmental improvement after 1989 have not materialized, some positive changes discussed above have already been recorded. One of the reasons for this progress was the willingness of new governments to comply with EU standards and to make the needed adjustments required for such compliance. The reduction in sulphur dioxide emissions can also be partially attributed to the decline in the production of lignites in the countries of Central Europe. Calculations show that, for example in the Czech Republic, it decreased by about 26% between 1991 and 1997 (The Europe... 1999; Eastern Europe 1994). Compared to the peak value recorded in 1984, production ten years later has declined by almost one-third (OECD 1991; EIU Country Profile 1996a). In Poland the decline in brown coal production from 1991 to 1997 was less at around 12%. In some countries it was due to a temporary overall reduction in industrial activity during the transition period. At the same time, it was also related to post-socialist economic restructuring aimed at the reduction of lignite consumption and introduction of technological improvements.

The overall environmental improvement was greatest in those countries of the Black Triangle that could afford to allocate substantial finances to this area, primarily in Germany: this was due to the reunification of the two German states in October 1990. Various surveys carried out recently (e.g. Brueggemann and Rolle 1998; Brueggemann and Spindler 1999) have observed a decline in the amount of harmful emissions, wet and dry deposition. The main changes after unification included the reformation of industry, particularly de-industrialization, reduction in stock farming, closure or modernization of many factories, reduction in the capacity of lignite-fired power plants, substitution of lignites by other fuels (petroleum and natural gas), a better removal of dust from flue gas, etc. In general, the causes of emission reduction in the Black Triangle countries have included more efficient use of energy sources and the introduction of de-sulphurization systems.

Generally, positive changes during the post-socialist period have been greatest in the countries that have achieved more in terms of socio-economic restructuring. For example, this was the case in Poland, which has chosen a more radical way to economic transformation. When the reunification of the two German states took place, its eastern neighbour had started a year of 'shock therapy' for its economy. Polish government ministers were heard to complain: 'Why cannot we have an affluent western Polish state to pay for our economic therapy?' (Kennedy 1993). However, despite the lack of external help, the latter country enjoyed the earliest economic revival and growth: GDP in 1999 is forecast to be 121% of the level recorded in 1989 (The Economist 20 November 1999). This is the highest increase in Central Europe and the FSU states. Post-socialist socio-economic changes here were the main prerequisites for the improvement in various spheres, including environmental.

However, along with positive developments, there is some evidence of lack of progress. Unfortunately, in general, during the post-socialist period there has been only a limited shift to a post-industrial type of economy with a lesser share of polluting industries that has been a typical trend in the developed countries of West Europe. The share of industry in the economic production structure in the countries of Central Europe in the mid-1990s still accounted for over 50%, while services sector made up only one third. This proportion can be compared with 32% and 65% respectively in Sweden (*Zelyoniy Mir* 25, 1996).

It appears that, as in the states of the FSU, in the countries of Central Europe the reduction in the overall pollutant load related to industrial production has been negated by an increase in the role of the transport sector as the main source of pollution. According to the second assessment report from the European Environment Agency, in 1995 transport accounted for 60% of the total pollution. An increase in the lead concentrations were recorded in some Central European cities (Europe's... 1998). As in future transport is expected to increase considerably, transport-related emissions are likely to rise. The envisaged phasing-out of leaded petrol would, in the long run, solve the problem of lead pollution but many other problems may persist.

A widespread shift from public to private transport is currently causing more urban congestion and uncontrolled parking in cities which during the socialist period were not designed to accommodate large numbers or private cars. For example, in the former GDR there was a 50% increase of traffic between 1991 and 1995 (Brueggemannn and Rolle 1998). The road network is expanding while in some countries, such as Poland, this trend is accompanied by a contraction of more environmentally friendly means of transport such as rail.

The major reason for the lack of a radical breakthrough in ecological improvement has arguably been the inability of most countries to integrate environmental considerations into other policy areas in the course of their economic and political transformation. As noted by Klarer and Francis, 'most decision-makers in the region do not understand the significance of this basic integration of environmental and economic policies and therefore do not see the need for it' (Klarer and Francis 1997, 17). Expectations that the new democratic countries could learn from the past mistakes and successes of the West have not come true.

While in the immediate aftermath of the Velvet Revolution a number of decisive steps related to pressing environmental concerns have been taken, for example in Czechoslovakia (Wolchik 1991), during the post-socialist period environmentalism has generally lost its strength and appeal to politicians. In most countries of the region ecological goals reduced in priority compared to economic and social issues. It should be noted that while the strategic objectives of all Black Triangle countries to harmonize their environmental legislation with EU regulation remained on the agenda, environmentalists failed to demonstrate an essential link between the economic transformation policies and their long-term environmental implications. The potential role of NGOs in the region has not been realized to its full potential (Matthews and Saiko 1994).

There is still no comprehensive environmental information on the magnitude of ecological damage and its effects on human health. Although there was a major improvement in the availability of ecological data via the publication of annual state-of-the-environment reports in most countries, the system of environmental monitoring in the region is still insufficient. In some areas, however, there was progress: in Poland the water quality monitoring service has improved since 1989. In addition to the grave 'socialist heritage' in the hot spots of the region, some new environmental challenges have appeared. For example, these are related to the encroachment of industry into the remaining relatively untouched natural landscapes and protected areas.

As noted by Fodor (1994), the latter areas present an attractive alternative for the location of new industrial enterprises, particularly, when compared to the existing heavily polluted regions that will require decades and billions of dollars of investment for rehabilitation. However, the second assessment report on the state of the European environment has also revealed that in contrast to severely degraded areas in Western Europe, 'the expected lower cost of providing adequate protection for the large almost undisturbed natural areas that can be still found in the eastern half of Europe should be seen an opportunity and challenge for the whole of Europe to maintain the natural value and function of these areas as significant parts of Europe's natural capital' (Europe's... 1998, 7).

CHAPTER 7

Kalmykia: the first anthropogenic desert in Europe

To say that catastrophe has befallen this enchanted world of grass and light would understate the case.

(Bingham 1997)

7.1 Introduction

This chapter examines the history of the creation of the first anthropogenic desert in Europe in the Republic of Kalmykia (or Khalmg Tangch, in the Kalmyk language) located on the north-western coast of the Caspian Sea and in the south of European Russia (Figure 7.1). The example of Kalmykia differs from many other case studies in this book, because environmental catastrophe in this republic has ensued not due to a complex of factors but as a result of a single type of human impact i.e. ecologically unsound agricultural activities. The physical geographical background is represented by dry steppe and semi-desert landscapes with specific features. The area of Kalmykia is 76,100 sq km which is bigger than the Republic of Ireland, or Belgium and Denmark taken together.

Kalmyks have many unique features compared to other European nations. They belong to Mongoloid race; their language pertains to the Altai family of Mongolian branch and they are the only Europeans who profess a form of Tibetan Buddhism. It is believed that Ghenghis Khan was a forefather of the Kalmyk people. Probably no other nation in the former USSR has suffered so many adversities from the political system. During the Soviet period the people of Kalmykia were twice deported from their native land! Their history is a long tale of miseries and misfortune. Meanwhile, in the relatively recent past, at the beginning of the 20th century free-roaming Kalmyk nomads enjoyed the beauty of their blossoming steppes and wisely mastered an ecological harmony with their fragile environment.

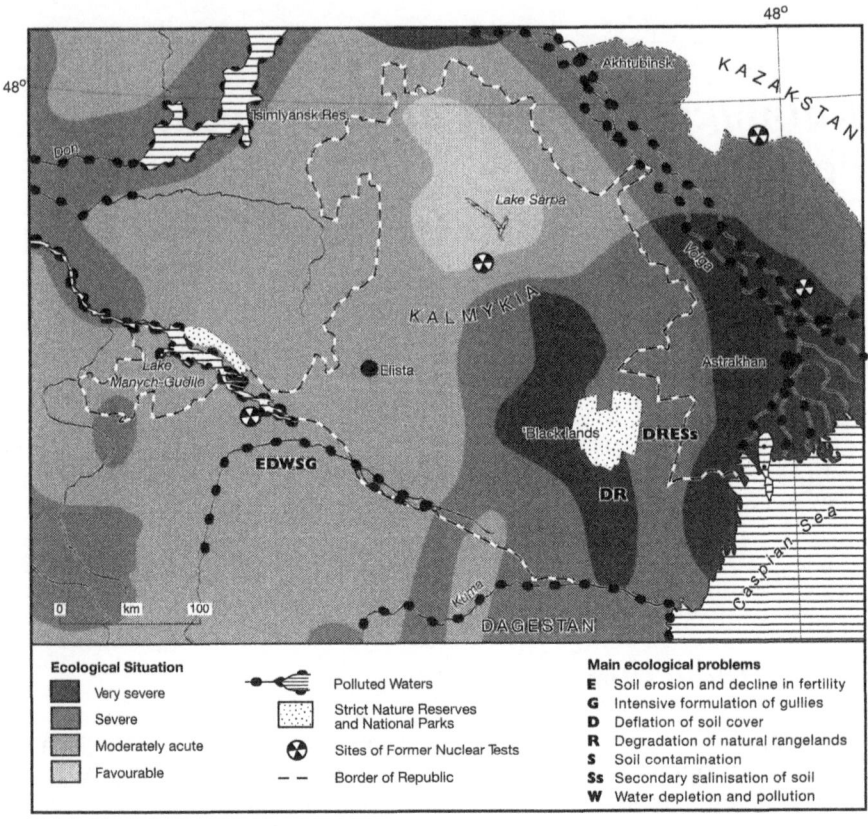

Figure 7.1   The present ecological situation in Kalmykia (adapted from Ekologicheskaya... 1999)

Over the second half of the 20th century, severe desertification processes in Kalmykia caused by overgrazing and cultivation of natural pastures in the so-called 'Black lands' have completely degraded one million ha of the formerly highly productive rangelands. By the 1990s almost 80% of the territory of the republic had been affected by desertification, and 13% had been transformed into a true desert. While in 1990 Kalmykia was not even mentioned in the USSR's 'State Report on the Status of the Environment in 1988' in the section on land resources (Sostoyaniye... 1990), in 1991 it was officially included in the list of 13 'ecological disaster areas' of the Russian Federation (Gosudarstvenniy... 1992). On 1 August 1993 Kirsan Ilyumzhinov, the President of Kalmykia, declared a state of ecological emergency in the Republic.

Until recently, the environmental problems of Kalmykia have been rather poorly covered in the Western scientific literature. Probably the first scientist who wrote about the formation of the first desert in Europe was Wolfson (in Stewart 1992). However, most publications related to this remote and isolated region of Europe are still mainly in the Russian language. Originally, desertification in Kalmykia was recognized within the framework of the UNEP/USSR project, 'Combating Desertification through Integrated Development', at the Centre for International Projects in Moscow, where the author of this book worked for 12 years. A monograph published under the aegis of the above project included an overview of Kalmykia's problems (Zonn 1986). Later, when UNEP implemented a global assessment of the status of desertification, Kalmykia was featured as an example of a FSU region which had experienced severe desertification (Dregne *et al.* 1991; Rozanov 1990).

Reviews of the ecological problems in Kalmykia appear in the yearly state reports on the status of the environment of the Russian Federation (Gosudarstvenniy... 1992–1998). The report for 1991 stated, 'The conditions of arid areas in the Russian Federation are impaired; their degradation and desertification are extensive in scale. The most prominent example of desertification is in the region of "Black lands" (Kalmykia) and Kizlyar (Dagestan) pastures, where a desert has formed and is progressively increasing now' (Gosudarstvenniy... 1992). Recently an overview on the ecological problems in Kalmykia edited by Zonn and Neronov was published in Russia (Zonn and Neronov 1995).

7.2 Physical geographical setting

7.2.1 Geographical location and relief

The Republic of Kalmykia is located in the south-east of the Russian plain on the western shore of the Caspian Sea close to the Lower Volga crossing the steppe, semi-desert and desert landscape zones (Figure 1.3). The republic stretches for 448 km from north to south and for 423 km from east to west. The length of the shoreline with the sea accounts for 110 km. It consists of three main landforms: the Circum-Caspian lowland, which lies in the east of Kalmykia, occupies the greatest area and consists of the 'Black lands' (called 'Chornye Zemli' in Russian) and the Sarpa lowland in the north (Figure 7.1). The latter plain is a depression with low-lying topography for the most part located below sea level. The lowland's highest sections are elevated only by 15–20 m above sea level.

The Yergeni upland with a maximum altitude of 222 m is located in the west. It is the oldest part of Kalmykia and forms a watershed between the Circum-Caspian lowland and the basin of the Lower Don river. A narrow elongated Kuma-Manych valley stretches along the southern border of Kalmykia. From the mid-Tertiary period it was repeatedly flooded by sea water which formed a strait connecting the

Caspian and Black Seas. The two main rivers of Kalmykia, Kuma and Manych flow along this valley. In the north-east the republican boundary comes close to the Volga river.

7.2.2 Climate and hydrology

Climate is extremely important for the formation of landscape in all arid environments, and it is largely responsible for the limitations and vulnerability of natural ecosystems. Kalmykia has an extreme continental climate which features a high degree of aridity and temperature extremes. The level of continentality sharply increases from west to east. The climate is hot and dry with long summers and short but sometimes rather severe winters. Mean January temperatures range from −8 degrees C in the south to −10 to −12 in the north with absolute minimum sometimes reaching −35 to −40 degrees C. In July mean temperatures vary from +23 to +26 degrees C with the maximums as high as +40 to +44 degrees.

Average annual precipitation is quite low at 210–340 mm, with 75% occurring during the warm part of the year. Only in the west does precipitation reach 400 mm in places. Normally winters have little snow cover or no snow at all. Droughts are a natural part of climatic pattern and lack of moisture is the main limiting factor of plant life and landscape evolution. It also determines the degree of their vulnerability to human impact. The probability of dry years in the region reaches 88–95%. By its level of aridity, Kalmykia is the driest region of Russia and is inferior only to deserts of Central Asia. The most arid section of Kalmykia classified as a 'desert' by some scientists is located in the extreme south-east of the republic. Generally, aridity increases from north-west to south-east. The greater part of eastern Kalmykia is classified as 'arid' (Vinogradov et al. 1995). The 'Black lands' stretch across the dry steppe and semi-desert zones. They are called such because in winter they are not completely covered with snow and look 'black' compared to the surrounding territories.

High atmospheric temperatures and low humidity often create conditions for generation of hot and dry winds called 'sukhoveys' ['sukhovey' means dry breath in Russian]. In summer up to 120 days have sukhovey winds (Sandjiev 1998). Wind speed can reach values of 15–20 m per second; temperatures rise to +35 degrees C while relative air humidity falls to 10–20%, initiating dust storms. In 1969 one dust storm which originated within the 'Black lands' reached Poland. According to remote sensing data, in 1984 a storm from the same region reached France. These storms accelerate the effect of wind erosion if land has been cleared of vegetation. Strong winds blow away organic matter from the upper humus horizon leaving the soil open to further degradation.

The **hydrological** system in Kalmykia is poorly developed because the Republic is arid and water deficient. The surface runoff is quite limited. Rivers flowing to the Manych valley dry up during the hot summer.

Shallow lakes with mineralized water are common in the Caspian Lowland and Kuma-Manych depression. They include Lake Manych-Gudilo, Lake Sarpa, etc. Short water streams which flow down from the Yergeny upland form semi-drying lakes or inland deltas. In recent geological history, these lakes were closely connected to the Caspian Sea basin. This connection is at present reflected in the groundwater pattern.

The depth of groundwater and salinity changes towards the coastal strip from 6 to 10 m depth with salinity of 15–30 g/l on the Yergeni upland to the depth of 2–3 m with the salinity level of 40 g/l near sea bays. Fresh water lenses, located at a depth of one to four metres below the ground, are common under the sandy soil and subsoil of the 'Black lands'. When tapped by wells for watering cattle, this shallow groundwater is rapidly depleted and replaced by mineralized or saline water.

7.2.3 Vegetation and soil cover

Vegetation cover comprises dry steppe and semi-desert types of communities. In the north typical and dry steppes are composed of European feather grass (*Stipa pennata*), sheep's fescue (*Festuca sulcata*), meadow and spear grasses (e.g. *Poa pratensis*), slough grass (*Beckmannia*) and wormwood species (*Artemisia lercheana, A. austriaca* and *A. pauciflora*). The canopy cover of these steppes accounts for 35–40% of the total area, and plant productivity amounts to 300–400 kg/ha. By the present time, the typical steppes of northern Kalmykia have been almost completely cultivated. Dry steppes are most common, occupying 40% of the republic's area. The dry steppes of the 'Black lands' have historically been the most valuable natural fodder base of Kalmykia. Until the Second World War their plant cover was very diverse and rich and included a combination of feather grass with sod grasses and gramineous associations, wormwoods and semi-shrubs (*Kohia prostrata*), ephemeral perennials (*Poa bulbosa*) and annual plants (*Lepidium perfoilatum, Ceratocephala falcata*).

To the south-east, dry steppes are replaced by semi-deserts which consist of gramineous communities and wormwoods with an increasing share of xerophytic shrubs and grasses. Plant canopy cover ranges between 30 and 35%. The proportion of ephemeral species increases in the vicinity of the Caspian Sea. Semi-deserts are located in the Yergeni upland, the eastern part of the Manych valley and in the south-east of the Circum-Caspian region. Along with psammophytes on sandy soils and grasses, the vegetation composition also includes halophytes and thickets of *Tamarix ramosissima*. *Agropyron desertorum* and *A. fragile* are also common in this landscape. Halophyte vegetation is found on saline sandy soils. Reeds and meadow grasses are common in depressions, while trees, mainly willow, elm and ash, grow only in the gullies of the Yergeni upland.

The **soils** of Kalmykia are formed under conditions of insufficient moisture, so leaching is periodically lacking within the dry steppes or has a pulse regime in the semi-deserts (Table 1.1). Within the steppe zone, the chestnut type of soil is more common. Chesnut soils have developed in the northern part of Kalmykia on the Yergeni upland and flood plain terraces of the Manych valley. They occur in complexes with 'solonetz' and meadow chestnut soils in micro-depressions. In general, the soil cover has a mosaic structure which is dependent on the bedrock and the type of microrelief.

Another typical feature of arid soils is their relatively high degree of natural salinity which can be partially attributed to the nature of their substratum: marine sediments of the former bed of the Caspian Sea, which since the mid-Tertiary period have undergone repeated transgressions and regressions. Within the most arid south-eastern part, the lack of continuous vegetation cover limits the development of humus, so their subsoil horizon is relatively poor in organic matter. The high evaporation rates during hot summer cause an enrichment of the upper layer of soil with soluble salts. Therefore, the so-called 'solonchak' soils with excess of salts and 'solonets' with excess of sodium are quite common. Both types of saline soils often have compacted surface crusts.

It was recently estimated, that over 75% of soils in Kalmykia have critically low humus levels and they are also poor in potassium contents. A trend of sharp decline in fertilizer application was recently evident throughout Russia: by 1996 the amount of fertilizers provided to farms was ten times less than needed. Only 9.1% of the necessary amount of mineral and 18.4% of the needed organic fertilizers were actually applied in 1995 (Gosudarstvenniy... 1996).

7.2.4 Overall vulnerability of landscape

Both steppes and semi-desert landscapes are very vulnerable to human impact. As seen from Figure 1.4 the whole republic lies within a zone where the natural ecosystems have a very low degree of resilience. Four main factors are responsible for its high sensitivity to human impact: (a) lack of moisture, (b) severe heat, (c) strong winds and (d) high salinity, aridity being the most important. Due to high seasonal aridity, plant growth during dry summer practically ceases, thus making pastures particularly vulnerable to overgrazing. This can be seen, for example, in the seasonal change in the pastures' overall carrying capacity. In autumn it amounts to 1.8 million head of sheep; in winter and spring it goes down to 1.1 million and during dry summer it declines to only 290,000 head (Trofimov 1995). Therefore, the dry steppes of the 'Black lands' are particularly sensitive to human impact. In addition, the discontinuous plant cover in the Kalmyk semi-deserts is a significant factor in their vulnerability.

Another important aspect of landscape sensitivity is related to the fact that the soils are underlaid by the ancient Caspian sands, which sometimes lie only 5–20 cm below the surface and thus can be easily eroded (Saiko and Zonn 1997). In addition, natural salinization of subsoil and compaction of fine-textured soils are responsible for a high overall susceptibility to secondary salinization if proper drainage conditions are not provided. In general, dry steppes and semi-desert landscapes represent marginal high-risk lands which require special care and sound management.

7.3 Historical perspective and agricultural development

The history of agricultural development in Kalmykia is directly linked to the state of the overall economy of this basically rural republic. The natural resource potential of Circum-Caspian lowland is limited to horse breeding, sheep and cattle production for meat and maintenance of wildlife habitat. The availability of lush natural semi-shrub and grass pastures, a long and warm growing season, and established traditions of cattle production favour the pastoral type of natural resource use. In 1991, out of a total area of 7.61 million ha, agricultural lands occupied 77.3%, including 11.9% of cultivated lands, 1.7% of hay lands, and 63.7% of natural unimproved pastures.

The history of agricultural development in Kalmykia can be subdivided into seven periods which featured major changes in the agricultural policies and practices and which were responsible for environmental degeneration in the republic and particularly, of the 'Black lands'.

(1) Before 1913.

Historically, the Kalmyks practised nomadic livestock breeding, as did their predecessors the Sarmatians, Huns and Scythians. The Kalmyks had learned that the earth had its own sacred laws which should be observed in order to maintain a stable equilibrium. Since the 12th century transhumance pastoral economy had become widespread. As a part of their yearly routine, in winter they took their herds to graze in the 'Black lands', which were green all year round, and in summer these were left to rest and regenerate.

Rich steppes provided enough forage for mixed herds of thousands of fat-tail sheep, horses, the local breeds of Kalmyk cattle and camels without endangering the thousands-head herds of endemic sheep-like antelope known as 'saiga' (Plate 13). Saiga is the only European antelope and it is very well adapted to arid climate. The trampling effect of saiga and of the comparatively flat-hooved fat-tail sheep was not destructive on steppe grasses. The status of pastures fluctuated due to climatic reasons: in dry or cold years it declined but herders would normally move to unaffected pastures to let natural self-rehabilitation of degraded areas take place.

Plate 13 A unique antelope of Europe: young saiga in the steppes of Kalmykia

Active economic development of the steppes in Kalmykia started in the middle of the 19th century. To control the increased pressure on rangelands, a special system of traditional relations was developed: steppe congresses of herder representatives were convened each autumn to allocate pastures for the coming winter season. In 1913 about 300,000 people lived in Kalmykia; animal population consisted of about 700,000 sheep and goats, 165,000 cattle, 71,000 horses and 21,000 camels (Table 7.1). The use of mixed herds was another ecologically sound traditional practice which allowed animal pressures to be spread more evenly and maximum use to be made of the available resources. It also yielded the highest animal production per unit area while preserving the long-term stability of pasture productivity (Vinogradov 1993). Therefore, despite the substantial total animal pressure on pastures in 1913, their carrying capacity was not exceeded. Kalmykia of that time served as the 'animal farm' for the whole of Russia and also supplied meat and wool to Persia and Turkestan (Saiko and Zonn 1997). By the end of this period limited enclosed winter keeping of animals was also practised.

The cultivation of dry steppes, which are of naturally marginal use for farming in Kalmykia, was always a very risky and poorly yielding enterprise. Therefore cultivation of crops was limited to small plots on the western steppes. In 1913 farming was practised on an area of only

Table 7.1 Change in animal pressure on pastures in Kalmykia during the 20th century (000 head)

Year or period	Sheep and goats	Cattle	Horses	Camels	Total or average	Used carrying capacity, %
1913	710	165	71	21	967	40
1923	140	97	5	2	241	15–20
1925	252	174	11	7	444	25
1931–1945	1000	220	100	15	1335	40–50
1946–1957	1800	137	80	10	2000	80–100
1958–1972	2200–2800	351	20	7	2900	300–430
1980	3100–3500	348	35	4	3700	500
1986–1987	2400	ND	ND	ND	2600	300
1991–1995	1500	250	20	ND	1770	250

Sources of data: Trofimov (1995); Zonn (1995b); Zonn (1995c)

110,000 ha. Wheat, oats and millet were cultivated there but yields were rather poor and fluctuated from year to year. According to Zonn (1995b), crop yields of wheat, oats and millet averaged 380–455 kg per ha falling to 100–200 kg in dry years.

(2) 1914–1923.

The First World War marked the beginning of an era of crucial hardship for the Kalmyk peoples. The first blow was made by the Tsarist government which authorized the requisition of some privately owned livestock for army needs. The second disaster for Kalmyks came with the civil war after the October Revolution when they supported 'White' Cossacks and were mercilessly repressed by the Bolsheviks. The population fell to 135,000 people by 1926. In addition, Bolsheviks expropriated horses for the Red Army and other livestock. The associated decline in animal population and a change in its structure by 1923 is shown in Table 7.1. These calamities did, however, reduce the pressure on pastures which amounted to only 15–20% of their carrying capacity in 1923.

(3) 1924–1945.

In 1924 the animal population in Kalmykia started to recover and rapidly increased before the Second World War. This was primarily due to the policy of forced collectivization of lands and the creation of collective and state farms, and the associated collectivization of livestock. This policy also involved a shift to large herds and a gradual transition to settled animal husbandry which ultimately disrupted the seasonal use of pastures. A prevailing agricultural pattern during this period represented a monoeconomy based on extensive animal husbandry.

Another important change during this period involved a transition from meat to wool production. The Bolsheviks dramatically changed the structure of pastoralism by promoting new breeds of animals, which were harmful for the fragile environment of dry steppes. They shifted the focus of local pastoralism to raising merino sheep and a fine-fleece Karakul (Astrakhan) sheep. In this way flat-hooved fat-tail sheep were gradually replaced by new breeds with sharp hooves which had an additional devastating effect on pastures.

The third calamity for Kalmykia occurred during the Second World War, when, after alleged collaboration with the Germans, 90,000 Kalmyks were collectively exiled to Siberia and Kazakstan in 1943. The human death toll was almost 50,000. They were not allowed to return to their lands until 1958 and as the indigenous population was expelled, traditional land management and conservation methods were abandoned.

(4) 1946–1957.

The main changes in agricultural policies after the Second World War were related to the need to revive agricultural economy destroyed by the war and to increase meat production. This was aimed to be achieved at the expense of the establishment of large state and collective farms. From 1954 pastures were permanently attached to farms stimulating farmers to use them continuously; a mechanized agriculture was introduced. Diversification of the agricultural economy in Kalmykia comprised a transition from a monopastoral economy to a mixed one: the cultivation of pastoral lands for fodder production was initiated. In accordance with this requirement, some high-risk marginal pastoral lands were cultivated. However, unsurprisingly low yields resulted in farmland being abandoned with further deflation and blowing out of soil. Gradual intensification of pastoralism was achieved not through an increase in meat production output from each animal but by a multiplication of livestock numbers. Poor pastoral management lacking rotation practices and protection measures was typical of this and the following periods.

The newly arrived Russian population was not aware of the specific nature and vulnerability of Kalmyk lands and lacked the traditional knowledge of local herders. They began using the 'Black lands' for grazing all year round; by the end of this period the total animal population increased by 50% compared with the pre-Soviet period. The sheep numbers more than doubled and the pastures' carrying capacity was used in full. Mechanized hay harvesting on natural rangelands placed an additional pressure on dwindling forage reserves.

(5) 1958–1972.

In 1958 Kalmykia became an autonomous republic. At the beginning of the 1960s, a large-scale campaign aimed at further intensification of agricultural production was initiated by Nikita Khrushchev who aimed to surpass US levels. The local authorities were encouraged to fulfil and over-fulfil the targets in animal production set by the yearly and five-year plans. Throughout the Soviet era, farmers would typically comply with all instructions given by the state and local authorities.

At the same time, Soviet development of animal husbandry required a continuous increase in the livestock population. An exponential increase in animal pressure on pastures and land degradation followed. The rise in animal population numbers was primarily achieved by an expansion of the sheep population, which trampled the delicate plant cover with their sharp hooves. By the end of this period the sheep population had expanded to over two million while the total animal population approached three million head. By this time, camel population had shrunk to one-third of the pre-Soviet period. This change adversely affected the state of pastures as camels eat weeds, including prickly plants, which improves the quality of rangelands. Furthermore, they do not trample vegetation or disturb soil cover with their flat hooves.

The average pressure on pastures in the 1960s–1970s exceeded their potential carrying capacity by three to four times (Table 7.1). According to Trofimov, in some years the surplus reached 6.8 times (Trofimov 1995). This intensification of sheep breeding was not supported by an improvement and increase in the available fodder resources. An additional impact on natural rangelands was associated with a transition to sedentary animal husbandry from transhumance pastoralism at the end of this period. A further policy of agricultural diversification involved more cultivation of high-risk pastoral lands for fodder and cereal production which accelerated wind erosion and ultimately had an adverse effect on soil fertility. In addition irrigated farming, which was introduced during this period, lacked proper drainage and thus initiated processes of secondary salinization of soils.

(6) 1973–1989.

This period featured exceptionally high animal pressures on pastures unsupported by an available fodder base (Table 7.2). It also comprised a further expansion of pasture cultivation and irrigated farming for rice growing which was not provided with proper drainage. Starting from 1973, permanent animal farms were established within the area of the 'Black lands'. According to Reznikov (1995) this was the main cause for the great acceleration in desertification processes. The structure of herds further changed: over the Soviet period, the proportion of large animals declined from 40% to 6.7% (Bananova 1993). The number of camels decreased by more than five times by 1980.

Table 7.2 The legacy of agricultural development and ecological change in Kalmykia

Period	Legacy of agricultural development	Ecological situation	Stage in ecological degeneration
Before 1913	Nomadic and transhumance pastoral economy; diverse herd structure; seasonal use of pastures; limited enclosed winter keeping of animals; very limited farming	Ecological equilibrium; fluctuations of pastoral status due to climatic causes; self rehabilitation of degraded areas in natural pastures	Satisfactory situation
1914–1923	Agricultural decline due to social calamities and wars; reduction in animal population; private ownership of livestock; its requisition by Tsarist and Bolshevik powers; change in animal structure; reduction in horse and camel numbers; seasonal use of pastures	Revival and blooming of natural pastures due to reduced animal pressure	Satisfactory situation
1924–1945	Collectivization of livestock and lands; creation of collective farms; shift to large herds; gradual transition to settled animal husbandry; monoeconomy based on extensive animal husbandry; disruption of seasonal use of pastures; shift from meat to wool production; replacement of fat-tail sheep to merino sheep and karakul (astrakhan)	Increase in area of trampled pastures on sandy soils; decline in manure amount; beginning of processes of soil compaction and erosion; increased surface runoff	Strained situation

Table 7.2 Continued

Period		
1946–1957	Creation of large state or collective farms; permanent attachment of pastures to farms; mechanization of agriculture; transition from monopastoral economy to mixed; cultivation of pastoral lands for fodder production; gradual intensification of pastoralism by an increase in livestock numbers; poor pastoral management	Gradual decline in pasture productivity; initiation of desertification; change in area of trampled pastures from 8% to 32%; formation of shifting sands on area over 100,000 ha; total desertified area – 5% by end of period	Critical situation
1958–1972	Transition to multi-sided economy; continuous increase in livestock; intensive animal husbandry unsupported by fodder resources; permanent use of pastures; cultivation of pastoral lands for fodder and grain crops; introduction of irrigated farming	Accelerated degradation of landscapes; precipitous decline in pasture productivity; activization of desertification process; expansion of trampled pastures area to 50%; active deflation and expansion of shifting sands area to 250,000 ha; total desertified area – 37%	Critical situation/ecological crisis
1973–1989	Intensive animal husbandry unsupported by available fodder base; cultivation of pastures for grain crop growing; irrigated farming for rice-growing	Acceleration of desertification; catastrophic decline in pasture productivity; shift to xerophyte and halophyte plants; area of trampled pastures reaches 100%; shifting sands area reaches 380,000 ha; secondary soil salinization and waterlogging in irrigated lands; total desertified area – 80%	Ecological crisis/ecological catastrophe
1990 to present time	Transition to the market economy; limited privatization of lands; weakening of state controls over agriculture	Anthropogenic desert has been created on area of 1 million ha; acceleration of salinization and waterlogging in irrigated lands	Ecological catastrophe

Furthermore, the official statistics did not account for sheep kept individually or illegally, which amounted to another 2 million head added to the state estimate of 3.5 million. According to Zonn (1995c), the peak of 3.7 million was reached in 1980. Therefore, the actual total numbers of sheep were well over 5 million head. Official estimates also suggested, that by the end of the 1980s, the carrying capacity of the 'Black lands' was exceeded by 2.5 times (Trofimov 1995). Thus the actual animal pressure on pastures could be four to five times greater than their carrying capacity. In addition the available biological reserves of fodder in the 'Black lands' were grazed by 90–100% instead of the optimum norm of 50–60% (Reznikov 1995). This, in turn, resulted in overgrazing and, ultimately, in the decline of the potential carrying capacity of pastures by two times over the last 35 years. The lack of fodder caused a massive famine in 1986–1987 when 800,000 sheep died (Zonn 1995c).

1990 up to the present time.

The post-Soviet period in Kalmykia featured a transition to the market economy and general weakening of state controls over agriculture. The latter circumstance and the catastrophic decline in pasture productivity which occurred over the previous two periods have resulted in a gradual reduction in the total sheep population to approximately half of its peak value in 1980 (Table 7.1). This can also be attributed as the reason for the overall decline in the standard of living in Kalmykia which has stimulated population to kill their animals and sell meat to be able to satisfy other essential needs. The privatization of lands in the early 1990s had a very limited effect on the overall grave situation.

An important event at the end of the previous period was the adoption in 1989 of the 'General Scheme of Desertification Control in the "Black Lands" and Kizlyar pastures' designed for the period up to 2000. This involved a set of actions for stabilization of shifting sands and phytoreclamation of pastoral lands which had slightly improved the ecological status in the former region. However, the implementation of rehabilitative measures was halted by the lack of financial resources for environmental needs that is typical of post-Soviet Russia. The legacy of agricultural development and the associated ecological changes are summed up in Table 7.2.

7.4 Ecological consequences and recent changes

As a result of overgrazing and the cultivation of high-risk marginal pastoral lands desertification processes have ensued. Desertification is 'land degradation in arid, semi-arid and dry sub-humid areas resulting from various factors, including climatic variations and human activities'. This latest, internationally negotiated definition was adopted by the United Nations Conference on Environment and Development (UNCED), Rio de Janeiro, Brazil, in June 1992 and included in the UN Convention on Desertification. Desertification is made apparent through various land degradation processes.

Degeneration of pastures is the most common type of desertification process in Kalmykia. It is expressed in the decline in the total biomass of the pastoral vegetation, its biological productivity, change in the vegetation composition and its structure. Deterioration of vegetation in the 'Black lands' is associated with degradation of the soil cover which includes the destruction and blowing out of humus subsoil and the formation of shifting sands or 'barchans' on sandy soils. On loamy soils it is expressed in the compaction of the soil surface and accelerated runoff. The evidence of these changes is ultimately reflected in the decline of the overall carrying capacity of the Kalmyk steppes and a large-scale development of shifting sands which has brought about the formation of an anthropogenic desert over the last two or three decades. Another type of desertification process is related to mismanagement of irrigated lands in the rice-growing areas of Kalmykia which has resulted in the development of secondary salinization and the waterlogging of irrigated lands.

7.4.1 *Degeneration of pastoral vegetation*

Before 1923 the pastures in Kalmykia were generally in a state of ecological equilibrium. Between 1913 and 1923 a decline in animal pressures encouraged a revival and blooming of pastures. Before the Second World War, despite the fact that the overall carrying capacity of pastures was not exceeded, an increase in the total numbers of livestock increased the area of trampled pastures on sandy soils. This was followed by the processes of soil compaction, accelerated water and wind erosion and increased surface runoff.

In general during the post-war period, degeneration of pastures was caused initially by overgrazing and trampling of rangelands by the rocketing numbers of sheep and cattle, particularly within the 'Black lands'. Favourable conditions for pasture ecosystems are determined by their resilience to exploitation, the ability for self-rehabilitation and reproduction after disturbance. As mentioned above, dry steppe pastures of the 'Black lands' are particularly vulnerable to anthropogenic impact, therefore their continuous over-use, which has exceeded carrying capacity, was the major reason for this degradation. According to Zonn (1995b), two additional indirect causes contributed to pastoral degeneration: (1) a period of prolonged drought which lasted during the 1960s and (2) the regression of the Caspian Sea during the 1970s.

Vinogradov (1993) noted that overall natural fodder yields during the late 1950s were still relatively favourable. They ranged from 600–800 kg/ha on spear grass and white wormwood pastures on loamy soils to 200–350 kg/ha on sandy soils with sod grass and wormwood pastures. According to Zonn, the average winter yields on the 'Black lands' declined between 1960 and 1972 from 210 kg of dry weight per ha to only 70–90 kg/ha (Zonn 1995b). By the 1980s, the total yields of pasture fodder plants had declined by 2.2–2.1 times compared to the 1950s level while the number of sheep had increased by 1.6–1.8 times. On loamy pastures the forage yield declined to 200–300 kg/ha and to 100–200 kg/ha on sandy pastures. As a result of continuous grazing and seasonal fluctuations of carrying capacity, summer fodder deficits became quite common.

Trampling of pasture vegetation is the major problem throughout the 'Black lands'. Between 1946 and 1958 the area of trampled pastures increased from 8% to 32%. Precipitous decline in pasture productivity and the activization of desertification processes followed during the 1960s through to the 1980s. In 1972 already half of all pastoral lands were over-grazed, while by 1975 this proportion had risen to 60%. By 1986 all pastures were classed as trampled while severely trampled rangelands comprised three-quarters of the total area (Reznikov 1993; Trofimov 1995).

The decline in productivity paralleled the change in the vegetation composition of rangelands towards an increase in the proportion of xero-phyte and halophyte plants and a reduction in the share of perennials and palatable species. According to Myalo and Volodina (1995), the first stage of digression was expressed in the disappearance of feather and gramineous grasses which have been replaced by sod grasses and worm-woods. Trofimov (1995) stated that between 1958 and 1986 the share of gramineous pastures decreased by ten times from 7.5% to 0.7% of the total 'Black lands' area. At the same time, white wormwood (*Artemisia lercheana*) communities increased from 1.4 to 5.7%.

During the second stage, white and black wormwood (*A. pauciflora*) started to dominate in the composition of plant communities while most sod grasses completely disappeared. Xerophyte ephemeral plants (*Poa bulbosa, Lepidium perfoliatum* and *Ceratocephala falcata*) became wide-spread. Halophyte species, e.g. *Anabasis aphylla*, became predominant on saline soils. The third stage of pasture digression involves a complete dominance of annual and ephemeral plants. Groups of black wormwood became dominated by saltwort (*Salsola*). The vertical and horizontal structure of plant communities became simpler, while vegetation cover became discontinuous. In essence, the degradation cycle was expressed through a transition to a more arid type of landscape in the following sequence: gramineous steppes–dry steppes–semi-deserts–deserts. Partially, this change could be attributable to the increased climatic arid-ity which coincided with the initial period of pastoral degeneration during the 1960s.

Pastures were particularly overgrazed around settlements and other places of livestock concentration such as enclosed sheep pens with roofs, wells and other watering points. Many pastures were degraded not only due to excessive stock pressures but also because persistent grazing occurred at the 'most sensitive' time of the year, that is at the start of the growing period, in the absence of favourable conditions for normal plant reproduction and growth. This kind of inappropriate grazing practice became so common that natural fodder yields varied not in accordance with a specific season, but with the degree of vegetation degeneration.

7.4.2 Wind erosion

Accelerated wind erosion or deflation is the second most common type of desertification process, especially hazardous in the parts of the 'Black lands' that are underlain by sand. These processes were initiated by an increase in animal pressure on pastures in the course of the post-war agricultural campaigns. By the end of the 1950s degradation of natural vegetation and wind erosion on the semi-stable sand soils in Kalmykia had resulted in the formation of shifting sands in an area exceeding 100,000 ha. During the next 15 years active deflation and blowing out of subsoil from trampled pastures expanded the area of moving sand dunes to 250,000 ha. By the end of the Soviet period, shifting sands encompassed an area of 380,000 ha (Table 7.2). By 1995, according to official estimates, the area of shifting sands accounted for 7.4% of the total area of the 'Black lands' (Gosudarstvenniy... 1996).

Wind erosion is spread not only on sandy soils but on other fine textured soils e.g. loamy soils as well. Typically strong 'sukhovey' winds which can reach 20 m/sec, blow out fine particles from soil and sometimes uproot plants with shallow root systems. Accelerated wind erosion processes originated not only within pastures but also on cultivated farmlands, creating local centres of deflation with the consequent formation of aeolian land forms. It should be emphasized that all attempts to compensate for the declined productivity at the expense of growing fodder on ploughed rangelands resulted in a vicious circle, because the animal pressure on the remaining pastures multiplied accordingly, thus contributing to further degeneration of the vegetation. In addition the cost of grown fodder was always higher than that naturally produced on rangelands.

The location of the main areas of shifting sands and the nucleus of desertification within the 'Black lands' are shown in Figure 7.2. This illustrates that only a narrow band of lands along the western border of the 'Black lands' are not prone to deflation. Shifting sands cover an area of about half a million hectares while unstable sands have an approximately similar size. Within this region sand dunes have encroached on pastures, roads and villages, many of which have been abandoned.

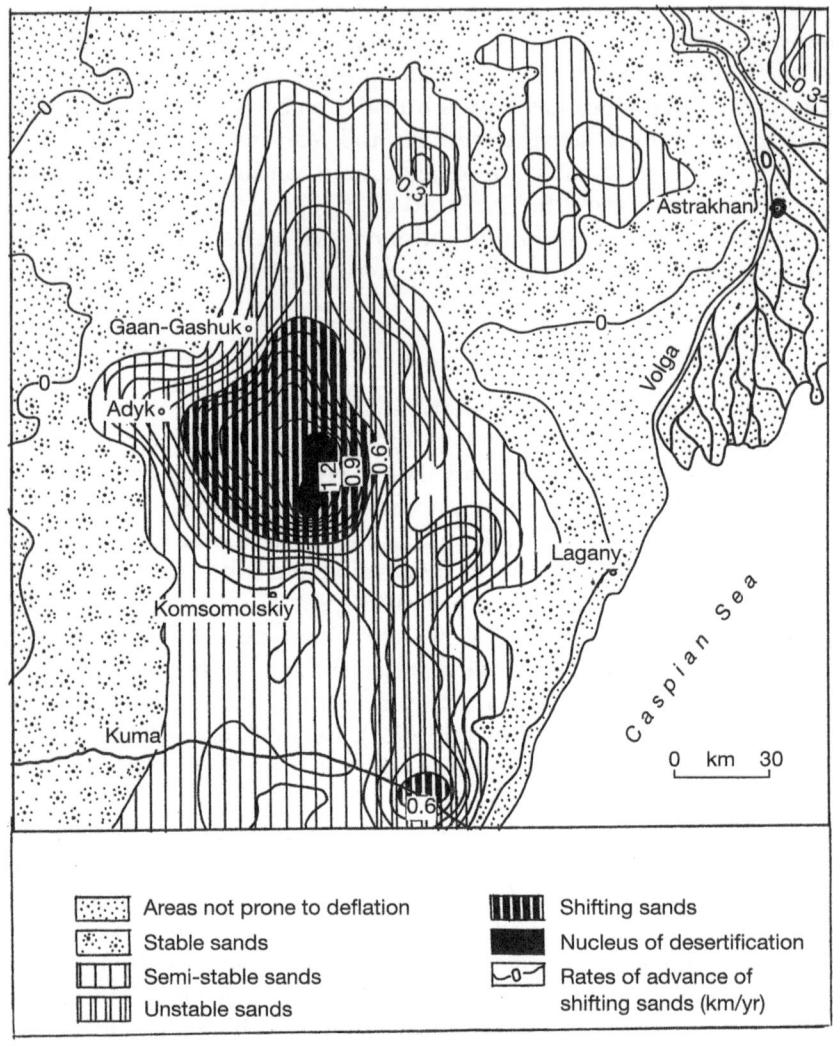

Figure 7.2 Desertification in the 'Black lands' of Kalmykia (after Gosudarstvenniy... 1995)

7.4.3 Soil salinization and waterlogging

In general, the processes of secondary salinization, alkalization and waterlogging are due to mismanagement of irrigation. But as has been mentioned above, Kalmyk soils are particularly susceptible to salinization. Unsurprisingly, this process emerged almost immediately after the introduction of irrigated farming in the republic. By 1970 irrigated lands in Kalmykia comprised an area of 124,000 ha. Construction of a drainage system lagged well behind the development of the irrigation network and, as estimated by Zonn (1995), at the beginning of 1993 only one-fifth of all irrigated lands was provided with an adequate system of drainage.

As a result of poor or totally lacking drainage on naturally saline soils, widespread ecological problems ensued. In the rice-growing fields of the Sarpa lowland an estimated 67% of all irrigated lands were subjected to the processes of secondary salinization and waterlogging. Other estimates suggest that practically all irrigated lands were affected to different degrees. According to Bananova (1993), during the whole period of irrigation in Kalmykia, the total area ruined by the above processes reached approximately 106,000 ha.

7.4.4 Impact on the saiga population

Major changes which occurred in the agricultural economy of Kalmykia had a consequent effect on the population and the natural habitat of the saiga. Before the 19th century they freely roamed the vast open steppes which stretched between the Volga and the Don rivers. Their numbers were estimated at hundreds of thousands or more, and herds of 6000–10,000 animals were common. A reduction in population numbers started in the 19th century mainly due to over-hunting but also because of natural causes such as very cold winters and famines (Bliznyuk 1995). Their population dwindled well into the 20th century, when the habitat of these relict animals shrank and was limited to the area of the 'Black lands' only. By 1919 they were almost completely extinct and hunting of saiga was banned by the Soviet government. As discussed above, the pre-war period featured relatively low total animal pressures on pastures and saiga population started to recover quite rapidly.

By 1940 they were already commonly seen in many parts of Kalmykia, and by 1948–1949 the saiga's habitat covered 50,000–60,000 sq km while the population reached 100,000 head. The role of the 'Black lands' in their restoration was paramount. Their numbers kept increasing and by 1958 the saiga population had reached its peak of 811,000 head, while their grazing area expanded to 120,000 sq km. From 1951 the hunting of saiga was authorized and in some years the share of killed animals amounted to 40–45% of the total population (Bliznyuk 1995). The hunting quotas were scientifically based and strictly controlled; therefore an increase in numbers continued until 1978, when the total saiga population was estimated at 715,000 head.

Since that time however, competition with an exponentially increasing population of domestic animals, a consequent decline in the quality and productivity of pastoral vegetation along with the expansion of cultivated lands, have dramatically reduced the saiga habitat and affected its

population. Construction of a system of irrigation canals for rice growing had a particularly adverse effect because large areas became
inaccessible for animals. By 1987 the total saiga population fell by almost
five times to 143,000–168,000 head while their habitat area shrank to
38,000 sq km. During the following five years it further decreased to
26,000–29,000 sq km.

Recognizing this situation, in 1990 the government of the former
Soviet Union established the 'Black lands' biosphere strict nature reserve
with an area of 121,902 ha to protect their habitat and restore the natural
steppe landscapes (Figure 7.1). However, when the USSR dissolved and
the transition to the market economy started, the unfortunate saiga confronted a new challenge. Poaching has become a major threat as these
animals are valued for their horns: the cost of three pairs of saiga horns at
the current world price is equivalent to the cost of 10–12 mink skins.

7.4.5 Other ecological problems

Along with specific issues associated with unsound agricultural policies, Kalmykia is affected by the ecological problems related to
industrial and urban development which are typical for other regions of
the former Soviet Union. According to the Russian State Report on the
Status of the Environment in 1994, the most urgent problems for
Kalmykia include integral protection of land resources to prevent further desertification and to raise fertility of soils; plus the provision of an
adequate supply and **quality of water** for drinking, agriculture and
industry (Gosudarstvenniy... 1995). A serious problem is associated with
water pollution due to the lack of treatment of wastewater. Although the
total amount of discharged wastes decreased between 1991 and 1995
from 57.8 million cubic m to 44.6 million, in the latter year *all wastewater*
was discharged *untreated* compared to 40% of the total volume being
treated in 1991. This fact was clearly responsible for the recent decline in
the quality of drinking water which is normally taken from surface
water sources.

Air pollution has become an increasingly important issue. The structure of atmospheric pollution from stationary sources consists of 52%
from oil extracting enterprises, 21.3% from energy industries, another
21.3% of pollutants are generated by construction enterprises and the rest
comes from meat processing plants. The situation here is somewhat similar to that with water pollution. Despite an overall decrease in the
amount of generated polluting substances between 1991 and 1995, there
was a 55% increase in the volume of emitted pollutants. In addition, over
the last decade the share of car transport in overall pollutant emissions
has increased dramatically, reaching 80% (Gosudarstvenniy... 1996).

7.4.6 Overall ecological situation

Table 7.2 illustrates the main stages in ecological degeneration. The overall ecological situation in Kalmykia, and the 'Black lands' in particular, was satisfactory until the mid-1920s when zoogenic and other pressures on the landscapes were low compared to their ability of self-rehabilitation. With a continuous increase in the livestock population after that time, before the start of the post-war agricultural policies the status of the physical environment became strained due to an increase in an area of trampled pastures on sandy soils. The turning point in landscape degeneration can be related to the start of large-scale permanent pressure on the fragile dry steppes. This was associated with the establishment of big collective and state farms and the permanent allocation and use of pastures. Therefore, during the period between 1946 and 1957 the overall ecological situation became critical: the maximum potential carrying capacity was reached and various desertification processes started. During this period the proportion of pastures with degraded vegetation reached one-third. It has been estimated that by the late 1950s approximately 5% of the area of Kalmykia had become desertified.

The situation remained critical during the 1960s with the further intensification of animal husbandry and reached crisis point when half of the natural vegetation cover of pastures in the 'Black lands' and over one-third of soil cover became severely degraded: by the beginning of the 1970s the total desertified area of the Republic increased to 37%. The ecological crisis continued during the 1970s and the 1980s when the degeneration of landscapes was greatly accelerated and they reached the irreversibility threshold. By the late 1980s the ecological situation had become catastrophic, with the total desertified area reaching 80% of the territory of Kalmykia. The dry steppe landscapes had virtually turned into semi-deserts, while semi-deserts became deserts. According to some experts (Reznikov 1995), in 1984–1986 the proportion of desertified lands was even greater at 94.6%.

This ecological situation coincided with the 'glasnost' period and the population of Russia became, for the first time, aware of its gravity. By the beginning of the post-Soviet period, for the 'Black lands' the situation was clearly catastrophic: an anthropogenic desert of 1 million ha has been created there. This amounts to almost 13% of the whole territory of Kalmykia. This desert features a settled combination of barren shifting sand expanses of up to 7000 ha each which alternate with patches of unstable and semi-stable sands.

Table 7.3 shows the present status of desertification in Kalmykia. It shows that the total desertified area at present amounts to over 5.9 million ha which amounts to 77.7% of the total territory of the Republic. This value includes over 2 million ha of moderately desertified lands, almost 1.9 million of severely and 1.5 million ha of very severely

Table 7.3 The present status of desertification in Kalmykia (000 ha)

Desertification type	Desertification stage*				Total area	% of area in Kalmykia
	Slight	Moderate	Severe	Very severe		
I. Zoogenic desertification: pastoral degradation						
1. Overgrazing	359.2	1806.6	1556.9	175.0	3897.7	51.2
2. Wind erosion	–	43.1	–	298.3	341.4	4.5
3. Water erosion	196.3	51.0	–	40.5	287.8	3.8
Total:	555.5	1900.7	1556.9	513.8	4526.9	59.5
II. Zoogenic and mining-induced desertification						
Overgrazing and land degradation from extraction of mineral resources	4.6	106.9	298.7	495.6	905.8	11.9
III. Technogenic and farming-based desertification						
1. Land alienation (settlements, industry and transport)	–	–	–	345.7	345.7	4.5
2. Wind erosion (under rainfed farming)	–	–	–	62.8	62.8	0.8
3. Salinization of lands (under irrigation)	–	–	17.0	58.1	75.1	1.0
Total:	–	–	17.0	466.6	483.6	6.3
Grand total:	560.1	2007.6	1872.6	1476.0	5916.3	77.7

*Source of data: Zonn (1995a)

degraded lands. The most common process of desertification is over-grazing which affects over half of the area in Kalmykia. Wind erosion comes second with 5.3% of Kalmyk territory. Processes of water erosion and salinization jointly affect another 4.8%. It should be noted that the State Report on the Status of the Environment in the Russian Federation in 1992 gave an even higher estimate for the area of very severely degraded lands: 1.8 million ha (Gosudarstvenniy... 1993). According to Zonn (1995b), the desert consumes about 50,000 ha of pastures annually. In 1992 the 'Black lands' were officially designated as an 'ecological disaster area' (Gosudarstvenniy... 1992).

7.5 Socio-economic implications and health issues

In Kalmykia the vegetation cover of pastures provides not only the fodder for animals but also the quality of human habitat for its farmers. As shown in Table 7.1, between 1913 and 1980 the total sheep population increased from 0.7 million head to 3.5 million. The swelling of animal population numbers was not accompanied by an increase in meat productivity. It was estimated that in 1913 one sheep yielded more than 30–40 kg of meat (Trofimov 1995), during 1954–1960 the output declined to 25–30 kg per head (Zonn 1995c) whereas in 1985 meat productivity was down to only 7–9 kg (Zonn 1995b). Thus, the quality of meat has also declined which has had negative implications for its marketability in the 'free market' system of the post-Soviet period. In this way, the deterioration in the character of animal husbandry has had an impact on the viability and the health of the local population. An overall decline in pastoral productivity attributed to the degradation of the ecological situation serves as a clear index of environmental degeneration.

Under the Soviet regime, the rangelands were increasingly exploited for animal production, and a part of these pastures was converted to cropland, completely ignoring the established natural resource management systems, traditions and customs of the native Kalmyk population. The ecologically unsound agricultural development of the fragile dry steppes in combination with an advance of industry and transportation over the past 70 years have completely ignored the possible ecological consequences of the over-exploitation of natural resources. Furthermore, it has undermined the economic basis of this agrarian republic.

In addition, major health implications in Kalmykia have been linked to the low standard of living, poor quality of drinking water and air pollution. Health implications can also partly be linked with the set of problems discussed above: (a) the secondary salinization of soils and, subsequently, increased mineralization of surface and ground water in irrigated lands, and (b) the increased air pollution with dust and sand particles from the anthropogenic desert. All the above points provide sound grounds for classifying the stage in environmental degeneration of the 'Black lands' as an *environmental catastrophe.*

7.6 Prospects for the future

As mentioned above, recognizing the importance of the 'Black lands' as the main fodder base of Kalmykia, the 'General Scheme of Desertification Control in the Black lands and Kizlyar pastures' was formulated in 1986 and adopted in 1989. The latter pastures are located in Dagestan and are also severely degraded. Although many attempts had been made to improve pastures over the previous 20 years, on this occasion, it was the first scientifically based, holistic and systematic approach for optimization of agricultural management. The scheme was designed for the period up to 2000 and, within Kalmykia, covered an area of 3.6 million hectares. The main target of this scheme was to halt and ultimately reverse desertification processes, at the same time restoring the natural fodder basis of animal husbandry in the Republic.

The programme of action included agro-technical, forestry, phytoreclamation and water management works in the 'Black lands'. In particular, it involved erosion control, stabilization and phytoreclamation of 661,000 ha of shifting sands, radical improvement of pastures by sowing fodder crops on an area of 440,600 ha and the development of a water supply system over entire pastoral lands. A set of measures for the enhancement of pastoral management comprised enclosing pastures and using them in a controlled rotational manner. Pastures most prone to deflation, which exceeded 600,000 ha, were meant to be used only during the winter period. However, the main problem in implementing these measures was that these pastures were overstocked with animals. Livestock had to be removed for any positive results to be achieved. Unfortunately, no adjacent region could accept the additional pressure from these animals due to inadequate fodder resources.

There has been some progress in the ecological situation particularly in the early years of implementing the scheme. The overall area under reclamation by 1993 accounted for over 300,000 ha. According to the Russian Ministry of Ecology and Natural Resources, the total area of very severe desertification decreased from 376,000 ha in 1985 to 278,000 ha in 1994 (Gosudarstvenniy... 1995). The rates of advance of the shifting sands were reduced by 50–60%, thus within 3–5 years desertification processes were slowed down or even halted. Reznikov stated that by October 1994 100,000 stabilized and reclaimed sands had been returned to use for grazing (Reznikov 1995). Despite these measures, as was shown earlier, almost 80% of the republic, including the major part of the Kalmyk 'Black lands', is still affected by the processes of desertification.

It should be emphasized that already during the Soviet period scientifically based methods for shifting sand stabilization and pasture improvement had been developed, for example at the Volgograd Research Institute of Agroforestry and Phytoreclamation (VNIALMI) and at the Institute of Desert Research in Ashgabat, Turkmenistan. Implementation of these technologies, which was normally carried out on a limited scale, sometimes led to amazing results. Within the reclaimed pastures of the 'Black lands' the annual productivity reached two tonnes of dry fodder weight per ha which is several times greater than the yields from natural dry steppe pastures.

However, it is difficult to be optimistic about the ultimate effect of this scheme for two reasons. Firstly, the strained economic situation in the Russian Federation does not allow it to divert the substantial financial investment needed for the improvement of the ecological situation in this relatively poor agrarian region. Secondly, reclaimed pastures are returned to the same farms and managers who are responsible for their poor conditions and nothing is in place to stop them from ruining them again. As Reznikov stated, 'The effectiveness of all implemented measures ... directly depends on the "management culture" of animal husbandry. If reclaimed pastures are not legally protected by the state, and if the current anarchy and lack of control in their use persist, all efforts will remain as "the labour of Sisyphus"' (Reznikov 1995, 92).

Therefore, unless the whole agrarian policy in Kalmykia is changed to incorporate the main conservation and management principles which would account for the natural vulnerability of the physical environment, restored pastures will inevitably revert to the same state. Furthermore, a desertification control programme should become an integral part of the economic and social development of the Republic.

The transition to the market economy has not basically changed the practices of the Soviet period, nor can it change the mentality of the people living on this land. In the opinion of Bingham (1997), the current management system is governed by a drive to get the most out of pasture dominated by annual grass. Initially cattle are allowed to consume the lushest growth and then sheep are left to graze as long as it takes to clean up all that remains. He is convinced that 'such a system of continuous grazing by scattered herds will guarantee that grasses will NEVER regenerate and degradation continues' (Bingham 1997, 14). The short-term approach and maximum immediate profit orientation of the 'new Kalmyks' do not provide much hope for a drastic improvement of the situation.

The deep-rooted causes of the lack of progress on a larger scale are due to the inability of the current agricultural system to manage land properly. The Soviet system of land tenure and control collapsed and the previous practices which functioned well before the Bolsheviks came to power cannot be reconstructed in a completely different environment.

The introduction of private farming has failed to bring about a change to a more caring attitude to land, at least for as long as the Russian Land Code does not provide a farmer with a right to sell his land.

The reason for a strong opposition to such a change is the existence of a coalition majority between the Communist and the Agrarian Party in the Russian Parliament known as the 'Duma'. Furthermore, contemporary representatives of the Agrarian Party in the Duma are, for the main part, the former chairmen of collective farms who always hated the idea of letting people own their land. Of course, private farming is not a panacea for land degradation and there is no guarantee that, even if this right is granted, people's attitudes would change dramatically in the near future.

According to the forecast made by Vinogradov (1993), if the pastoral economy continues to be governed by the current inappropriate agricultural policies, by 2030 the 'Black lands' will completely vanish from the face of the Earth giving place to an anthropogenic desert of three million ha.

CHAPTER 8

The Volga basin: a massive degradation of ecosystems

> The Volga is not only the principal river of Russia. It is a symbol of Russia, its pride.

> (Yanshin and Melua 1991, 7)

8.1 Introduction

Not many West Europeans realize that the Volga, stretching from north to south for 3531 km, is Europe's longest river. Russians warmly call it 'Mother Volga', appreciating its historically great importance for the well-being of the whole country. The Volga drainage basin occupies 62.2% of the territory of European Russia or almost 13% of the whole of Europe, accounting for an impressive area of 1,358,000 sq km (Figure 8.1). This area is two and a half times the size of France. The Volga basin concentrates the main industrial and agricultural potential of Russia. Over 40% of the country's population live in this region. During recent decades the whole drainage basin of the Volga, the former industrial heartland of Russia, was struck by a multitude of problems which reached such a level that massive degradation of ecosystems ensued. In 1992 the Middle Volga and Lower Kama regions were included in the list of 'ecological disaster areas' of the Russian Federation (Gosudarstvenniy... 1993).

This enormous territory features an unrivalled diversity of geographical landscapes from its northern to southern borders, a range that includes the taiga, mixed forest, broad-leaved deciduous forest, forested steppe, steppe and semi-desert. Unsurprisingly, the Volga river system manifests a tremendous influence on the overall 'health' of the environment in this part of the Russian Federation.

The mean inter-annual discharge of this great river amounts to around 236 cubic km which ultimately enter the Caspian Sea. Exploitable ground

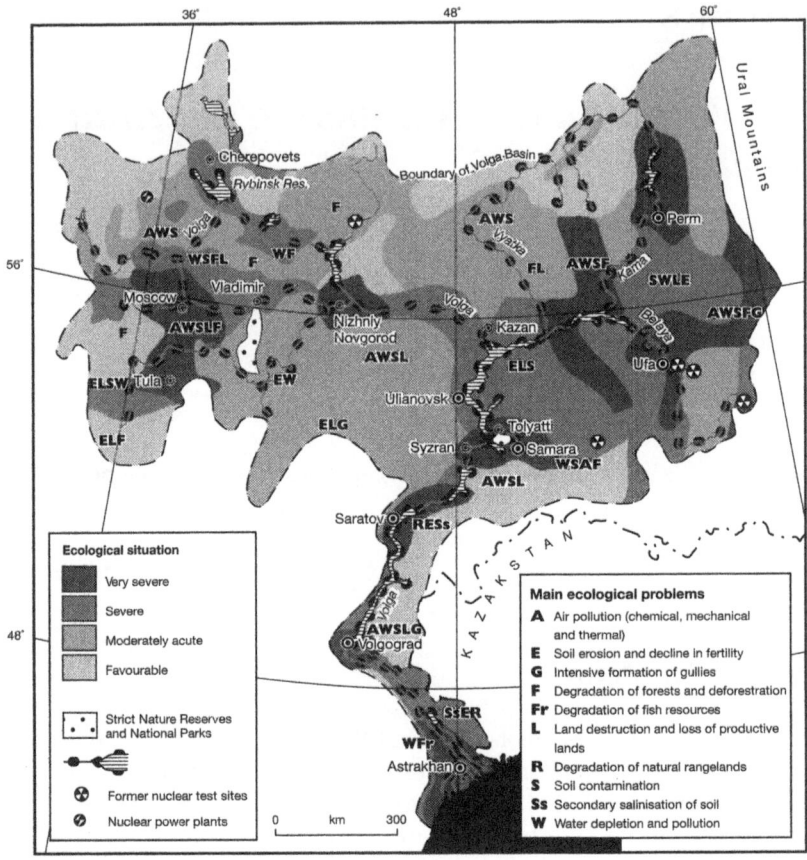

Figure 8.1   The present ecological situation in the Volga drainage basin (adapted from Ekologicheskaya... 1999)

water resources within the Volga basin are also quite significant accounting for 39.7 cubic km per annum. In 1997 almost one-third of all water consumption and 40% of all discharged polluted wastes in the Russian Federation fell within the Volga basin (Gosudarstvenniy... 1998a). The Volga provides 85% of the surface runoff into the Caspian Sea via its extensive delta and thus determines the water and salt balance in the sea as a whole. Moreover, the quality of water in the Volga delta which has an immense area of about 19,000 sq km, to a large degree influences the ecological situation in the northern Caspian Sea.

The catchment of the Volga encompasses parts or total territories of eight republics and 30 oblasts (provinces) of Russia and two administrative regions in Kazakstan. Considering the huge influence the river manifests over economic activities in the adjacent areas, the entire territories of the above units are included in the concept of the 'Volga economic region', although, for example, only 0.07% of the territory of Kalmykia is actually located within the Volga drainage basin (Vozrozhdeniye... 1996).

It can be noted that problems of the Volga basin did not find an adequate coverage in the English language scientific media. Substantial factual data on the main economic regions of the basin are included in one of the first comprehensive reviews of the state of the environment in the former Soviet Union prepared by Mnatsakanian (1992). In the Russian scientific media extensive studies on the region were published recently in a two-volume edition prepared as background documents of the Federal Programme 'Revival of the Volga' (Vozrozhdeniye... 1996, 1997). A substantial amount of information has also been collated from Russian State Reports on the Status of the Environment published annually between 1992 and 1998 (Gosudarstvenniy... 1992–1998).

8.2 Physical geographical setting

Along its course across the vast Russian plain from its source to its mouth the Volga crosses several landscape zones (Figure 1.3). It originates in the European taiga; from Kazan to Samara it passes through forested steppes; then through the steppe zone until Volgograd and further south through semi-deserts and deserts. In conformity with climatic differences and inherent combinations of heat and moisture, soil and vegetation cover vary in the corresponding zonal pattern from north to south.

8.2.1 Geographical location and relief

The drainage basin of the Volga is confined between 62 and 45 degrees North and between 36 and 62 degrees East. It stretches from north to south for 2500 km and from east to west for around 2300 km. In the west it almost reaches the state border with Belarus and in the east it is bordered by the Ural mountains. As can be seen in Figure 8.1, it covers extensive areas of the central and southern parts of the enormous Russian plain which also extends into East Europe. The main part of the catchment is located within the Russian Federation and only a minor left-bank section of the Lower Volga pertains to Kazakstan. The river originates on the Valdai upland, nearly 500 km north of Moscow. Until Kazan city the Volga's course takes an approximately latitudinal direction, which then changes to longitudinal.

The topography of the drainage basin is relatively uniform and consists of alternating low-lying plains, the most notable of these being the Upper Volga and Circum-Caspian lowlands. This circumstance creates favourable conditions for removing air pollutants. Streams and rivers normally originate on the uplands which, for example, include Valdai, Smolensk–Moscow and Middle Russian uplands and also in the foothills of the Ural mountains. The Middle Volga's right bank is elevated between Cheboksary, the capital of the Chuvash republic, and Volgograd, while its left bank is low-lying. Absolute altitudes throughout the basin generally vary between 100 and 300 m, although they can be as low as 30–40 m within the Circum-Caspian lowland and as high as 400 m in the upstream reaches of the Belaya river.

8.2.2 *Climate and hydrology*

Due to the vast area of the basin the **climate** of the Volga basin is extremely diverse and changes greatly from north to south; with its predominantly flat relief, the Russian plain experiences a pronounced moderating influence from the Atlantic Ocean and in particular from the warm North Atlantic current. The latter determines the formation of humid atmospheric masses, warm in winter and cool in summer. This explains why winters in the European part of Russia are less severe and summers are less torrid than in Siberia. At the same time, the winter weather is also influenced by the cold polar and Siberian air, while the summer weather is affected by hot continental and tropical air masses from the south, as the low Ural mountains do not present an insurmountable barrier for any of them. Interaction between different atmospheric masses creates frequent alterations of rather unstable weather conditions within the Russian plain.

Mean January temperatures within the Volga basin range from –9 degrees C along the western border to –15 to –16 degrees C along the foothills of the Ural mountains. Two different climatic regions are distinguished: northern and southern. The boundary between the two types of climates lies along the line of balance between heat and moisture: to the north precipitation is greater than potential evaporation, to the south it is less. The northern type of climate is dominated by a cyclonic activity with the prevailing westerly direction of winds. This is the wettest part of the Volga basin where the value of mean annual precipitation varies from 700 mm along the western border to 600 mm in the east. Winters in this climatic type are cold and snowy, summers are cool and cloudy with average July temperatures not exceeding +20 degrees C even in the south.

The southern type of climate is characterized by a greater variability of weather conditions due to the greater influence of continental atmospheric masses. Anti-cyclonic activity is more frequent and stable and generates less rainfall e.g. annual precipitation along the Middle Volga varies between 400 and 300 mm. Winters in the southern type of climate are shorter and warmer than in the north, with thin and brief snow cover. Summers here are long and warm or even hot in the south-east of the basin. Mean July temperatures range between +21 degrees C near

Samara and +25 degrees C closer to the Volga delta. Aridity to the south-east of Volgograd sharply increases and droughts become part of the normal climatic pattern.

In the deltaic region of the Volga, the climate is the most arid and continental in the Russian plain because humid polar atmospheric masses rarely reach that far south. Besides, when they do so, the quality of air considerably changes after a long journey across the continent. In summer hot air from Central Asia frequently reaches this area forming the local weather patterns. Temperatures can vary from a minimum of –4 degrees C to a maximum of +45. Mean January temperatures range between –5 and –7 degrees C with an absolute minimum of –28 degrees C. Mean July temperatures exceed +25 degrees C with an absolute maximum of +38 degrees C. The annual precipitation in the delta is very low and varies from 160–180 mm at the Caspian coast to 240–310 mm in the north of the deltaic region (Finlayson *et al.* 1993). At the same time, potential evaporation ranges from 280 mm over unflooded meadows to 1400 mm over swamp areas (Ust'yevaya... 1998).

The **hydrological** network of the Volga drainage basin incorporates a dense system of over 150,000 water courses longer than 2 km which includes 2600 rivers. The total area of the small rivers catchments amount to 45% of the whole drainage basin. The Volga has two main tributaries – the Kama on the left and the Oka on the right. The Vyatka and the Belaya rivers are the two major tributaries of the Kama. Due to the unrivalled diversity of physical geographical conditions throughout the basin the river's hydrological regime varies in different sections of its course. There are even more variations in the hydrological conditions of its tributaries located at great distances from each other. Nevertheless, there are also some common features.

Snow cover, rainfall and ground waters are the main sources feeding the river system. Water supply of the Oka is composed of the thawed water 65%, rainfall 20% and ground waters 15%. The Kama water supply structure is 57%, 18% and 25% accordingly (Vozrozhdeniye... 1996). The hydrological regime of the Volga depends on the patterns of seasonal discharge of the two main tributaries. Most of the seasonal water distribution in both tributaries occurs during the spring period, when flood is high and prolonged and least – during summer.

Before construction of the Volga–Kama cascade of reservoirs, floods would continue for up to two months and the natural amplitude was highest in the Kama mouth reaching 17 m. Annual flood levels reached 10 m at Nizhniy Novgorod and 12 m near the Kama mouth and had a high sanitary value. The natural purification system, which was marked by a rapid river current and high oxygen content, did not create conditions for accumulation of organic and mineral substances and bacterial blooms. However, construction of artificial reservoirs has smoothed out the inter-annual range of fluctuations and has had various adverse implications for the environment.

8.2.3 Vegetation and soil cover

Owing to the huge area covered by the Volga basin, its **vegetation** cover is extremely diverse. Out of 248 types of vegetation found within the European part of the former Soviet Union, 105 or 42% are represented here (Vozrozhdeniye... 1996). Almost all territories with natural grass cover have already been cultivated, while the remaining vegetation is in various stages of degradation. Forest covers approximately 35% of the territory compared to the Russian average of 45%. The overall forested area amounts to 94.2 million ha, 21% of which comprises the most important forests with water conservation functions. Throughout the basin, forests are spread very unevenly ranging from 0.2% in Kalmykia and 2% in the Astrakhan oblast to 60–70% in the Perm oblast.

Composition of the forest zones is more diverse than in the Asiatic part of the FSU, because both European and Siberian species are intermixed within the East European plain. The eastern boundaries of some broad-leaved West European tree species, e.g. oak (*Quercus robur*), ash (*Fraxinus excelsior*) and maple (*Acer campestre*), run across the Volga basin. At the same time, it contains the western boundary of some coniferous Siberian species e.g. Siberian firs (*Picea obovata* and *Abies sibirica*), Sukatchev larch (*Larix sukaczewi*), Siberian stone pine (*Pinus sibirica*), etc.

It is fascinating to view the sequence of various geographical zones travelling from the north to the south across the Volga drainage basin. In the north, the mild and wet climate and adequate drainage provide favourable conditions for the growth of fir species which predominate in the European dark-coniferous **taiga** zone. The dense shade cast by fir trees prevents shrubs and grasses from inhabiting these forests. Therefore, ground cover is normally composed of mosses. Patches of forests composed of pines and small-leaved deciduous species, mainly birch (*Betula verrucosa*) and aspen (*Populus tremula*), are common, the latter particularly in the southern taiga subzone. Phytomass values range between 100 and 350 tonnes per ha from north to south and plant biodiversity is greater than in the Siberian light-coniferous taiga – 400 to 700 species per 100 sq km (Table 1.2). Soil formation in this geographical zone is represented by podzolic processes with predominance of leaching of chemical elements, which is limited in the north by excessive moisture, and in the south by peat formation under the well-developed grass cover of the forested steppe and steppe zones. Therefore zonal soil types of the European taiga are podzolic soils (Table 1.1).

Mixed forests cover the western part of the central Volga drainage basin including the Moscow and the Nizhniy Novgorod oblasts. However, they do not spread further east than Kazan where taiga is directly replaced by forested steppes. Pure oak groves are common in the mixed forests along with the European taiga. There is also a whole range of transitional types of forests: fir forests with the upper tier of broad-leaved species, fir forests with oak species in the shrub tier and to the south of the zone oak groves mixed with fir species. The value of phytomass in this zone increases to 300–400 tonnes per ha, while plant biodiversity reaches 600–700 species per 100 sq km. The annual biological production is also greater than in the taiga and amounts to 12–20 tonnes per ha compared to 4–12 in the former zone. Although mixed forests are generally more resistant to human impact than the taiga forests, a long history of economic development in European Russia has transformed them to a greater degree. Thus the most common type of anthropogenic landscapes can be described as patches of forest among vast cultivated lands.

The zonal type of soil is peaty-podzolic: processes of leaching occur along with accumulation, typical of more southern geographical zones (Table 1.1). This is explained by a specific combination of heat and moisture because the line of neutral moisture balance coincides with the southern boundary of the forest belt. To the south of this line the value of potential evaporation is greater than the value of annual precipitation. It is also an important geobotanical and soil boundary: to the south of this line woods are composed only of broad-leaved species and they are split by areas of grass steppes. Podzolic processes of soil formation give way to accumulation of peat and a more intensive formation of humus, which further become predominant in the forested steppe and steppe zones culminating in the creation of 'chernozems' or black earth type of soils.

The **forested steppe** geographical zone stretches in a broad belt across the Middle Volga from the Ural mountains in the east to the Don river's drainage basin in the west. A combination of broad-leaved deciduous forests with grass steppes forms the typical image of this type of landscape, the most widespread in the whole of southern Russia. Complex ecosystems of this zone are quite resistant to external impact due to a combination of heat and moisture which provides optimal conditions for the growth of trees, shrubs and grasses.

Plant biodiversity reaches its maximum of 1100 species per 100 sq km (Table 1.2). The phytomass value is lower than in the mixed or broad-leaved deciduous forests and amounts to 100–300 tonnes per ha. At the same time, annual biological production reaches its highest values in this and the steppe zone at 18–20 and 18–25 tonnes per ha accordingly. Forests are normally multi-tiered: the upper tier is composed of oak, ash, maple and elm; the second of pear and apple trees. Lower tiers are made up of shrubs such as hazel nut (*Corylus avellana*), spindle-tree (*Euonimus verrucosus* and *E. europaeus*), honey-suckle (*Lonicera xylosteum*), etc. In contrast to the taiga, mosses are lacking in the ground layer and are substituted by a diverse grass cover. Grey forest soils are formed under the forested steppe vegetation.

Steppes replace the forested steppe zone to the south and conditions here become governed by the lack of moisture. Continentality of climate increases in this geographical zone and aridity determines the conditions for vegetation growth. Trees and patches of woodland are found only along the river valleys and various grasses dominate the landscape. From north to the south the grass steppe subzone is replaced by gramineous steppes and then by the dry steppes. The main species found within grass steppes are *Adonis vernali, Trifolium montanum, Myosotis suaveolens* and *Stipa joannis*. Gramineous steppes are most common and are composed of various feather grasses (*Stipa capillata, S. lessingiana, S. sareptana* and *S. stenophylla*) along with *Fectuca sulcata* and *Koeleria gracilis*. To the south, in the dry steppes, vegetation cover becomes discontinued, then sparse, and more arid species of grasses and shrubs join the gramineous to form xerophytic plant communities e.g. *Agropyrum desertorum, Artemisia lercheana, A. pauciflora, Pyrethrum achilleifolium* and *Kochia prostrata*. The share of ephemeral plants also increases to include tulips (*Tulipa*). Plant biodiversity in grass steppes is extremely high at 800–900 species per 100 sq km but the value of phytomass is significantly lower than in the forested steppes, varying between 15 and 30 tonnes per ha. Phytomass becomes even lower in the dry steppe subzone where it is reduced to 8 tonnes.

Increased aridity also defines the direction of soil formation and explains the predominance of accumulation over leaching. In the north of the steppe zone where dryness is lowest, rich grass cover had contributed to the formation of the most fertile soil type – 'chernozem' or black earth soils with a powerful humus horizon. These are subdivided into three subtypes: podzolic, typical and southern chernozems. Further to the south, they give way to poorer chestnut and brown soils of the **semi-desert** geographical zone and on a limited area to alkaline–saline soils called 'solonets' and 'solonchak'. The latter soils become more common in the **desert** landscapes along with grey desert soils and have scarce xerophytic and ephemeral vegetation cover. Values of phytomass and biological production in both zones are very low, at 5 to 30 tonnes per ha and 2–8 tonnes per ha per year accordingly. Plant biodiversity reaching 200–400 species per 100 sq km is comparable to that of the forested tundra zone (Table 1.2). Large areas along both banks of the Volga closest to the delta are composed of sands. Vegetation cover of the northern Circum-Caspian region is in sharp contrast with the lavish vegetation of the Volga's delta. Up to 50% of the upper delta is dominated by meadow plants; meadow and reed vegetation are common in the middle part, while swamp plants prevail in the lower delta.

8.2.4 Overall vulnerability of landscapes

The overall vulnerability of various landscapes which alternate from north to south differs accordingly. In general, to the north of the line of the equal balance of heat and moisture, high susceptibility to human impact is due to lack of heat, while to the south it is due to lack of moisture. The mixed forest zone and forested steppes which cover vast areas of the western and central parts of the Volga basin have an overall high natural resistance to human impact and accordingly a relatively high potential for rehabilitation, while the dry steppes and semi-deserts of the Lower Volga are much more vulnerable to anthropogenic influence.

Due to the moderating influence of the Atlantic Ocean and also to the lack of permafrost, the ecosystems are more stable than in a similar type of landscape to the east of the Ural mountains – for example, the European taiga is more diverse in species composition and has a more complex forest structure than the Siberian taiga. This relatively high potential resilience of ecosystems in the European taiga, mixed forest and forested steppe zones which cover extensive areas of the northern and central parts of the Volga basin can be seen on the FSU map of the vulnerability of natural landscapes (Figure 1.4). Another favourable feature of the Volga basin is a low-lying relief; this does not create a barrier which could prevent polluted air from being dispersed by the alternating atmospheric masses. To continue the comparison with other case studies, there are no phenomena similar to the 'temperature inversions' of the Arctic latitudes.

The vulnerability of ecosystems increases from the central part of the basin to the south–south-east and is greatest in the semi-deserts and deserts of the Lower Volga. If human impact terminates, normally steppes more easily recover from anthropogenic disturbance than forests: duration of secondary succession in gramineous steppes is 35–40 years compared to 100–200 years in oak groves and 120–150 years in the dark coniferous taiga (Sokhraneniye... 1997). In case soil is heavily degraded and/or anthropogenic impact persists, rehabilitation of dry steppes also presents a major problem as in the case of Kalmykia (see Chapter 7).

However, the pressures of anthropogenic origin throughout the Volga basin has had an unmatched duration, scope and magnitude of impact due to the long-term history of human settlement within the Russian plain, and in the central part of the region in particular.

8.3 Historical perspective and economic development

Historically, the Volga river has had an unsurpassed importance for Russia, which initially was called Rus', by connecting it with Byzantium, and later, linking Rus' with Central Asia, Persia and India via the Caspian Sea. From the 13th century onwards it became 'an artery for the land of Russia'. For more than eight centuries it has remained the focal route of trade and economic life within the European part of the country.

The Volga has always played an exceptional role in the economy of Russia. The rich fish resources and the fertile lands on the banks of this great river have provided food and livelihoods for vast areas stretching along the Volga. Many large industrial centres in the heartland of Russia were able to develop successfully owing to her part in the exchange of raw materials and produce with other parts of the country. The appearance of steam boats in the 1840s added to the value of this powerful river. However, every year powerful spring flooding created a whole range of problems, so only with the construction of a series of reservoirs on the Volga was her wild nature tamed and value significantly increased.

At the beginning of the 20th century the main importance of the Volga was still due to the river's transportation and navigation functions. However, in the 1930s, under the 'Great Volga Scheme', construction of a number of major hydraulic complexes on the river was started. Implementation of this project had considerably improved the navigation conditions and gave a mighty impulse to the start of large-scale industrial development in the region on the basis of cheap and abundant hydroelectric energy.

In the 1930s a nationwide 'electrification' campaign was pursued by the Soviet government. This was associated with the implementation of many large-scale hydroelectric energy projects. On the Volga three hydroelectric power-stations (HEPS) were constructed before the Second World War, including Rybinsk station at a capacity of 330,000 kilowatts. During the 1950s four more major hydroelectric stations were put into operation. The largest of them were Kuibyshev (renamed as Samara in the 1990s) and Stalingrad (renamed as Volgograd after Stalin's death) stations (Plate 14), which at that time were the world's most powerful at a capacity of 2.3 million and 2.5 million kilowatts respectively. In 1972 Cheboksary HEPS at a capacity of 1.4 million kilowatts was built in the capital of Chuvash republic. At present, the total energy capacity of the Volga's HEP system is 11.7 million kilowatts and its electric energy is used within an area of around 3 million sq km (Vozrozhdeniye... 1996). At the same time, the Volga continues to maintain an exceptional transport role by linking the industrial heartland with the Middle Volga, the Ural mountains and the Caspian region.

Historically, the easily accessible and abundant natural resources of the Russian plain, and the Volga region in particular, attracted diversified economic development and urbanization. The prioritized development of the military complex, heavy and energy industries within a limited number of large industrial centres was common practice under the planned economy of the command administrative system of the former USSR. It was normally accompanied by a simultaneous retardation in development of the peripheral branches and regions, small and medium-sized towns. The Soviet approach to industrial growth focused on enlarging such enterprises which were normally high consumers of energy, water and material.

As a result of this kind of development, the Volga region now accommodates a large number of enterprises of ferrous and non-ferrous metallurgy, chemical and petrochemical industries, which have

Plate 14 One of the former Soviet giants, the Volzhskaya dam at Volgograd, has cut off almost 90% of the sturgeon's spawning grounds

expanded without taking the existing resource limitations into account. The maximum concentration of industry is in the Central Economic Region which includes Moscow city and oblast and 11 other oblasts. Another feature of the region is a large share of enterprises related to the military-defence complex. Moreover, disregard for the quality of the environment, generally representative of the Soviet period, has contributed to the creation of a significant share of polluting or otherwise harmful industries and enterprises. Many of them are still using backward or outdated machinery and have polluting technological processes. In most administrative units of the Volga region the degree to which machinery is worn out exceeds 40% (Vozrozhdeniye... 1996).

All the above characteristics, evident by the end of the Soviet period, are still present to a large degree and determine the current problems of the Volga basin. The economic decline during the period of transition to the market economy had an immense effect on all branches of industrial production. In 1997 the share of unprofitable enterprises in the country reached 60.4% compared to 15.3% in 1992 (Sodruzhestvo... 1998b). Between 1994 to 1997 the overall index of industrial production in the Russian Federation ranged between 56 and 52% as compared to the 1990 level (Sodruzhestvo... 1998a). In 1994 this value was even lower than 50% in 22 administrative units within the Volga basin with the worst economic indices being in the Russian textile industry centre, Ivanovo oblast, which is located 320 km north-east of Moscow. Another badly affected region is Bryansk oblast, situated 380 km to the south-west of the capital and also noted in this book in the context of radioactive contamination from the Chernobyl accident.

On the whole, during 1990–1995 the overall industrial production decline of over 65% was recorded in the two above-mentioned oblasts. Reduction ranging between 60% and 65% was registered in four oblasts of the Volga basin including Moscow oblast, Chuvash and Mariy El republics (Vozrozhdeniye... 1997). This fall hit most severely the high-tech, defence and machinery production, textile and food processing industries, and agricultural production. For example, some estimates show that production at defence industries, despite the introduction of large-scale changes, declined by six to eight times since 1989 (Vozrozhdeniye... 1996). More successful were regions specializing in the extraction of raw resources, particularly oil and gas, energy production and metallurgy industries.

Agricultural restructuring and changes in land tenure have as yet failed to bring in the much needed increases in the productivity and yields to enable farmers to invest in the long-term fertility of soils in most regions. On the whole, agricultural decline in the Russian Federation amounted to one-third of the Soviet level (Vozrozhdeniye... 1997). However, the situation is far from uniform in various regions of the Volga basin: the more fertile chernozem lands in the south appear to be less affected by the decline than the northern areas. The only exception within the Volga basin, where agricultural production has not shrunk since the Soviet period, is the republic of Tatarstan which enjoys a massive investment support of agriculture by the regional government.

A general slump in productivity, which has been recorded during the last decade in the majority of old and new land tenure forms of collective enterprises, has been somewhat compensated by a productivity increase on private farms and family allotments. A devastating grain yield of 1998 of 47.8 million tonnes, which was mainly due to the prolonged drought conditions, has aggravated the overall situation in agriculture.

8.4 Ecological consequences and recent changes

The main specific feature of ecological problems in the Volga drainage basin is the complex and multi-sided nature of the anthropogenic impact. Analysis of the current state of the environment in the Volga region can only be all-inclusive if a catchment approach is applied, since various anthropogenic influences within the drainage basin will ultimately tell on the quality of water in the Volga river. It can be further shown that at present all components of the basin's physical environment, i.e. surface and ground water, vegetation and soils and even local climate, have already been affected by human actions. The main reasons for environmental destruction in the basin comprise excessive water consumption, large-scale hydro-technical construction, generation of energy, industrial and communal water use, irrigation, timber rafting, over-hunting and fishing, violation of the status of protected territories, etc. Furthermore, these causes often operate in combinations, particularly within urban areas.

8.4.1 Changes in the water supply and hydrological regime

In the course of many decades of intensive industrial development, the overall water consumption within the Volga basin kept increasing and ultimately peaked in the 1980s, when water withdrawal exceeded 46 cubic km per annum, or almost one fifth of the total river's flow. In the 1980s some experts believed that to protect the Volga ecosystems and fish resources only 4–5% of the annual discharge or 10–12 cubic km could be withdrawn safely (Doklad... 1989). It was also estimated that between 1946 and 1986 the total reduction in the annual discharge of the Volga amounted to 35 cubic km.

By the late 1990s, it was planned to increase the annual uptake to 70 cubic km, which was seven times greater than recommended. The main objective for this expansion of water use was associated with irrigation of steppe and dry steppe agricultural lands in the Lower Volga. Meanwhile, the average productivity on irrigated lands during the late 1980s did not exceed that on most efficient rain-fed farms, which is evidence that, by that time, the southern Volga region has already reached a water economic catastrophe.

By the 1980s the available water supply in some parts of the Volga basin was already depleted. For example, excessive use of water resources has led to the formation of depressions in Kursk region. These 'craters' have a radius of 25 to 70 km. Additionally, from 1946 to 1986 the total annual flow of the small rivers in the Volga basin was reduced by 20–30% and the upper reaches of dozens of thousands of minor rivers, streams and springs retreated from their original sources by dozens of kilometres (Doklad... 1989). Approximately one-third of the small rivers and streams in the Volga basin have disappeared.

During the post-Soviet period, economic recession and the reduction in the area of irrigation both contributed to an overall decrease in the amount of water withdrawal in the Volga basin; however, this was not a uniform inter-annual trend. Thus in 1992, water intake went down to 37.3 cubic km or approximately one-third of all Russian freshwater withdrawal (Gosudarstvenniy... 1993). Three years later, in 1995 it increased to 44 cubic km with a value of irrecoverable consumption amounting to 23 cubic km (Gosudarstvenniy... 1996)! By 1997, the volume of withdrawn water decreased again and reached a record minimum of only 28.6 cubic km or 12% of the river flow. The overall water consumption also diminished between the 1980s and 1997, but it was not uniform throughout the basin as illustrated by Table 8.1.

Table 8.1 Freshwater use in selected regions of the Volga basin, 1985–1997

Water user	Freshwater use (cub km)									1997 as % of 1985*
	1985	1990	1991	1992	1993	1994	1995	1996	1997	
Russia	101.0	96.2	95.4	90.0	84.1	77.1	75.8	73.2	70.2	69.5
Astrakhan oblast	2.9	2.3	2.4	2.3	2.2	2.0	1.9	1.7	1.5	53.2
Moscow region	5.5	5.6	5.7	5.9	5.9	5.6	5.6	5.4	4.9	88.8
NizhniyNovgorod oblast	1.6	1.6	1.6	1.6	1.5	1.5	1.5	1.4	1.3	84.9
Perm oblast	2.0	3.1	3.2	3.0	2.9	2.6	2.4	2.4	2.3	115.0
Samara oblast	1.7	1.6	1.8	1.7	1.5	1.3	1.5	1.2	1.1	65.8
Saratov oblast	2.6	1.6	1.8	1.7	1.4	1.1	1.2	1.2	1.2	47.8
Tatarstan Republic	1.5	1.4	1.4	1.3	1.2	1.2	1.2	1.1	1.0	66.1
Volgograd oblast	2.3	1.7	2.0	2.0	1.5	1.5	1.6	1.4	1.3	56.4

Source of data: Rossiyskiy… (1998)
* Calculated by data in million cubic metres

While in the whole of the Russian Federation this decline, compared to 1985, amounted to 30.5%, in Saratov oblast the reduction was greatest at 52.2% and in Perm oblast consumption has actually even *increased*, reaching 115% of the 1985 level. In other administrative units the amount of this reduction varied from 12–15% in Moscow and Nizhniy Novgorod oblasts and over 40% in Astrakhan and Volgograd oblasts. Some of this reduction was achieved by an increase in the share of secondary used and recycled water which reached 51.1 cubic km or 82% of the total water use in 1997 (Gosudarstvenniy... 1998a).

There is an expert opinion that until 1940 the river had a 'natural' type of hydrological regime and discharge and between 1940 and 1955 it was only 'relatively changed' (Doklad... 1989). During the period from 1956 to 1960 most hydraulic works were completed and, as a result, significant changes have occurred in the natural depth and route of the river bed which was largely straightened out. Consequently, after 1961 the hydrological regime of the Volga river system became 'completely regulated', which had a number of significant implications.

Construction of hydroelectric power stations was accompanied by the creation of almost 300 artificial reservoirs on the Volga and its tributary, Kama, with a total volume of about 110 cubic km. The largest of them is the gigantic Kuibyshev reservoir which covers an area of 5900 sq km and has a volume of 56 cubic km. Its length exceeds 500 km with a maximum width of 27 km. An approximate scheme of major water reservoirs in the Volga-Kama system is given in Figure 8.2. The diagram shows that some of these water bodies are so huge that they cross the boundaries of several administrative units. Consequently, *de facto* the Volga has been transformed from a mighty free-flowing river into a chain of successive water reservoirs. At present, the proper term for the Volga river system would probably be '*the Volga river–reservoir system*'.

It has been estimated that after completion of the construction works, water exchange within the drainage basin slowed down by 12 times. Before hydro-construction, water from the river source at Rybinsk reached Volgograd in 50 days and during flood periods in 30 days. By the beginning of the 1990s, it would only pass the same distance within 450–500 days (Doklad... 1989). The above anthropogenic cause in its turn, increased the natural vulnerability of aquatic ecosystems in the Volga river basin.

Moreover, construction of huge reservoirs has changed climatic conditions in the adjacent territories. It has been revealed that these water bodies contribute to the proliferation of droughts within a distance of 10–30 km from their coastline. In winter, cold atmospheric masses formed over the extensive ice surfaces which cool the surrounding territories and delay the beginning of the growing period by 10–30 days. It is estimated that 16–24 million ha or 13–18% of the basin's area are affected by these adverse climate changes (Doklad... 1989).

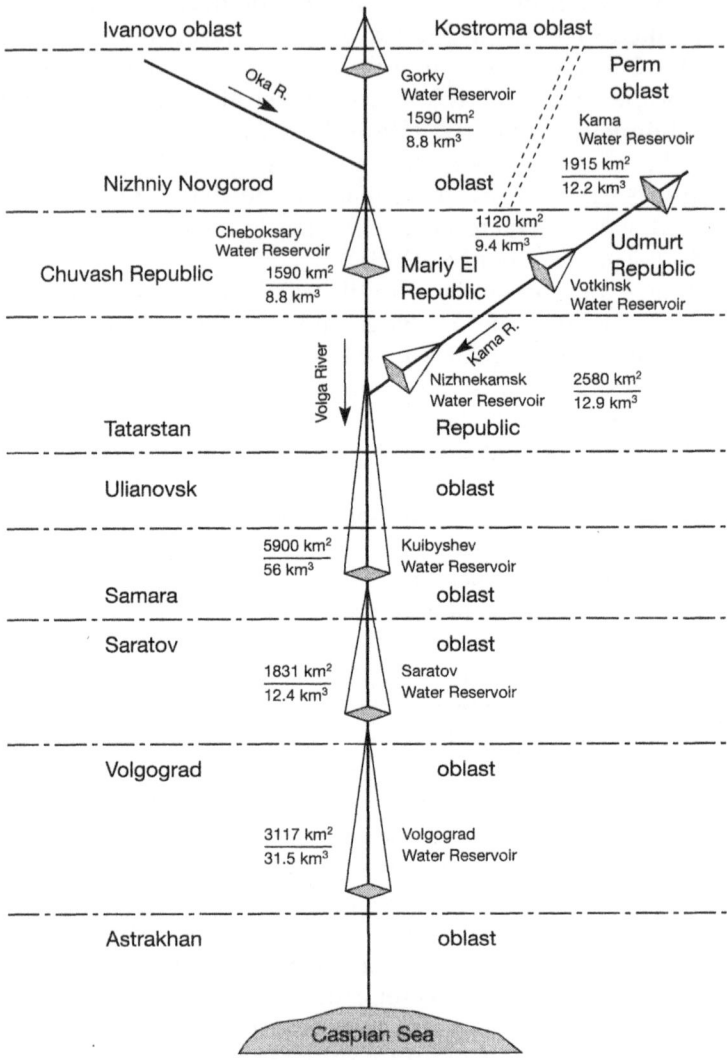

Figure 8.2 A scheme of the Volga 'river-reservoir system'

8.4.2 Decline in the water quality

The long-term technogenic impact on the environment and intensive industrial, urban and agricultural pressures have turned the Volga river system into the collector of all wastes generated in the region. The average annual toxic pressure on the Volga ecosystems is approximately five times greater than in other regions of the Russian Federation. As a result of this, the water in the Volga and its tributaries has become inadequate for the dilution and neutralization of even the 'standard treated' waste-waters. The overall self-purification system in the Volga basin was greatly undermined during the 1960s–1970s due to the radical change in the hydrological balance and had become almost completely destroyed by the late 1980s, owing to the enormous burden of discharged pollutants and a consequent decline in the oxygen level.

The Volga river itself is being polluted right from its upstream reaches where a large number of industrial plants are located. Every town and industrial enterprise situated further down the river, adds to this poisonous load which has an extremely complex chemical composition. Additionally, pollutants discharged into the multitude of small streams and rivers ultimately enter the Volga tributaries and find their way into the main stream. Furthermore, agricultural discharge from the farms lying along the Volga banks adds to the 'cocktail' with its own share of mineral, organic and toxic pollutants. If this poison was not partially neutralized in the water reservoirs and went directly to the lower Volga and the Caspian Sea, all aquatic life there would have been completely destroyed by now.

At present, the Volga basin accounts for over one-third of all water pollution in the Russian Federation. During the late 1980s about 12 cubic km of polluted industrial, agricultural and domestic wastes were annually discharged into the Volga River system. It was estimated that for dilution they would require about 200–240 cubic km or almost all the river flow (Doklad... 1989). The post-Soviet period has been marked by an overall decrease in the amount of polluted wastes reaching the water courses as illustrated by Table 8.2. However, the share of the Volga basin in the total amount of polluted wastes in the Russian Federation remained unchanged at 37–39% of the country's total.

Table 8.2 Pollutants discharged into the Volga basin, 1994–1997

Pollution index	1994	1995	1996	1997
Total volume of polluted wastes, cubic km	9.6	9.2	8.7	8.7
Oil products, 000 tonnes	6.8	5.6	4.1	3.7
Suspended particles, 000 tonnes	257.0	195.0	198.7	169.5
Sulphates, 000 tonnes	1344.0	1139.0	996.0	1557.0
Organic compounds, 000 tonnes	156.0	150.0	142.0	126.0
Total nitrogen, 000 tonnes	12.0	12.7	9.6	8.2
Total nitrites, 000 tonnes	4.7	4.7	6.9	4.3
Zinc, 000 tonnes	0.6	0.5	0.4	0.4
Aluminium, 000 tonnes	5.5	5.6	7.3	8.1
Mercury, kg	61.0	38.0	26.0	23.0

Sources of data: Gosudarstvenniy... (1995, 1996, 1997, 1998a)

The overall pollutant load in 1997 comprised almost 1.6 million tonnes of sulphates, 169,500 tonnes of particulate matter, 126,000 tonnes of organic compounds and around 4000 tonnes of oil products and nitrites. It also included 15,000 tonnes of iron, 8100 tonnes of aluminium, 4000 kg of zinc and 23 kg of mercury. There is a clear evidence of the decline in the amount of pollutants during the 1990s and particularly if compared to the 1980s; for instance, the amount of organic compounds had decreased since the 1980s by three times. Table 8.2 also shows that between 1994 and 1997 the volume of sulphates and aluminium increased.

An important point is that, despite a reduction in the overall amount of generated wastewaters, *the quality of water sources did not improve* (Gosudarstvenniy... 1997). This surprising phenomenon was examined in detail by experts involved in the preparation of 'The Revival of the Volga' national programme. They revealed that the actual volume of pollutants reaching the Volga river system also includes a substantial amount of pollutants originating within urban areas, sites of industrial enterprises and agricultural lands which enter water bodies with surface or shower runoff. It is estimated that at present over half of the Volga's overall annual flow is polluted with excessive concentrations of oil products, phenols, ammonium and nitrates, while concentrations of copper and zinc exceed the maximum permissible levels all year round (Vozrozhdeniye... 1996).

The situation is worst in the minor rivers and streams in the upper reaches of the Volga's tributaries. For example, in 1994 water in most of the Upper Volga tributaries was qualified as 'dirty', 'very dirty' and 'extremely dirty' (Gosudarstvenniy... 1995). One survey carried out in 1993 revealed that all small rivers within the sphere of influence of Cherepovets were polluted with oil products in concentrations ranging from 3 to 43 maximum permissible concentrations (MPC). In the same year, values of polychlorine diphenyl contents in some river sections were three magnitudes (1000 times) greater than the WHO recommendations for their concentrations in the clean water (Gosudarstvenniy... 1994).

Cherepovets has a long-standing bad reputation for the quality of its environment. The major threat is associated with the 'Ammofos' chemical complex. Already by the late 1980s, 16 million tonnes of concentrated toxic wastes including hydrofluoric and other strong acids are kept in the adjacent ponds within an area of 200 ha. The rates of waste accumulation are so high, that the protecting dam had to be raised by four metres every year (Yanshin and Melua 1991). Accumulation ponds in another district of Cherepovets contain heavy concentrations of deadly arsenic. Loudspeaker systems were installed around these ponds to keep birds off these lethal 'water bodies' but still many of them die.

A catastrophic situation with water pollution was recorded in the town of Chapayevsk located 45 km to the south-west of Samara. Until the end of the Second World War, wastes from the town's main defence complex, which produced such hazardous chemical weapons as mustard gas, phosgene and lewisite were being discharged untreated directly into the Chapayevka river which then enters the Kuibyshev water reservoir. Since 1947 this enterprise has been retooled to the production of agricultural pesticides and other chemicals. Accumulation of extremely toxic and highly persistent chlororganic compounds or dioxins in soils, which was occurring for many decades, remains one of the major environmental problems. Concentrations of these toxins reached 400–700 MPC. Even after the closure of this enterprise in 1987, average concentrations of dioxins exceeded the maximum permissible level by 20 times (Vozrozhdeniye... 1997).

Another major environmental threat comes from technical accidents. Dzerzhinsk, one of the most polluted towns in Nizhniy Novgorod oblast, is a centre of the chemical industry. Over the last six years there have been 60 emergency situations associated with toxic and hazardous substances, some with lethal implications (Vozrozhdeniye... 1996). After one technical accident in January 1987 in Cherepovets, the town's integrated metallurgical works discharged untreated waste directly into the Rybinsk reservoir. Heavy metals are also notable for their extensive impact, particularly on human health. By the late 1980s, concentrations of copper in the Lower Volga already amounted to 7–15 MPC (Sostoyaniye... 1990) and the situation has not improved since then. Volgograd and Samara oblasts have areas where ground waters are contaminated with heavy metals classified as 'extremely dangerous' such as mercury and beryllium.

The quality of water in the Volga reservoirs remained very poor over recent decades. While in general water reservoirs play an important positive role in the accumulation and neutralization of pollutants, when their assimilative capacity is exceeded, stagnation conditions in the reservoirs can contribute to a fast accumulation of organic and mineral substances, algae blooms and catastrophic pollution of their waters. Toxicity in the bottom layers of water reservoirs greatly increases. Additionally, algae abundant for example in the Kuibyshev reservoir, sink to the bottom after death and enrich the water with carbon dioxide and hydrogen sulphide which makes it deadly for any living organisms. When this water is discharged further into Volgograd reservoir and reaches the Lower Volga it poses an extreme hazard for all aquatic life.

By the 1980s water pollution by chlororganic compounds called 'dioxins' had already created major problems in the Saratov and Kuibyshev water reservoirs, where the maximum permissible concentration of chlororganic compounds was exceeded by dozens of times. It was perceived as the main reason for massive fish deaths in the above reservoirs in June 1988 (Doklad... 1989).

Furthermore, in some water reservoirs secondary processes of 'self-pollution' had already started in the 1980s. For example, in the Kuibyshev reservoir industrial, domestic and agricultural wastes have been accumulating for over 35 years as bottom sediments. At present this sediment load pollutes water in the reservoir with phosphorus and other chemicals. Now, even if new pollutant discharges are stopped, self-pollution would proceed. This secondary pollution from toxic substances accumulated in the sediments at the bottom of the reservoirs and their eutrophication is now evident in many parts of the drainage basin.

During the 1990s the quality of water in all major reservoirs was classified as 'polluted', 'very polluted', 'dirty' or 'very dirty' (Gosudarstvenniy... 1992–1997, 1998a). In the early 1990s average concentrations of phenols in most water reservoirs reached 2–5 MPC, oil products 3–10 and copper compounds 5–6 MPC. In 1996–1997 the worst quality of water was recorded in the Gorkiy water reservoir near Nizhniy Novgorod, in the Cheboksary and Kuibyshev reservoirs. The former was polluted with oil products at 5–31 MPC, the Cheboksary reservoir with formaldehyde at 7–8 MPC and copper at up to 70 MPC, while in the Kuibyshev reservoir concentrations of oil products amounted to 18–20 MPC near Ulianovsk and levels of phenols and copper were more than ten times higher than the permissible norm.

An integral index of the total pollutants load is the amount which reaches the river delta. A long-term survey, carried out in the Volga mouth area (Ust'yevaya... 1998) has revealed that there were major fluctuations in the annual values of various pollutant contributions but most, for example, the amount of petroleum hydrocarbons in the upper part of the delta ranged from 22,000 tonnes in 1979 to 161,000 in 1988 and organic materials from 619 tonnes in 1977 to 3496 in 1988. Most pollutants showed a clear trend of an increase by the end of the 1980s.

The average total sediment load in the deltaic area during the 1950–1955 period reached 13.6 million tonnes. After the construction and filling up of major reservoirs this value significantly decreased to only 4.5 million tonnes per year between 1971 and 1977. However, after 1978 the annual solid discharge increased again due to an overall rise in water consumption within the Volga basin and a subsequent activation of erosion activities in the river bed. During the period between 1978 and 1993 the average annual sediment load amounted to 8.6 million tonnes per year. An additional input of organic material contributed to eutrophication of the Volga delta and the northern Caspian Sea. A study (Lychagina et al. 1998) has also revealed increased concentrations of heavy metals in the bottom sediments of the delta.

8.4.3 Air pollution

Annually 10.6 million tonnes of pollutants are emitted into the atmosphere within the Volga basin, which amounts to 27% of all air pollution in the Russian Federation. In 1997 the list of 33 Russian cities with the highest permanent levels of pollution included the Volga region's cities of Cherepovets, Lipetsk, Moscow, Samara, Saratov, Syzran, Tolyatti and Ulianovsk, all shown on the map of the state of the environment in Figure 7.1 (Gosudarstvenniy... 1998a).

It would be interesting to note that during 1989–1992, despite an overall reduction in atmospheric pollution in Russia, in Lipetsk and Saratov average concentrations of various pollutants *increased* due to the more frequent instantaneous discharges into atmosphere related to technical accidents which is associated with the lack of rhythmic work of enterprises (Gosudarstvenniy... 1995). Data on the dynamics of atmospheric pollutant emissions from stationary sources between 1992 and 1998 for the first five of the above-mentioned cities are presented in Table 2.2. It shows that in both years the situation was worst in Cherepovets and Lipetsk which emitted over half a million tonnes of pollutants each. A decline in overall industrial production is reflected in the total reduction by 1998 of the total pollution amount, which accounted for 74.8% of the 1992 level in Cherepovets and for 66.6% in Lipetsk. In Samara and Saratov the decline in emissions was even more obvious reducing to only 31% in the latter city.

However, at the same time, atmospheric pollution from transport dramatically increased during the last decade reflecting an increase in the number of private cars and decline in the quality of used petrol, which explains why both cities were included in the list of the worst polluted towns. Unfortunately, data on the amount of pollutants emitted by vehicles is not available for the majority of the Russian towns, so this point is based on the analysis of the situation in Moscow city which is discussed in detail in Chapter 4.

The most common pollutants include particulate matter, sulphur and nitrogen dioxides, benzopyrene, formaldehyde and phenol. In 1997 in many industrial centres of the region the maximum permissible concentrations of these pollutants were exceeded. For example, in Perm contents of benzopyrene in the atmosphere reached the value of 13.3 MPC, in Tolyatti the level of formaldehyde was 20 times more than permitted and in Saratov the concentrations of xylol exceeded 77 MPC! In the same year, the Astrakhan oblast had Russia's highest background atmospheric concentration of cadmium, which is extremely dangerous for human health. The Astrakhan gas condensate and sulphur production complex shown in Plate 15 is the major air polluter in the delta region of the Volga. From the start of operation this industrial 'monster' had practically no treatment facilities. Certain improvements were made later, after large-scale damage had already occurred.

8.4.4 Degradation of vegetation and soil cover

Historically the main transforming human factor within European Russia has been deforestation to make way for pastoral development and cultivation of extensive fertile lands. An extremely prolonged duration of

Plate 15 Construction of this Astrakhan gas condensate complex near the Volga delta had a dire impact on its ecosystem

impact had created a typical anthropogenic type of landscape within all the geographical zones in the central and southern parts of the basin. During the 19th and the 20th centuries, however, nature and landscapes throughout the Volga region have undergone probably the greatest transformation since the commencement of economic development. At present the total forested area accounts for only 10.4% of the territory of the Volga basin, which means a five times decline over the last 100 years. Steppe landscapes in the southern part of the basin have been almost entirely ploughed up.

In the late 1980s, prevailing agrolandscapes lacked trees in water catchment areas and within fields, which left them totally unprotected from wind and water erosion. It was estimated that the woodland around small rivers was 2–3 times less than was needed for conservation purposes. For example, the Volgograd oblast, which requires over 400,000 ha of wooded areas for soil protection, had only 170,000. Annually only 4500 ha of new afforested areas were created at the end of the Soviet period, which meant that the adopted agroforestry programme in this oblast would not be implemented until the year 2040! It should be emphasized that, more recently, afforestation rates have further declined owing to a number of factors, including the lack of finances. Another major problem is the continuing disregard for past-cultivated lands which need urgent rehabilitation and improvement. Approximately 10–12 million ha of such lands are found within 60,000 abandoned or destitute villages and settlements in the Volga region.

One of the adverse implications of the construction of a reservoir system was the flooding of fertile lands due to the predominantly lowland relief of the Volga's flood-plains. The total value of land losses has been estimated at 21 million ha, which comprised 11.0% of cropland; 31.4% meadows; 7.4% pastures; 36.8% forest and woodland; 13.4% other uses (Vozrozhdeniye... 1996). One of the most priceless losses was that of the 1.2 million ha of the best flood-plain meadows in the region, which used to provide cheap fodder for animal stock.

Along with agricultural lands, numerous settlements have been flooded, particularly in the course of the construction of water reservoirs within the most densely populated part of the basin. More than 2500 rural and 96 urban settlements, 200 industrial enterprises, over 380 km of railroads and 1620 km of motorways have been removed and flooded. Approximately 700,000 people had to be resettled to make way for the construction of the Volga-Kama cascade (Doklad... 1989). It was estimated by the latter source that all the revenue received during the last 30 years from the energy production on the Volga-Kama hydroelectric power plant system had not covered these immense losses.

During the last 30 years losses from soil erosion account for 10 million ha or 7% of the basin's area. In the forested steppe zone 60–70% of humus is lost from the soils; in the taiga zone 50–60%. The most fertile chernozem soils in the basin have lost one-third of their humus contents. In the Central Economic Region, the humus layer of soil is lost completely in an area of one million ha, and lost by 50% on four million ha. The rates of fertility losses within rain-fed lands are rapidly accelerating. Despite the seriousness of the situation erosion control activities have been substantially curtailed during the 1990s particularly in the Saratov oblast.

Desertification processes are active in Volgograd and Astrakhan oblasts and they increasingly threaten the forested steppes of the Middle Volga and southern steppes of the basin, particularly in Saratov oblast. In the zone of deficient moisture, particularly on the irrigated farms of Astrakhan, Volgograd, Samara and Saratov oblasts the level of ground water is rapidly rising due to excessive watering norms and inadequate drainage. Waterlogging and soil salinization adversely affect the state of 741,000 ha or 15% of all irrigated lands there (Gosudarstvenniy... 1998a).

Additionally, erosion of the banks of the Volga and its tributaries is a serious problem. Subsequently, bank erosion of the water reservoirs has become a real disaster. Although planners forecasted their coastline retreat by dozens of metres, in reality it has regressed by hundreds. Furthermore, catastrophic landslide processes are common on the slopes of the Cheboksary reservoir between Nizhniy Novgorod and Cheboksary; of the Kuibyshev reservoir between the Kama mouth and the Zhiguli mountains; of the Saratov reservoir between Syzran to Khvalynsk, and of the Volgograd reservoir between Dubovka and Kamyshin. Whole sections of bank slopes, including forests, fields and villages, have slid down into the reservoirs. By the late 1980s, already more than 70,000 ha of productive lands had been lost from destruction of river banks.

Within irrigated areas, the soil geochemistry has drastically changed. Salinization, waterlogging and flooding are threatening the irrigated lands of the Lower Volga. By 1987, in the two oblasts of Saratov and Volgograd alone over 300,000 ha of land had been degraded and needed urgent rehabilitation measures. In the 1980s, the overall area of lands in the Volga region which had lost their productivity due to permanent waterlogging, flooding and adverse change in soil and vegetation cover was estimated at 3.5–5 million ha (Doklad... 1989).

Air pollution often affects the quality of soil cover through precipitation and thawing of snow cover. In 1997 the level of rainfall mineralization, mainly with sulphur ions, in the Central Economic Region increased by 1.5 times as compared with 1995–1996 (Gosudarstvenniy... 1998). Soil contamination with sulphur compounds affects almost the entire Volga basin with average annual sulphur deposition densities exceeding 400 kg per sq km. Areas located adjacent to the main industrial centres e.g. Perm, Samara or Nizhniy Novgorod, are polluted at levels ranging between 600 and 1000 kg per sq km. The worst situation is found around Moscow and to the east of Tula where pollution values exceed 1500 kg per sq km (Atlas... 1998). Soil pollution with benzopyrene is greatest around major urban centres as transport is its main source. The largest affected territory in the Volga basin covers an area of several hundred kilometres along the axis between Moscow and Tula. Other polluted locations include territories around Lipetsk, Nizhniy Novgorod, Samara, Astrakhan, Volgograd and other major cities of the region.

Contamination of soils with such dangerous pollutants as heavy metals and dioxins is also evident throughout the drainage basin of the Volga. For example, in Ivanovo oblast pollution of the soil by lead and copper is ten times greater than the permissible norm. Additionally, maximum levels of contamination with oil products in this region reach 9–56 MPC. In 1997 Moscow and Samara oblasts were particularly severely polluted with pesticides (Gosudarstvenniy... 1998a). Additionally, lands have become burdened by the load of solid wastes which has rapidly increased and presents a major disposal problem. The total amount of solid wastes that accumulate each year within the Volga basin is about 55 million tonnes including 10.5 million tonnes of toxic substances. In Astrakhan, only 2% of the 170,000 tonnes of highly toxic wastes produced annually is currently utilized.

Furthermore, in several areas radioactive contamination associated with the Chernobyl accident, radioactive waste disposal and past nuclear tests presents a major problem. As mentioned in Chapter 5, the Bryansk region in the south-west of the region is particularly severely contaminated with caesium-137. The current threat in the Volga basin is aggravated by the fact that the existing and planned nuclear power plants are often located in densely populated regions, on fertile lands, in the upstream reaches of rivers and in territories subject to landslides or which are tectonically unsafe.

8.4.5 Decline in fish stocks and biodiversity of terrestrial ecosystems

As a result of all the above-mentioned influences, the natural aquatic ecosystems within the Volga river system have been completely transformed; some people are of the opinions that they have been entirely destroyed by the construction of the system of dams and reservoirs. Currently, these so-called 'anti-ecosystems' (Doklad... 1989) are heading towards permanent degradation, while the secondary self-pollution processes which are evident in many reservoirs confirm the above point. It appears that this situation may ultimately culminate in the emergence of a large-scale environmental catastrophe.

For decades pollutants have accumulated in the Volga river system, but the ecological threshold in the aquatic system was probably reached when the Astrakhan gas condensate plant was constructed (Plate 13). In 1987 this complex emitted around one million tonnes of sulphur dioxide into the atmosphere; this was even accompanied by human losses. Afterwards, acidity in the Lower Volga sharply increased and it is considered that when water becomes acid, the effect of toxic substances on organisms multiplies by many times (Yanshin and Melua 1991). Through the food chain, pollutants ultimately accumulate in fish influencing their health and reproductory functions. For example, it was recently revealed that pollution of water, bottom sediments and aquatic organisms have resulted in the development of oncological diseases in fish (Gosudarstvenniy... 1998a).

Fish stocks have been adversely affected due to a number of reasons. Large-scale hydroelectric construction has radically transformed the hydrological and thermal regimes of the river system, cut off many spawning grounds of transitory fish and worsened conditions for survival and reproduction of many fish species. For example, migratory herring species have completely lost their spawning grounds (Vozrozhdeniye... 1996). Due to the continuing ecological degradation and pollution, accumulation of pollutants in fish muscles and roe, loss of fish habitats and over-fishing, an overall reduction in the fish catch was recorded throughout the Volga–Kama system. The total annual commercial fish catch on the Volga before construction of the Volgograd dam amounted to 215–220,000 tonnes, while in 1988 it totalled only 7000 (Doklad... 1989).

Commercial fish stocks in the polluted water reservoirs have suffered a particularly dramatic decline as illustrated by Table 8.3. Our calculations show that between 1989 and 1997 the total fish catch in the Saratov reservoir decreased by 2.3 times, in the Kuibyshev reservoir by 2.6, in the Rybinsk reservoir by 3.2 and in the Volgograd reservoir by 4.5 times! The most valuable fish species in the Volga and the Caspian Sea is the sturgeon (*Acipeseridae*) which has also undergone a significant decline in stocks due to pollution, hydro-construction and poaching (see Chapter 9). The latter was a major new challenge throughout the Volga basin. It

Table 8.3 Decline in fish catch in selected Volga reservoirs, 1989–1997

| | Total fish catch (000 tonnes) | | | | | | | | | 1997 as % |
	1989	1990	1991	1992	1993	1994	1995	1996	1997	of 1989
Kuibyshev reservoir	6.0	5.5	5.2	4.4	3.2	2.6	3.2	2.8	2.3	31
Rybinsk reservoir	3.2	2.3	2.6	2.3	1.2	1.8	1.4	1.2	1.0	44
Saratov reservoir	1.8	1.9	1.9	1.4	1.1	0.8	0.8	0.7	0.8	22
Volgograd reservoir	5.0	4.2	3.6	2.3	1.8	1.4	1.0	0.9	1.1	38

Sources of data: Vozrozhdeniye... (1996); Gosudarstvenniy... (1997, 1998a)

should be emphasized, that *decline in fish stocks continues* despite a reduction in the recorded amounts of discharged pollutants. This, again, could possibly be explained by the fact that the resistance threshold in aquatic ecosystems has been already passed and rehabilitation is impossible without positive human intervention.

The biodiversity of the terrestrial ecosystems within the Volga basin has been dramatically affected as well. Destruction of animal habitats, over-hunting, introduction of alien species and widespread pollution, particularly with toxic chemicals within the Volga basin have adversely affected biodiversity which is minimal despite a wide range of zonal types of landscapes stretching across the basin. For example in the Perm oblast only 12 species of amphibians, 22 species of reptiles and 100 species of mammals are now being recorded (Vozrozhdeniye... 1996).

8.4.6 Overall ecological situation

Figure 8.1 illustrates the contemporary ecological situation in the Volga drainage basin. It demonstrates that there are very few areas with a satisfactory ecological status while the majority has moderately acute, severe and very severe conditions. These conditions appear particularly critical in the Moscow, Tula and Nizhniy Novgorod oblasts of the Upper Volga, in the Middle and Lower Kama, the Samara and Saratov oblasts of the Middle Volga, and in the Volgograd and Astrakhan oblasts of the Lower Volga.

This complies with the results of a multidisciplinary survey of the ecological conditions within the Volga basin (Vozrozhdeniye... 1996), according to which this region can be divided into three groups of territories by degree of ecological degradation. The first group includes ecological disaster areas, some of which are already in a catastrophic state. It can be further subdivided into two subgroups: 'old industrial' and 'new industrial' regions. The former consists of Tula, Moscow, Ivanovo, Vladimir and the Nizhniy Novgorod oblasts on the Upper Volga. The first two of these are the worst within the Volga basin by ecological status. Samara oblast comes third worst within the basin, and together with Tatarstan republic they form the 'new industrial' subgroup on the Middle Volga. The second group with unfavourable conditions comprises three republics and five oblasts including the Chuvash republic and Ulianovsk oblast.

The map in Figure 8.1 also shows the main ecological problems which vary throughout the region. Water depletion and pollution appears to be the most common problem in all parts of the drainage basin; air pollution and soil contamination are most severe near the major industrial centres. Deforestation is a serious issue in the northern areas, while soil erosion and loss of productive lands are common in the central and southern parts of the region.

It was mentioned earlier that natural landscapes of the Volga basin have a relatively high potential ability to recover from anthropogenic disturbance. However, it should be reiterated that human impact within the Volga basin is characterized by great complexity, spatial distribution, enormous magnitude and an exceptionally long-term duration. Unsurprisingly, the total environmental influence was and remains enormous. All the main components of natural landscapes including water, soil and vegetation cover and even climate, as well as whole landscapes, have already been adversely affected. Consequently massive degradation of ecosystems is currently evident throughout the drainage basin.

Investigations carried out by the Volga ecological expedition of the USSR Academy of Sciences in 1983–1988 (Doklad... 1989) have revealed that an ecological catastrophe had occurred in the Volga drainage basin. In the early 1970s there were 15–20 scattered areas of acute pollution throughout the Volga basin. By the late 1980s these degraded sites had spread to form a stable 'ecological disaster zone' in many parts of the region. As the above source evaluated, only 20% of the ecosystems in the Volga basin remained in the 'normal' state; 40% were in the 'critical' and 40% in the 'catastrophic' state (Doklad... 1989).

The analysis of the contemporary ecological situation in the Volga basin made in this chapter further supports the view of Zeev Wolfson (Boris Komarov), the Soviet expert who was the first to openly discuss in Western scientific literature the gravity of the Soviet environmental inheritance. Speaking of the dangerous ecological situation in the Far North and the south of Russia, he stated that 'there is good reason to speak of the formation of an *united front of ecological degradation between Scandinavia and the Black Sea*' (Wolfson 1992, 63).

8.5 Socio-economic and health issues

In 1994, 59.6 million people lived in the Volga region. This part of the Russian Federation has the highest population densities in the country. The urban population inhabits 600 cities or half of the Russian total. Seven of them, Moscow, Nizhniy Novgorod, Samara, Perm, Kazan, Ufa and Volgograd have a population of over one million people. The share of rural population is 22.7%. The population is distributed unevenly with 83% living within the Middle Volga in four administrative units – the Tatarstan and Chuvash republics; Samara and Saratov oblasts. Population density is highest in the Moscow region at 331 people per sq km. The Lower Volga also has high population densities but concentrated along the narrow coastal band. The periphery of the region, which includes the upstream areas of the Volga and Kama, have few large cities, low total population and population densities and a low share of urban population.

Excessive concentration of industry in the largest cities of the Volga have adversely affected the environment inside the cities and in adjacent territories, creating ecological disaster areas. They also exert a negative impact on the social conditions of urban dwellers, such as housing and employment problems, which are compounded by an acute problem of rural out migration. Out of 60,000 villages abandoned during 1979–1989 in the Russian Federation, two-thirds were located within the Volga region (Vozrozhdeniye... 1996).

Various health problems are evident within the Volga basin, which are at least partially linked with degradation of the environment. To start with, poor quality of drinking water has been responsible for many health problems throughout the Volga basin. At present, 85% of all drinking water comes from surface water sources, none of which throughout the region conforms to the Russian sanitary-hygienic and microbiological standards for drinking water quality. Water reservoirs are the main sources of industrial and communal water supply, hence their pollution has had a multitude of adverse health implications. Furthermore, over 80% of all water intake points within the Volga region, particularly those located further away from the main Volga river course, lack treatment and disinfectant facilities. Consequently, water is often contaminated with pathogenic micro-organisms which generate massive infectious diseases.

In 1994 a health survey in the Saratov oblast revealed some major problems. Only 12% of children and teenagers were recognized as healthy, while 44% had various chronic illnesses (Gosudarstvenniy... 1995). An extraordinary situation has arisen in the Lipetsk oblast, where breast milk in women contains 20 times higher levels of dioxins when compared to even such highly polluted Volga cities as Dzerzhinsk and Ufa (Gosudarstvenniy... 1998a). A recent study into child health carried out in Lipetsk has revealed neurological illnesses and deviations related to the impact of heavy metals in 38.3% of children (Gosudarstvenniy... 1996).

It has been estimated that in the major industrial centres like Perm up to 40% of pathological health changes are related to the quality of air, drinking water and food products, soil contamination and specific living conditions. Generally, children are more susceptible than adults to ecologically caused diseases. For example, a general health survey revealed that air pollution contributed to 37% of the total illness frequencies in children compared to 10% in adults. The same value for respiratory diseases was even greater at 41% (Gosudarstvenniy... 1996). A study carried out in 1993 in one of the small towns in Tula oblast have recorded an even higher rate of respiratory diseases at 50% (Gosudarstvenniy... 1994).

In Chapaevsk and Samara increased frequencies of tumours, blood and nervous illnesses and genetic abnormalities were recorded in all age groups of children. In the former town, where the environment is in a catastrophic state, health problems can be directly attributed to ecological factors. Increased frequencies of illnesses of the endocrine, respiratory and intestinal systems, as well as skin diseases, were recorded in all age groups. The reproductive system in women has been also affected, and only 4% of all newborn babies can be considered healthy. Dioxins are found not only in water and soil cover but even in women's breast milk. Adults in this town suffer from the so-called 'Chapayevsk syndrome' – a pathological early ageing and intellectual degeneration (Vozrozhdeniye... 1997).

Adverse ecological conditions along with economic difficulties and social problems have had an ultimate adverse effect on demographic trends. Starting from 1990–1991, increased death rates accompanied by low birth rates have resulted in negative values of natural growth throughout the Volga basin. While in the whole of the Russian Federation the first year of recorded negative growth is 1992, in some oblasts of the Volga region e.g. the Tula and Ivanovo oblasts, the decline started even earlier, in 1980 and 1985 respectively (Rossiyskiy... 1998). All previous analysis of health and the demographic situation within the Volga basin gives evidence that an *overall environmental crisis* is occurring throughout the basin with *catastrophic ecological situations* in many regions.

8.6 Prospects for the future

As shown above, numerous significant health implications have already ensued in many parts of the region with an overall subsequent unfavourable effect on demographic trends. The multitude of ecological problems have had a negative secondary impact on the economy of the region in many cases by increasing production and processing costs. Therefore, it appears that the overall environmental situation within the Volga drainage basin can be with due reason evaluated as a large-scale environmental crisis deepened by the grave economic situation in many parts of the region.

The catastrophic ecological situation within the Volga basin was created during the 1960–1980s but did not receive a comprehensive and critical assessment until the end of the Soviet period (Doklad... 1989). Conclusions based on this report were focused on the need to change priorities from industrial and hydraulic construction to agriculture and fisheries development. However, little has changed for the better during the recent post-Soviet period.

Lack of material and financial resources, plus difficulties in selling produce in most administrative units of the region, have made enterprises temporarily stop or curtail production. Therefore, an overall production decline has decreased the amount of some pollutants in wastes reaching the environment. At the same time, rising inflation and general insecurity about the future have made enterprises abandon more complex, high-tech and resource-saving technologies and economize on installing and using waste treatment equipment and ecologically sound technologies.

The transition period has had a number of economic implications which have affected not only the state of the environment in the Volga basin, but also the ability of experts to study and cope with the ecological problems. One adverse consequence of the overall lack of financing during the last 10–15 years has been a reduction in the number of hydrometeorological stations and posts along the Volga, and particularly in the river delta. Moreover, the monitoring capabilities in the basin have dramatically declined and the execution of some detailed measurements, e.g. measuring pollution in the delta arms, has been abandoned (Ust'yevaya... 1998).

At present, hopes for the rehabilitation of the ecological situation are centred around the implementation of the 'Revival of the Volga' National Programme which is designed to cover a 15-year period and was adopted on 24 April 1998. The concept of this programme was earlier approved by the Russian government in February 1996 (Gosudarstvenniy... 1996). Only 12% of its cost will come from the state budget, while the rest will be covered by the local budgets of the 38 oblasts and republics comprising the Volga region. One of the mechanisms for financing this programme includes the introduction of payment for the use of water and water bodies which has been lacking so far. This could help to transform the typical wasteful attitude towards 'free of charge' natural resources which dominated throughout the Soviet period.

This comprehensive two-stage programme has a whole range of economic, social and environmental goals and is ultimately aimed at the sustainable development of this vast territory. This is expected to be achieved through an improvement in the ecological conditions as well as in the standard of living and overall health of the population, and an increase in economic production and reduction of the associated material and energy losses. Among more specific objectives are the following:

(a) in the course of the *first stage* – to reduce by 30% the volume of discharged polluted wastes into the water bodies; to decrease by 40% the use of drinking water for industrial purposes and to reduce by 20% the consumption of raw materials and energy per production unit;

(b) in the course of the *second stage* – to cease the discharge of polluted wastes into the water bodies; to reduce atmospheric emissions from stationary sources by 1.9 times; to complete the creation of the system of protected territories, increasing its area to up to 3% of the whole Volga basin; to double fish productivity in the Volga–Kama reservoirs (*Zelyoniy Mir* 18, 1998).

Unfortunately, it appears that the economic and financial crisis of 17 August 1998 has pushed back this ambitious programme by at least another five years. The unsatisfactory state of the implementation of this programme has been noted more recently (*Zelyoniy Mir* 21, 1999). Even the most urgently needed measures of the federal, basin-wide and inter-regional levels failed to be realized. These include reservoir improvement, coastal defence measures, protection from inundation and flooding, storage of toxic wastes and environmental monitoring.

CHAPTER 9

The Caspian Sea region: inter-state problems of a shared water body

From the late 19th century, the Caspian moves into the oil era of its existence and lives in two dimensions: a sea of water and a sea of oil.

(Zonn 1999, 36)

9.1 Introduction

The Caspian Sea is the largest closed inland water body on the planet; in many ways it is a unique phenomenon. With a surface area ranging between 363–390,000 sq km it is equivalent to the territory of Germany and the Netherlands taken together. The sea is located well below the average level of the world's oceans; at present it is more than 26 metres below average sea level. Its bed stretches from north to south for almost 1200 km and its width varies between 200 and 450 km. The total length of the coastline is almost 7000 km. Since 1991 the Caspian Sea has been shared by five littoral countries: Azerbaijan, the Islamic Republic of Iran, Kazakstan, the Russian Federation and Turkmenistan (Figure 9.1).

The Volga River basin forms nearly 40% of the territory of the Caspian's catchment area but it supplies over 80% of the total volume of its sea water. Sometimes the Caspian Sea is called a 'Hard Currency Sea' because of its wealth of the two kinds of 'black gold' – oil and caviar; 90% of the world's most precious sturgeon reserves are contained in the Caspian Sea. The above combination, however beneficial it may seem for the owners of these resources, could easily become a perilous 'alliance' in the new millennium in view of the expected substantial growth of oil production in the region. In the early 1990s this region was sometimes called the 'Kuwait of the Third Millennium'. Despite a major controversy concerning the actual oil reserves, with estimates varying widely between 17 and 250 billion barrels or 2.3–34.1 billion tonnes, there is no doubt of its key importance as a source of hydrocarbon fuel in the future.

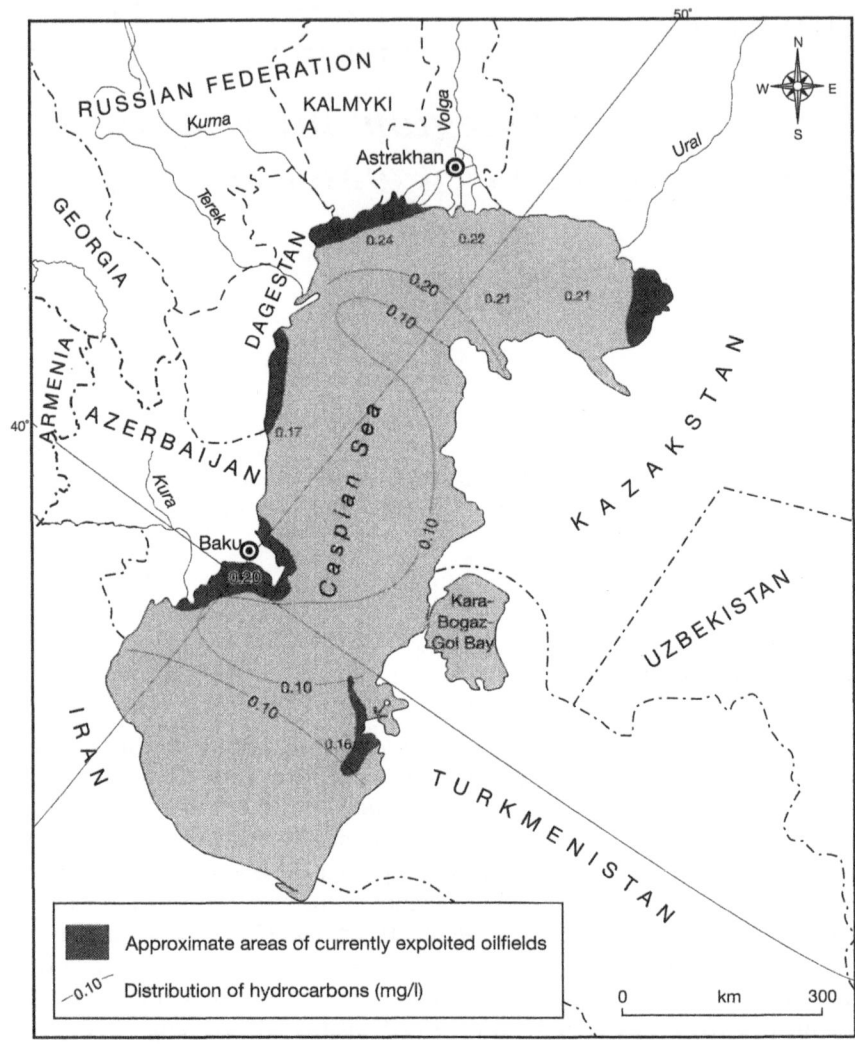

Figure 9.1 Political divisions in the Caspian Sea region, exploited oil-fields and marine pollution with hydrocarbons (data on hydrocarbons from Mekhtiev and Gul 1997)

However, oil pollution, which had already started to threaten the Caspian in the 19th century, could result in a major environmental disaster and prove fatal for the precious sturgeon stock.

Another 'puzzle' associated with the Caspian Sea is related to the fluctuations in its level. Its continuous decline between 1930 and 1977 has brought into existence a notorious scheme to divert the water from the Pechora, which flows north to the Kara Sea, to supplement the falling

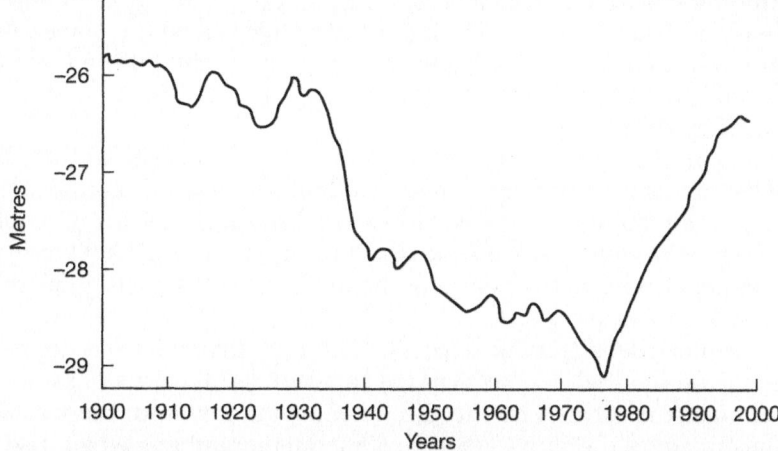

Figure 9.2 Change in the level of the Caspian Sea during this century (updated after Voropayev 1997 with amendments)

Caspian's level. From 1978 onwards the sea could be considered as a model area for the study of the effects of a possible future sea level rise in connection with climatic change. Most recently, in 1996, it seems to have 'changed its mind' again (Figure 9.2).

In the last decade the Caspian region has been facing new challenges stemming from the dissolution of the USSR and the resultant changes in the geopolitical situation. The interests of many more countries are focused on this region thus making it of global rather than of regional importance. The most significant question is the legal status of the sea, which has not been agreed upon, although the actual exploitation of the Caspian Sea reserves has already started.

The aim of this chapter is to focus on two major environmental issues: the continuing water pollution, particularly by oil and oil products, and a recent catastrophic decline in stocks of the sturgeon species. A whole set of various ecological problems confronting five littoral countries will be also outlined. Recent books on the problems of the Caspian Sea include works by Glantz and Zonn (Glantz and Zonn 1997; Zonn 1997a, b, 1999) and Kuksa (1994). Many articles focusing on this fascinating region have been also published over the last decade.

9.2 Physical geographical setting

The Caspian Sea is located in the inner part of Eurasia predominantly within the semi-desert and desert landscape zones. It is a major closed continental basin which lost its links with the oceans of the world some six million years ago. Nevertheless, it has features that are common of both seas and lakes which makes it the major lake-sea on the planet. It is an extremely dynamic water body and all components of the Caspian ecosystem, directly or indirectly and to a greater or lesser extent, are influenced by the river flow.

The area of the Caspian Sea accounts for 12% of its basin. In turn, the size of its catchment amounts to over 3 million sq km which is equivalent to 10% of the total area of the world's closed drainage basins (Golubev 1997). It stretches for 2500 km from north to south and for 1000 km from west to east, also partially including the territories of Armenia, Georgia, Turkey and Uzbekistan.

The sea fills a deep tectonic depression and its current level is approximately 26 metres below the average level of the world's oceans. Its northern coast is formed by the Circum-Caspian lowland; the western coast incorporates the ranges of high Caucasus mountains which rise to over 4000 m above sea level and are bordered by the Terek-Kuma lowland in the north and the Kura-Araks lowland in the south. The southern coast of the Caspian Sea is mountainous, while on the eastern shore uplifted plateaus alternate with low-lying surfaces.

The **climate** of the Circum-Caspian region is extremely diverse and climatic conditions of the northern and southern parts differ, particularly in winter. The high Caucasus mountain range serves as an important divide between the temperate and subtropical climatic belts and as a barrier for humid atmospheric masses from the Atlantic Ocean. In the north of the region, average temperatures in January range between –7 and –11 degrees C, while in the south they are much higher and amount to +8 to +10 degrees C. In summer this difference is significantly less and temperature values measure +22 to +26 degrees C in the north and +25 to +29 degrees C in the south. The range of precipitation is extremely high: from 60 mm per year in Turkmenistan deserts to 1600 mm in the Caucasus south-west. In the north, the average annual amount of precipitation is less than 300 mm and on the western coast it varies between 300–600 mm. The main feature of climate in the region is its high degree of aridity and even both lowlands adjacent to the Caucasus constitute semi-deserts.

The **hydrology** of the Caspian Sea is determined by the 'behaviour' of all inflowing rivers and by the economic activities in the drainage basin; therefore, its regime is extremely unstable. The surface water inflow into the sea is formed by the flow of the Volga, Terek, Sulak and Kuma flowing through Russia; the Ural river in Kazakstan; the Kura in Georgia and Azerbaijan; the Samur flowing along the border between Dagestan and Azerbaijan; small rivers of the Caucasus; and the rivers of the Iranian coast including the Sefidrud.

The hydrological balance is formed by the ingress of surface runoff and precipitation, and the loss of water through evaporation and outflow into the Kara-Bogaz-Gol Bay, which is now notorious as having been one of the major anthropogenic disasters of the last two decades. This bay has an area of about 12,000 sq km and before 1980 was linked to the sea by a narrow 11 km strait. Kara-Bogaz-Gol is located more than four metres below the Caspian Sea level. Annually 8–10 cubic km of the sea's water containing 130–150 million tonnes of salts enters the bay to be further evaporated creating a natural brine of various chemical salts, particularly mirabilite or Glauber salt and halite. This one-way link makes the bay the main 'demineralizing mechanism' and 'evaporating frying pan' of the Caspian Sea and a source of immense renewable wealth. However, in 1980 it was dammed in order 'to protect the level of the Caspian Sea from falling' as it was perceived as the main 'consumer' of the sea's water. As a result of this measure, by 1983 the bay had completely dried up, turning into a gigantic salt 'mirror' almost the size of Wales. This obviously terminated the occurrence of complex physical and chemical processes which sustained the natural renewal of valuable mineral resources. Ironically, the level of the Caspian Sea started to rise even before this 'saving' project was implemented. In 1984 a narrow passage was made through the dam to allow two cubic km of water to flow into what had become a salt lake. More recently, in 1992 the dam was broken down but the damage was already irreparable.

Water salinity in the northern part of the sea is the lowest at 1–8 mg/l, and in the middle and southern parts it increases to 12.6–13.2 mg/l. Low salinity level is one of the features which allows us to refer to the Caspian Sea as a major lake. In addition the Caspian water is poor in sodium and chloride salts and in sulphates. It has also been noticed that the main proportion of aboriginal fauna avoids high levels of salinity.

The range of **landscapes** found on the coasts of the Caspian Sea is quite diverse although they mainly tend to be more arid types: dry steppes, semi-deserts and deserts, which are described in detail in the context of Kalmykia and the Aral Sea in Chapters 7 and 10. A high degree of aridity determines the overall vulnerability of natural landscapes. However, as regards the Caspian Sea probably, even more important is the nature of its aquatic life which comprises many valuable species. One of these is a vulnerable Caspian seal which has a certain similarity with the Baikal nerpa. The total population of Caspian seals amounts to 440–450,000. The Caspian contains 120 species of fish, and 40 of them have commercial value. The sea is the only place on Earth which contains commercial stocks of the most precious species of the sturgeon family.

Sturgeons (*Acipeseridae*) are the source of the world's most delicious and expensive fish product – caviar. In the early 1990s about 65% of the sturgeon catch in the Caspian was associated with the Volga, including 97% of the Russian Sturgeon (*Acipenser sturio*), 66% of the Beluga or Giant Sturgeon (*Huso huso*), and 25% of the Stellate Sturgeon (*Acipenser stellatus*) called 'sevryuga' in Russian. Six species of sturgeon come to the Volga to spawn. The Beluga is the largest fish in the Caspian and can

Plate 16 Sturgeon versus oil: the main dilemma of the Caspian Sea

reach 4.2 m in length, weigh one tonne and reach 75 years of age. However, the average weight of the Beluga, which is becoming increasingly rarer in the Volga, is 70–80 kg. The Russian Sturgeon can reach 230 cm in length and weigh up to 150 kg although its average length is 80 to 150 cm (Plate 16) and average weight about 25 kg (Finlayson *et al.* 1993). 'Sevryuga' or 'star sturgeon' is the smallest of the three weighing from 2.2 kg to a maximum of 24.7 kg. These creatures have been under an increasing threat ever since the discovery of offshore oil reserves in Azerbaijan and the construction of a series of dams on the Volga river.

9.3 Historical perspective and economic development

Economic activity in the Caspian Sea basin is concentrated not only on oil wealth but on the overall industrial potential of the five littoral countries. These issues in regard to the largest state, the Russian Federation, have already been discussed in the context of the Volga river basin in Chapter 7. Intensive industrial development in all former republics of the USSR has always been associated with high natural resource and energy consumption and the generation of excessive amounts of waste discharged into the waterways or otherwise dumped into the environment. By the end of the Soviet period, the European part of the country was overwhelmed by a waste load it could not possibly assimilate.

Probably the oldest legacy of oil pollution and environmental negligence is associated with the development of oil industry in Azerbaijan. It dates back to the 19th century when huge oil reserves were discovered on the Apsheron peninsula and in the offshore zone near Baku. Industrial production of oil started in the 1860s and continued throughout the Soviet period. Until 1951 this was the top oil-producing region in the whole USSR. Discovery of immense oil and gas resources in West Siberia has shifted the centre of focus of the Soviet oil industry into the Asiatic part of the country. However, until the dissolution of the former Soviet Union in December 1991 Azerbaijan remained the centre of oil processing and produced the main share of oil-processing equipment. Recent discoveries of extensive oil and gas reserves in its offshore oilfield, assisted by foreign companies, has put Azerbaijan back in the focus of world attention and heated political debates.

Figure 9.1 shows the location of offshore oil resources within the Caspian shelf. However, the value of potential reserves are different for each country. According to various projections and estimates cited by Zonn (1999) it appears that, taking account of resources located only within the Circum-Caspian region, in terms of oil Azerbaijan is the richest of four former Soviet republics with 3.6 billion tonnes. Kazakstan's potential prospected reserves are only slightly less at 3.4 billion. Turkmenistan comes third with only 600 million tonnes and Russia fourth with only 400. It should be noted that along with oil, Turkmenistan has immense gas reserves. There are no reliable estimates for Iran's reserves although for the moment they appear as the smallest in amount. At the same time, proved reserves estimated at 1.4–2.2 billion tonnes are greater in Kazakstan compared to 0.5–1.5 billion for Azerbaijan.

After the dissolution of the USSR, the dependence of all former Soviet republics on oil and gas has dramatically increased because it became the main source of revenue and hard currency for most countries during the 1990s. For some states it is the only available means to revive the economy and protect sovereignty. According to various sources, in the late 1990s the share of oil and gas in total exports amounted to about 90% in Iran (Europa... 1995, 1999); 52.4% in Azerbaijan (EIU 1996e); 45.8% in Russia (Rossiyskiy... 1998); 43.5% in Turkmenistan (EIU 1996c) and 29.7% in Kazakstan (EIU 1996b). Five major deals related to offshore oil production over the following 30 years, with a total cost exceeding US$20 billion, were signed between September 1994 and January 1997 by Azerbaijan with international oil consortiums incorporating such well-known companies as AMOCO, British Petroleum, UNOCAL, etc.

It should be realized however that at the present time, the actual production of oil in the Caspian basin is still quite low. Table 9.1 illustrates the current values of oil production by four countries. It can be estimated that, for example, in 1996 the total production output of these countries was only 39 million tonnes. Kazakstan is currently the major oil producer

Table 9.1 Oil production in four littoral countries of the Caspian basin, 1990–1997*

Country	1990	1992	1994	1995	1996	1997
	Total production, million tonnes					
Azerbaijan	12.5	11.1	9.6	9.2	9.1	9.0
Kazakstan	25.8	25.8	20.3	20.5	23.0	25.8
Russia**	3.1	ND	ND	3.0	2.6	ND
Turkmenistan	5.7	5.2	4.4	4.5	4.3	ND
	Per capita production*, kilograms					
Azerbaijan	1754	1512	1281	1219	1205	1187
Kazakstan	1542	1523	1207	1232	1430	1624
Russia	3481	2686	2142	2071	2039	2075
Turkmenistan	1538	1290	1000	1000	947	ND

Source of data: Rossiyskiy... (1998)
ND - no data available
*total country's data
**data for Circum-Caspian region only, from Zonn (1999)

in the region. At the same time the value of its production output amounting to about 26 million tonnes is still less than one-tenth that of the Russian Federation which in 1996 exceeded 300 million tonnes. Most of Russia's oil comes from West Siberia. The latter fact explains very high per capita values of oil production in this country. The table also shows that oil production between 1990 and 1997 declined in all countries except Kazakstan.

There have been projections that in the early 2000s Kazakstan's annual crude oil exports would increase to around 70 million tonnes (*Moscow Times*, 20 January 1995) while the annual oil production from the new oil-fields in Azerbaijan would rise to 45 million tonnes (*Monitor – Jamestown Foundation*, 3 November 1995). It now appears, that these forecasts were over-optimistic. Modest estimates based on the current rates of production increase show that by 2010, the Caspian may become an oil-producing region comparable to the North Sea and its value will keep growing (Zonn 1999).

9.4 Ecological consequences and recent changes

The environmental problems facing the five littoral countries of the Circum-Caspian region are numerous and varied (Figure 9.3). The Russian Federation is confronted with the longest list of acute ecological problems; many of these were discussed in the previous chapter. They

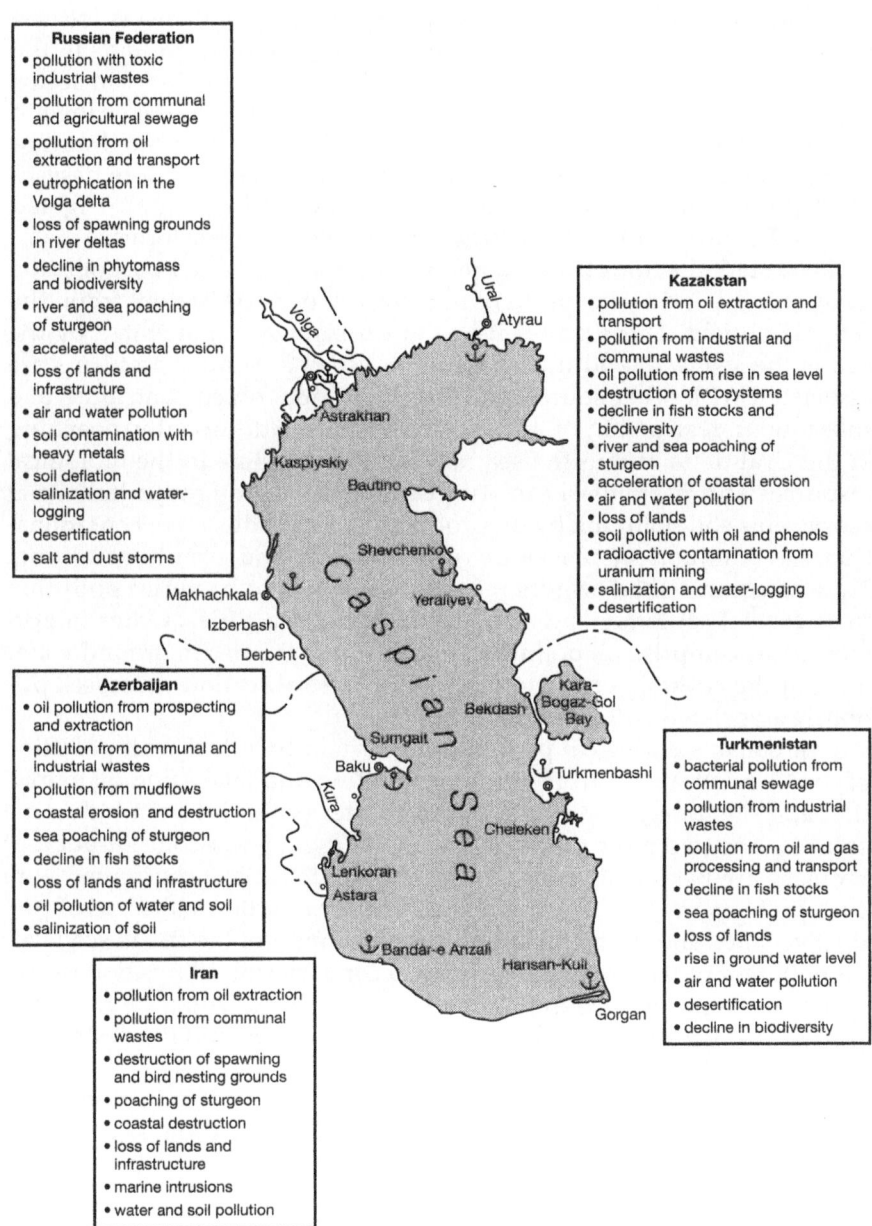

Russian Federation
- pollution with toxic industrial wastes
- pollution from communal and agricultural sewage
- pollution from oil extraction and transport
- eutrophication in the Volga delta
- loss of spawning grounds in river deltas
- decline in phytomass and biodiversity
- river and sea poaching of sturgeon
- accelerated coastal erosion
- loss of lands and infrastructure
- air and water pollution
- soil contamination with heavy metals
- soil deflation salinization and water-logging
- desertification
- salt and dust storms

Kazakstan
- pollution from oil extraction and transport
- pollution from industrial and communal wastes
- oil pollution from rise in sea level
- destruction of ecosystems
- decline in fish stocks and biodiversity
- river and sea poaching of sturgeon
- acceleration of coastal erosion
- air and water pollution
- loss of lands
- soil pollution with oil and phenols
- radioactive contamination from uranium mining
- salinization and water-logging
- desertification

Azerbaijan
- oil pollution from prospecting and extraction
- pollution from communal and industrial wastes
- pollution from mudflows
- coastal erosion and destruction
- sea poaching of sturgeon
- decline in fish stocks
- loss of lands and infrastructure
- oil pollution of water and soil
- salinization of soil

Turkmenistan
- bacterial pollution from communal sewage
- pollution from industrial wastes
- pollution from oil and gas processing and transport
- decline in fish stocks
- sea poaching of sturgeon
- loss of lands
- rise in ground water level
- air and water pollution
- desertification
- decline in biodiversity

Iran
- pollution from oil extraction
- pollution from communal wastes
- destruction of spawning and bird nesting grounds
- poaching of sturgeon
- coastal destruction
- loss of lands and infrastructure
- marine intrusions
- water and soil pollution

Figure 9.3 A scheme of environmental problems faced by littoral states in the Caspian Sea region

include those directly affecting the water quality in the Caspian and those developed throughout its terrestrial ecosystems which have an indirect effect. The former include direct pollution of the Caspian Sea with toxic industrial wastes and oil products; pollution from the discharge of domestic and agricultural wastes from settlements located along its coast; eutrophication and the loss of spawning grounds in the Volga delta; decline in biodiversity and fish stocks in the Russian sector of the Caspian Sea, and accelerated coastal erosion due to the change in sea level. The latter include air and water pollution within the drainage basin; contamination of soil by heavy metals; wind erosion from desertified lands e.g. in Kalmykia; salinization and waterlogging of irrigated lands; other processes of desertification and salt and dust storms.

Kazakstan's sector of the Caspian is seriously affected by pollution from the oil industry, particularly by oil transportation from the Tenghiz oilfield. These problems, as in Russia, were aggravated by the rise in the sea level and the flooding of oilfields. Similar problems are associated with other sources of direct pollution which contribute to a subsequent destruction of aquatic ecosystems. Additionally, poaching in the Ural delta and in the sea itself cause a decline in the biological resources and particularly in sturgeon stocks. Throughout terrestrial ecosystems soil pollution by oil products and its radioactive contamination are common in places of oil extraction and uranium mining. Turkmenistan's shelf waters record high levels of bacterial pollution from coastal settlements which lack treatment facilities. Other important issues comprise oil pollution; over-fishing; the rise in ground water level in the coastal zone; and large-scale desertification processes primarily associated with overgrazing.

Azerbaijan's ecological problems are primarily linked with pollution of the Caspian waters from the long-standing and increasing oil extraction and offshore prospecting: air and water pollution from industrial sources. The two most polluted cities with the highest concentration of industry are Baku and Sumgait. The country is further confronted with such problems as coastal erosion and frequent mudflows; the loss of productive lands and infrastructure due to the impact of the rise in sea level; oil pollution of terrestrial ecosystems; secondary soil salinization of irrigated lands; and decline in fish stocks. Iran is faced with a similar set of problems which also include marine water intrusions and the destruction of spawning and nesting grounds.

The ecological problems which will be discussed further in more detail include water pollution with river flow, eutrophication of the Volga delta, oil pollution and depletion of commercial fish stocks, primarily of sturgeon species.

9.4.1 Water pollution from river flow

The main sources of water pollution of the Caspian Sea are: surface river runoff polluted by discharge of inadequately treated or untreated industrial, domestic and agricultural wastes; exploration, production and transportation of oil from offshore and coastal oilfields; technical accidents at the coastal enterprises and in the sea; surface storm runoff which contains industrial and agricultural wastes and is particularly linked with the increased effects of the sea level e.g. arising from inundated oil and gas fields, industrial estates, settlements, etc. Surface river runoff is responsible for 90% of the total pollution load in the Caspian Sea (Gosudarstvenniy... 1995).

Table 9.2 shows the change in the total amount of polluted waste waters discharged by the littoral countries between 1991 and 1997. It is evident that although by 1997 the total amount of polluted effluents decreased by over 20%, the actual amount of pollutants reaching the sea is enormous. Russia's share was greatest at 91.3% in 1991 when official data was available from all countries of the region. The second largest amount of waste waters came from Kazakstan, followed next by Turkmenistan and Azerbaijan. The average flow of pollutants from each country between 1978–1992 was assessed by Mekhtiev and Gul (1997). Although Russia's total share is a similar size at over 84%, Azerbaijan's estimated contribution was 15.4% of all pollutants, while the remaining very minor load was almost equally divided between Turkmenistan and Kazakstan. The latter detail raises certain doubts due to the notoriously high level of pollution in the Ural river flowing from the heavily contaminated industrial regions of the Ural mountains.

According to these authors, Russia is responsible for almost 90% of phenols and over 80% of oil products, pesticides and metals. The contribution of Azerbaijan is over 15% of oil products, 32% of synthetic surfactants, 17% pesticides and about 9% of metals. The share of other countries is shown as insignificant (Mekhtiev and Gul 1997). The overall annual amount of pollutants is estimated at 124,040 tonnes. If unaccounted sources of pollution, discussed previously in the context of the Volga basin, are also included, the overall situation would appear as quite dramatic.

Table 9.2 Discharge of polluted waste waters by Caspian littoral countries, 1991–1997 (million cubic m)

Country	1991	as %	1992	1993	1994	1995	1997
Azerbaijan	254	1.9	229	255	262	134	179
Kazakstan	338	2.6	318	290	236	230	188
Russia*	12,016	91.3	11,593	10,162	11,000	10,400	9,800
Turkmenistan	290	2.2	ND	ND	ND	ND	ND
Total	13,152	100.0	12,430**	10,997**	11,788**	11,054**	10,457**

Sources of data: Okhrana... (1998); Okruzhayushchaya ... (1996); Rossiyskiy... (1998)
ND – no data available
* within Caspian basin only
** taking Turkmenistan's value as in 1991

The problems arising from the pollution of the Volga were discussed in detail in Chapter 8. Annually about 2.5 cubic km of untreated and 7 cubic km of insufficiently treated wastes are discharged into the Volga basin. Direct and indirect pollution of the river system over many decades has ultimately turned the deltaic area and northern Caspian ecosystems into a collector of various harmful pollutants. The water quality in the Volga mouth in the 1990s is greatly different to that recorded in the early part of this century. This is clearly illustrated by changes in the phytoplankton communities. Cyanobacterial blooms are now quite common, which is a clear evidence that the water quality has deteriorated over the past 50 years.

Concentrations of the most 'typical' pollutants of the Volga, that is oil products, zinc, copper and mercury compounds, suspended organic matter and pesticides, exceed their maximum permissible levels by 2–3 times throughout the Russian Caspian. In some sea sectors, the current level of pollution is greater than 5 MPC (Gosudarstvenniy... 1996). Additionally, the level of pollution increases due to an influx of pollutants from bottom sediments in the river deltas. Water in the western parts of the Northern Caspian is classed as 'dirty' while in other sections of the Caspian as 'polluted'. There is a steady tendency of an *increase* in the pollution level.

Other rivers flowing into the sea are also heavily polluted. For example, the Ural river has high concentrations of nitrogen, sulphates, chlorides, iron compounds and oil products. Its worst polluted tributary is the Blyava near Mednogorsk in the Ural mountains which in 1993 recorded copper contents exceeding up to 210 times the maximum permissible level! Concentration of zinc amounted to 37 MPC while the content of nitrogen was six times greater than the permissible level (Gosudarstvenniy... 1994). The extremely high content of chromium in Ural's tributary Ilek was noted as a source that affected the water quality of this river (Bridges and Bridges 1996).

The Terek river is connected with a very high level of economic activity: in 1993 water withdrawal in its basin amounted to 6.2 cubic km or 59.6% of the mean annual discharge. About 1.6 cubic km is lost in irrigation and 1.5 cubic km of polluted wastes are discharged back into the river. Only 23.3% of these effluents are treated. In the lower reaches this river is polluted with nitrogen which exceeds the standard by 3–9 times and copper by 2–7 times. In addition it contains excessive amounts of iron compounds, oil products, ammonium and suspended matter.

Eutrophication of the delta and the northern Caspian Sea has greatly increased since the late 1970s (Finlayson *et al.* 1993). On average, 30–40% of the nitrogen and phosphorus load comes from anthropogenic sources while in some years it increases to 60%. The amount of organic matter deposited annually in the delta has grown from 3.5 tonnes in the 1960s to 6–7 tonnes recently (Ust'yevaya... 1998). Thousands of tonnes of nitrates from farmlands reach the water system via surface runoff, particularly during the spring floods. In addition there are very high concentrations of pesticides widely used in rice-growing in the Lower Volga region.

Air pollution and pollutant deposition contribute to the overall pollution load. The majority of enterprises are not equipped with adequate treatment facilities. Normally 61–85% of solid particles are intercepted in Kalmykia and Dagestan as compared to the Russian average of 94%. Gaseous compounds are hardly intercepted at all, and in 1992 all discharged sulphur dioxide and other gaseous pollutants in these regions reached the atmosphere (Natsional'niy... 1995). Many of these pollutants are deposited in the aquatic surface of the sea in the form of wet or dry deposition, thus also affecting its water quality.

9.4.2 Oil pollution

Oil pollution of water and soil is considered separately because it is a very important issue for the Caspian Sea. It is already a serious problem throughout the sea and threatens water quality and the health of aquatic ecosystems. Oil and gas prospecting, extraction and transportation both in offshore and coastal areas present an ever-increasing danger. The equipment 'normally' applied in the former Soviet republics is old and sometimes obsolete while technologies are often harmful for marine life. An appraisal of the overall status of water quality in the Caspian at the end of the Soviet period was made by Mnatsakanian (1992). According to his estimates, throughout the aquatic surface the average level of pollution by oil products and phenols exceeded the maximum permissible concentrations (MPC) by 2–43 times. The worst polluted areas were near the Volga mouth, near the Ural River delta and along Azerbaijan's coast. In the latter section pollution levels around the Apsheron peninsula reached 10 MPC for oil and 18 MPC for phenol products.

During 1986–1990 the overall amount of oil and oil products polluting the Caspian Sea more than doubled from 62,000 to 169,000 tonnes (Kuksa 1994). Most of this pollution was associated with river flow; however, in 1986 85% of the total amount came from direct industrial discharges into the sea. In addition pollution by phenols, which in combination with water produce particularly toxic compounds, ranged between 367–1354 tonnes from both sources during the same period. This wide variation of values once again emphasizes the lack of reliable data which, along with the above-mentioned causes, has been due to poor analytical equipment available in the Caspian basin. As Kuksa stated, 'we obtain poorly representative data on marine pollution by heavy metals, chloroorganic pesticides and other toxic substances, and have only fragmentary information about processes of secondary pollution and exchange between toxic materials with bottom sediments and sea water. Data is also lacking on the accumulation of pollutants in aquatic organisms which can exceed the background values of water pollution by hundreds and thousands times' (Kuksa 1994, 95). An official source of information suggested that in 1996 average phenol concentrations in the northern Caspian exceeded the maximum permissible level by five times (Gosudarstvenniy... 1997).

As noted above, according to Mekhtiev and Gul (1997), the main sources of oil pollution are located in Russia and Azerbaijan while the shares of Kazakstan and Turkmenistan amount to 0.01 and 0.3%. However, this contradicts their own data on the distribution of hydrocarbons included in the map in Figure 9.1. It shows that the values of hydrocarbon content are very high both near the capital of Azerbaijan and in the northern and north-eastern sections of the Caspian Sea. It was estimated in 1993 that the average pollution levels near the coast of Kazakstan were four to six times higher than standard (*Izvestiya* 6 April 1993). This fact is not surprising taking into account the above values of the current oil production in these two countries.

In addition, data from Kazak sources (Sultangazin and Tsukatani 1995) suggest that the situation with oil pollution within the coastal zone of Kazakstan is no less serious. These authors noted that pollution is associated with the exploitation of over one hundred oil and gas deposits including the giant Tenghiz oilfield with recoverable reserves estimated at one billion tonnes of oil. Oil lakes and areas contaminated with diesel on the Mangyshlak peninsula cover an area of thousands of hectares. The ecological situation is especially hazardous in the region of the Karachaganak oil and gas condensate complex.

Extensive prospecting works in the shelf zones of all states carried out over the last decade have contributed to practically uncontrolled pollution throughout the aquatic area of the sea. The threat of oil pollution relates not only to the areas of offshore oil exploration and extraction or shipping across the Caspian Sea; it is also associated with its transportation by pipelines across the coastal zones. Overall oil losses in extraction, transport and processing amount to 2% of its total volume. During the Soviet period, seismic exploration for oil and gas reserves had already resulted in massive deaths of sturgeon by 1987 (Natsional'niy... 1995). Drilling works in the shelf zone and oil spills owing to technical accidents such as the one shown in Plate 17 present a particular hazard for marine life because, in this case, concentrations of pollutants may exceed the maximum permissible levels by hundreds and thousands of times and accumulate in the food chain, thus affecting the sea's potential bioresources.

It should be noted that currently there are many contradictions between the data on the level of oil pollution in each of the five states. A major reason for this is that before the dissolution of the FSU, environmental monitoring was based at the USSR's Academy of Sciences, while after 1992 each country has had to proceed on its own. For Turkmenistan, for example, this meant a start from almost zero and as the exchange of

Plate 17 Fire on the offshore rig near Baku, Azerbaijan: a major threat to the Caspian comes from its wealth of oil

information has nearly stopped, it is now very difficult to obtain comprehensive and objective information on the ecological situation. It is particularly hard because, due to the present economic problems, no country can afford to sustain a proper scientific fleet to monitor the pollution levels. Even in Russia, as was mentioned in the context of the Volga drainage basin (see Chapter 7), during the transitional period monitoring capabilities in the deltaic region have declined. This is reflected in the fact that pollution observations in the Northern Caspian in the last 10–15 years have been carried out only occasionally (Ust'yevaya... 1998).

9.4.3 Effects of human activities on marine life

Three main sets of reasons are responsible for the overall decline in fish resources over the last decades. The first is linked to the shrinkage of the area of fish spawning grounds in the course of hydraulic construction on rivers flowing into the sea. The second cause is the declining water quality in the Caspian Sea due to increasing water pollution from various sources. The third reason, which keeps growing in importance since the dissolution of the former Soviet Union, appears to be poaching of fish and primarily sturgeon for its precious caviar.

Before the construction of the series of dams, Beluga sturgeon used to have the longest migration route into the Volga tributaries, the Kama and Oka and even further upstream. The migration pattern of sturgeon to a great extent depends on the hydrological regime of the Volga. After damming of the Volga at Volgograd, only 400–430 ha or about 10–12% of the former 3600–4000 ha of the **spawning grounds** remained accessible to sturgeon. It is estimated that 100% of the natural spawning grounds of the Beluga sturgeon, 80% of the Russian sturgeon and 60% of the Stellate sturgeon was lost.

Furthermore, changes in the hydrological regime have caused a delay in the spawning of the Beluga and its reproduction has suffered. Previously, Beluga spent the first three months migrating up the river and reached a length of 36 cm and weight of 172 g. In the beginning of the 1990s their size only reached 9 cm and weight 4.2 g (Finlayson et al. 1993).

Problems with spawning grounds are typical not only for the Volga but also for the Terek and Ural rivers. In the Terek's drainage basin, excessive water withdrawal for irrigation and the lack of fish passage in the Kargalin hydraulic power system are further aggravated by insufficient water discharge into the lower reaches of the river. These changes have completely stopped sturgeon from entering the river for spawning. A similar situation is found in the Sulak river of Dagestan. The rivers in Dagestan have practically lost their spawning potential.

As a consequence of **water pollution** with toxic substances there have been reproduction problems in mature fish. Over the last five years reproductive functions in sturgeon have kept declining while the number of fish with complete fat degeneration of gonads increased by four times compared with 1991. Permanently high amounts of toxic substances in water accumulate in greater concentrations on the bottom, and via the food chain in aquatic organisms. Further up, they accrue in the internal organs and meat of fish which ultimately disrupts their reproductive function and causes the death of fry and adult fish and loss of gustatory qualities. The sturgeon, which lives much longer than other fish species, is particularly susceptible to pollution risks. In the past, systematic pollution had already induced mass sturgeon deaths. For example, in 1988 8496 sturgeon and in 1989 another 380 died downstream from Volgograd (Gosudarstvenniy... 1996). More recently, accumulation of toxic pollutants in fish muscles, particularly in the caviar producers which stay in the Volga over winter, has dramatically affected their reproduction potential. It has been estimated that 18% of all sturgeon females and 21% of Beluga females show signs of resorption of caviar. Adverse toxicological conditions in the river due to water contamination with heavy metals have also resulted in a disease that manifests itself in delamination of muscular tissue and weakening of egg shells, which affects the quality of caviar. A study of the health of Russian sturgeon revealed that they also suffer from disruption of ion exchange, protein-carbohydrate homeostasis and metabolism, liver dystrophy and kidney problems. Concentrations of pesticides including DDT in their liver and fat exceed the standard for food products by two to five times. In addition their liver contains cadmium, nickel, mercury, lead and copper in concentrations several times above the standard (Kuksa 1994). The above complex of pathological changes was diagnosed as a cumulative toxicosis with multi-system destruction. Analysis of physiological and biochemical indices gives evidence of the continuing depressed status of sturgeon and the Caspian seals (Gosudarstvenniy... 1994).

Oil pollution presents a specific threat to aquatic life. It is estimated that one gram of spilt oil products spoils 20,000 litres of water. An increase in the surface oil film to 0.1 mm which occurs at the value of one gram per sq m of oil, disrupts processes of gas exchange and threatens hydrobiotic organisms. Oil is toxic for all aquatic marine life: for fish, at concentrations of over 0.01 mg/l and for phytoplankton at over 0.1 mg/l (Zonn 1999). As can be seen from the map in Figure 9.1, contemporary concentrations exceed the threshold for fish by 10 to more than 20 times and are twice the toxic level for phytoplankton. Oil pollution primarily affects larvae and fry but it also has an adverse impact on the reproductive functions of adult fish. It has been noted that in combination with other pollutants, the toxic effects of oil significantly increase.

It is interesting that a concentration of 0.21 mg/l, which is common throughout the northern Caspian, is only four times greater than the maximum permissible concentration for drinking water. This fact confirms that marine organisms have a greater sensitivity compared to humans. The latter fact is also reflected in an exponential increase in fish diseases, mutagenic and pathological effects and an ultimate decline in their reserves.

An additional but probably more critical threat has recently arisen from the **poaching** of sturgeon for caviar and its delicious meat both to be sold unofficially abroad at the markets of Moscow and other big cities (Cullen 1999; *Izvestiya* 7 August 1999; *Zelyoniy Mir* 9 (303), 1999; Zonn 1997a, b, 1999). Frequently poachers hunt sturgeon only for its caviar. The so-called 'Fish Mafia' (*Izvestiya* 13 April 1999) protects an extremely profitable illegal sturgeon fishing operation in the Volga and the Caspian Sea.

During the Soviet period, the former USSR introduced the prohibition of sturgeon fishing in the open sea. Since that time, catching sturgeon has been carried out in the river deltas. It should be noted that sturgeons reproduce very slowly: the fish do not spawn for the first time until they reach the age of 20–25 years. However, as was noted by Sakwa, 'the disintegration of the USSR jeopardized Soviet environmental and conservation legislation as each republic sought maximum advantage for itself, leading, for example, to sea fishing of sturgeon in the Caspian, banned in 1962 by the Soviet Ministry of Fisheries' (Sakwa 1993, 263).

The continuing economic crisis in all the former Soviet states which border the sea and the lack of a joint agreement on the legal status of the Caspian Sea 'stimulate' the poaching of sturgeon both for local consumption and sales. As a result, poaching of mature fish has dramatically increased and is practically uncontrolled during the spawning period, not only within the Volga but also in the open sea along the coastlines of Russia, Kazakstan, Azerbaijan and Turkmenistan (Figure 9.3). The cost of top-grade Beluga caviar sells on world markets for up to US$1200 per kilogram. Bought in Azerbaijan, it costs only a small fraction of that offering huge profits for smugglers. The volume of illegal fishing is estimated to be, at a minimum, 70% of the total reproductive sturgeon stock. More recently, there have even been piracy acts recorded in the open waters of the Caspian Sea (*Caspian Sea Bulletin* 2, 1999).

Poachers target the female fish and thus undermine the reproductive potential and structure of fish stocks. The latter has also showed a decline in the share of less valuable species and the predominance of 'sevryuga' in the fish catch. The structure of other fish stocks has also changed dramatically with an increase in the low value fish species (Kuksa 1994). While before the 1950s 80% of all the fish catch consisted of sturgeon species, by the mid-1980s only 20% accounted for valuable while 80% accounted for low-value fish. At the present time the latter species, such as sprats, make up almost all the catch in Azerbaijan and Turkmenistan (Zonn 1999).

The integral adverse anthropogenic impact on the Volga-Caspian ecosystem has culminated in the overall **decline in fish catch and their reserves**. While in 1932–1936 the average annual catch of such precious commercial fish as sturgeon, carp, Caspian roach, carp-bream, pike and herring in the USSR reached 400,000 tonnes, by 1951–1955 this had reduced to only 280,000. Furthermore, in the 1990s the decline became catastrophic as the total catch volume did not exceed 45–50,000 tonnes (Gosudarstvenniy... 1996). This decline was particularly dramatic for the sturgeon species. The most valuable species, Beluga, is becoming more and more rare, while smaller sized sturgeon species dominate in the fish catch (Plate 16).

The Volga-Caspian region historically had an outstanding role in the fisheries of the Russian Empire. In 1889 the total catch of sturgeon species was estimated at 106,800 tonnes. It accounted for 22% of the total fish catch of 491,200 tonnes (Rossiya... 1995b). As a result of all the causes discussed above, sturgeon catches during the 20th century has dramatically declined. On the Volga, the value of the catch shrank from a record of 39,000 tonnes in 1903 to 18.8 in 1913 (Brokgauz and Efron 1995), rising to over 20,000 in the 1930s; then it fell even lower to 4500 at the end of the Second World War (*Izvestiya* 7 August 1999). After an introduction of a raft of governmental measures which included the expansion of artificial fish hatcheries, the total sturgeon catch rose to 27,000 tonnes in 1977 but a further reduction was recorded in the 1980s. In 1983 only 16,800 sturgeon were caught (Gosudarstvenniy... 1995).

The most precipitous decline followed in the 1990s when state control over fishing in Russia has weakened. In 1993–1994 the total official catch figures were only 3200 and 2000 tonnes accordingly. In 1995 the sturgeon catch in the Volga-Caspian region reached 2176 tonnes and taken together with Turkmenistan and Azerbaijan quotas it amounted to 2830 tonnes (Gosudarstvenniy... 1996). Therefore compared to the pre-Soviet period, the amount of sturgeon fish catch *declined by almost 40 times!* In 1995 fishing regulations changed: the length of the summer banning period was extended to 15 June–31 August. Nevertheless, the total population of sturgeon in the northern Caspian continued to decrease, particularly the adult fish group. In 1996 the value of sturgeon catch dropped even lower to 1700 tonnes and in 1997 was down to its lowest value of 1300 tonnes (Gosudarstvenniy... 1997, 1998a). The most recent estimate suggested that in 1998 it had reduced further to about 1000 tonnes while the projected total catch for 1999 was only 639 tonnes (*Sunday Business* 10 October 1999).

During the 1980s, Soviet enterprises produced up to 2500 metric tonnes of black caviar per annum, while the Iranians produced 250 tonnes. However, according to recent assessments, in 1995 Russian enterprises produced only 70–90 metric tonnes of sturgeon caviar, and Iran only 200 metric tonnes (*Business World* 9 June 1995).

According to some estimates, the overall stocks of adult sturgeons in the Caspian Sea dropped from 142 million in 1978 to no more than 43.5 million in 1994 (Malyutin 1995 cited in Zonn 1999). An analysis of the situation in the Volga-Caspian region made in 1995 revealed that sturgeon stocks of reproductive age were *in a catastrophic state*. The efficiency of natural reproduction in the Volga was lower than the critical threshold due to a sharp decline of spawning sturgeons. The numbers of sturgeons arriving at the existing spawning grounds in the delta is less than its spawning potential. Pollution and over-fishing have coincided with the fact that at the present time fish spawned in low-yield years are reaching maturity. The current situation has become so alarming that the Russian National Report on the State of the Environment in 1996 stated that probably the *total banning of commercial sturgeon fishing* should be introduced (Gosudarstvenniy... 1997).

An additional threat to the environment during the period between 1978 and 1995 was associated with the **increase in sea level** by 2.4 m (Figure 9.2). Its causes and various implications have been discussed by Golitsyn (1997), Kasimov (1997), Svitoch (1997), Voropayev (1997) and Zonn (1997a, b and 1999). The most important environmental impact was associated with the flooding and inundation of the coastal oilfields in the littoral states. For example, according to Babaev (1997), in the latter state, oil and gas pipelines in Cheleken city have been completely submerged under water; in Turkmenbashi city the oil pipelines and oil processing plant's infrastructure are also threatened. Seven oil deposits have been already flooded in Kazakstan and another 28 are in direct danger of inundation (Zonn 1999). Flooding of oil extracting sites around Baku in Azerbaijan creates numerous oil lakes along the coastline. In addition, there are various dramatic socio-economic implications of the rise in the sea level; these will be discussed later in this chapter.

9.4.4 Overall ecological situation

Many scientists believe that the current ecological situation in the Circum-Caspian region is disastrous. For example, according to Azerbaijan experts, 'At present, pollution of the Caspian has reached catastrophic proportions' (Mekhtiev and Gul 1997, 83). It appears that the overall ecological situation can still be classed as a crisis while the state of the Caspian's sturgeon stocks, as was shown above, is in a catastrophic state. The main argument to support this is that, theoretically, if strict control is introduced and implemented over marine pollution and fishing throughout the sea's area it would be possible to reverse the current situation.

However, the current geopolitical situation in the region, lack of financial investments and adverse socio-economic circumstances do not give much hope for such a scenario. It seems that the threshold of irreversibility will be reached in the next century when the current rates of oil production will grow by dozens of times. Then the overall contemporary crisis will inevitably become an ecological catastrophe.

9.5 Socio-economic and political issues

There is still only relatively limited local evidence of health deterioration associated with ecological degeneration. Therefore we cannot classify the situation in the Circum-Caspian region as an 'environmental crisis' but rather as an *ecological crisis*. However, major disastrous socio-economic consequences have resulted from the rise in the Caspian sea level. These have primarily been caused by unsound economic development and colonization of the high-risk coastal zone during the low-level period of 1930–1977 (Saiko 1997). New settlements and industrial enterprises were placed there without due regard for the hazards involved. Within the Russian coast, in Astrakhan oblast about 400,000 ha of land was flooded including 12,000 ha of valuable irrigated lands, thus 10% of agricultural lands have been withdrawn from economic use (Zonn 1999). An additional 257,000 ha are in the zone of impact from the rise in sea level. In Dagestan, 150,000 ha of agricultural lands have been flooded and another 70,000 ha inundated in Kalmykia. In addition, frequent storms typical of the area often destroy settlements and infrastructure located inside the impact zones. In the spring of 1995 it was reported that in Kalmykia two people were killed, 900 evacuated and 441 homes were flooded as a result of unusually severe storm winds hurling sea water at the coast. In Dagestan 81,500 sheep and cattle died, 127 homes were destroyed and 500 homes flooded (*Guardian* 16 March 1995).

Emigration of people from these predominantly agrarian and already overpopulated regions followed. For instance, in Dagestan the total population of local national origin in 1979–1989 increased by 14.5% while outside the republic in Russia it increased by 31% (Natsional'niy... 1995). It should be noted that in cases of resettlement, the affected population is not properly compensated for losses in housing and property.

In Kazakstan the sea transgressed over an intensively used territory up to 70 km from the former shoreline in Atyrau oblast. The lower delta of the Ural has been flooded along with almost one million hectares of agricultural lands in the former oblast. Losses from the rise in the sea level in this country are estimated at US$150 million (Ozturk *et al.* 1997). Azerbaijan has the most populated coast in the Circum-Caspian region. Flooding here annually 'consumed' 10–20 m of the coast land coming very close to Sumgait city. According to Zonn (1999), the Kyzylgach strict nature reserve has been already flooded along with over 1000 ha of fertile lands in the Lenkoran' lowland. Over 10,000 houses in the coastal cities have been damaged and destroyed in the Islamic Republic of Iran as a result of the rise in sea level.

Environmental and economic problems in the Caspian region are closely linked to **political issues**, which have caused heated debates and even conflicts after the dissolution of the former Soviet Union. It should be emphasized that only effective inter-state co-operation between all littoral countries can help to take control of the current crisis in the ecological situation and to try to revert it. But the complexity of co-operation in environmental protection in this region stems primarily from the lack of agreement on the principal issue: the legal status of the Caspian Sea. This seemingly unimportant question has nevertheless a whole set of various implications, including those directly relevant to the protection of its biological resources.

The dispute is related to whether the Caspian should be regarded as an inland sea or an indivisible lake. Oil proves to be the stumbling-block on the way to a consensus in this matter. Before the dissolution of the former Soviet Union in December 1991, the legal status of the Caspian Sea was that of a shared lake, which implied common possession, management and utilization of its resources by the USSR and Iran or a 'condominium' regime. It was based on two agreements adopted in 1921 and 1940, which applied the principle of the equality of rights, and also on a number of bilateral treaties between the two countries concerning navigation and fishing (Vinogradov 1997). The sea was legally 'closed' to all third countries and their nationals. However, as Vinogradov noted, 'an exact definition of the legal status of the Caspian Sea has never been set out in any existing international instrument' (ibid., 56). Consequently such important issues as the exploitation of the sea's mineral resources and protection of marine biological resources needed more direct delineation.

If the Caspian Sea possessed the legal status of 'sea' it would be subject to a different legal regime stipulated by the 1982 United Nations Convention on the Law of the Sea (UNCLOS). In this case, each littoral country is assigned its own 'exclusive economic zone' which includes 12 miles of territorial waters and 200 miles of shelf, within which it has exclusive rights to prospect and extract mineral resources contained in the sea's bed. It also has an exclusive right to grant licences related to exploitation of these resources. The latter point has the most important political implications because the interests of more than 20 countries are focused on this region, but also because most littoral countries cannot afford to develop their resources without external investment.

After the demise of the Soviet Union and the emergence of sovereign states, five rather than two countries bordered the Caspian Sea in a new geopolitical situation and each had its own conflicting economic interests over the immense resources 'hidden' inside it. Unsurprisingly, individual attitudes to this issue often reflected the amount of offshore hydrocarbon reserves discovered within each country's 'sector'. From the beginning of the dispute, Iran and Russia with the least 'share' of Caspian oil deposits pursued the adoption of the legal status of 'lake' with joint use of all resources. Azerbaijan *de facto* applied the 'sea' approach via sector division by carrying out prospecting and extraction works and signing on 20 September 1994 the 'contract of the century' with an international consortium for a total investment value of US$8 billion. Kazakstan also took a 'sector' stand while Turkmenistan's position kept changing.

Further to this, a number of conflicts ensued between Azerbaijan and Turkmenistan; Turkmenistan and Iran; and, more recently, between Kazakstan and Russia in relation to the ownership of oil deposits located in the middle section of the sea. Another heated political issue is related to the routes of the main pipelines which will transport oil to Western markets. However, the scope of this chapter does not allow us to go into further detail regarding the pipelines controversy or the evolution of relations between the five countries in respect of the Caspian's legal status.

The important point is that although an overall consensus has not been still reached, there has recently been partial progress made on the way to solving this problem. On 6 July 1998, the Russian Federation and Kazakstan signed an agreement which envisaged the division of the sea bottom and subsequently its mineral resources into two parts, with the middle line drawn at an equal distance from the sea shores in accordance with the sea level mark of –27 m as at 1 January 1998. This document also implied that waters of the sea would be used jointly. Within the Russian Federation there is already strong opposition to this agreement which implied a principal 'change of mind' by the leadership of this country.

In addition to political complications at the decision-making level, there is an actual lack of local concern for the state of the environment which is explained by the serious economic difficulties of the transition period and low standards of living of the majority of the population. For all the former Soviet republics these problems include high inflation rates and a continuous decline in overall economic production. People living around the Caspian Sea are more concerned with their everyday struggle for survival. For instance, in 1996 Turkmenistan's population of the coastal region had serious flour and food shortages. They are 'distanced' from the problems of the decline in sturgeon stocks also because the majority of the population, excluding poachers, have almost certainly never tried either the precious fish or caviar. The cost of 100 g of caviar is about US$40 while for instance in Turkmenistan in 1994 an average monthly salary was estimated at US$30 (EIU 1996c).

9.6 Prospects for the future

Scientists still cannot agree whether the recent water 'regression' is a temporary phenomenon or the beginning of a new trend. It is expected that if the sea level continues to rise, in the Northern Caspian, part of the oil and gas fields within the current coastal zone will be flooded, thus increasing the total amount of oil products reaching the open sea. It is estimated that by 2005 within the Russian territory alone, eight towns and 105 villages with a total population of 197,000 people will be affected by inundation and flooding; in addition, 631,000 ha of agricultural lands may suffer as well as an area of 15 oil wells. Within the Russian territory the most serious problems are due to occur within the Dagestan coast where the main portion of deposits are located. In Kalmykia it is predicted that an increase to the –25 m level will completely cover the Kaspiyskoye oil and gas field (Gosudarstvenniy... 1995).

In Kazakstan several major oilfields will be flooded as well as 594,000 ha of agricultural lands and 30 settlements. In Turkmenistan 15 settlements, large animal farms, ports and most of the coastal oil and gas fields will be affected (Centre... 1995). It is also expected that the rise in the Caspian Sea's level may contribute to contamination of the shallow waters of the Volga delta with hydrogen sulphide, thus diminishing its productivity by 4–5 times and resulting in decline in sturgeon catches by up to 1.5 times (Gosudarstvenniy... 1995; Shlikhunov 1993).

Due to the overall decline in industrial production in the last few years there has been a minor decrease in the total pollutant loads. However, while in 1994 the overall decline in economic production in the Russian Federation amounted to 21% (compared to 1993), the total volume of air pollution decreased by only 12.7% and water pollution by 9%. This demonstrates clearly that countries affected by the economic crisis are inclined to economize on environmental expenditures. Normally wealthier countries tend to spend more on environmental protection; another reason for this is that in developed countries the value of human lives is much higher than in the countries of the former totalitarian regimes. In the poorer countries there seems to be a vicious circle between the state of the national economy and the health of the environment.

While the amount of the total revenue of the Russian Federation from exports of oil, gas and oil products in 1994 reached US$20.6 billion (Rossiyskiy... 1995), according to the Ministry of Ecology and Natural Resources, only US$1 million or 0.005% was spent on the construction of facilities and equipment for collection of oil products from the surface of rivers, ports and seas (Gosudarstvenniy... 1995). At the same time the total revenue from exports of all canned fish products, including caviar, was only US$27.6 million. It is clear that from the national perspective, sturgeon can easily be 'sacrificed' to gain higher oil outputs, even by the main caviar-producing country, Russia.

These facts also explain why environmental matters are given a very low priority in the national economic programmes. As was discussed earlier by the author of this book, the main obstacle to environmental enhancement in the Circum-Caspian region is due to the conflict of national priorities in all five countries (Saiko 1997). There is a contradiction between on the one hand the need for economic revival and the protection of sovereignty, and on the other hand, the necessity to divert financial expenditure towards environmental management and conservation. In this situation, the importance of ecological issues in the eyes of decision makers inevitably declines. It appears that the potential source of major funds for environmental improvement should be sought within the oil industry, which currently gets the greatest revenues from exports of this potentially dangerous pollutant.

It is also clear that the expected substantial increase in oil production will ultimately result in incomparably higher levels of oil pollution in the Caspian Sea and the neighbouring coastal regions at the expense of its biological diversity and the health of its ecosystem. The rates of oil extraction in the region are accelerating. For example, in Kazakstan the share of oil and oil products in the overall industrial production increased from 19.6% in the first quarter of 1999 to 31.2% in the second quarter (Kazakstan... 1999). Clearly, this exponential rise will continue in the early 21st century. It is doubtful whether the countries with collapsed economies will, for example, be able to afford expensive oil-collecting facilities to be used in cases of oil spills or other technical accidents. It is most likely that the 'interests' of sturgeon will be overlooked again. According to one recent source, the Russian State Fisheries Committee announced that the sturgeon was facing extinction (*Sunday Business* 10 October 1999).

It seems clear that to use and conserve the Caspian resources effectively it is essential to reach a consensus on the basic issue of the legal status of the Caspian Sea. Despite the recent agreement reached by Kazakstan and Russia, the overall status of the Caspian Sea remains unclear. The sturgeon does not 'recognize' state boundaries in its migratory routes, so its protection can be only provided when *all states agree* on the main principles of environmental protection, not only in the sea itself but also within the adjacent parts of the Circum-Caspian region. Unless such genuine co-operation is achieved, the expected expansion of oil production will result in a rapid deterioration of the sea's environment and create a **major ecological catastrophe** early in the new millennium.

CHAPTER 10

Desiccation of the Aral Sea : the hidden costs of irrigation

It is difficult to speak about the Aral Sea. It has died: one can now count the years of its death-agony and final disappearance.

(Reznichenko 1982, 110)

10.1 Introduction

One of the most notorious global environmental catastrophes is the desiccation of the Aral Sea: in 1960 this was the fourth largest inland body of water on the Earth with a surface area of 66,000 sq km and a volume of 1,090 cubic km. The Aral Sea basin is a large landlocked, predominantly desert region in Central Asia with a total area of 1.8 million sq km, which is equivalent to the combined area of France, Germany, Spain and the United Kingdom. This chapter aims to examine the causes of this disaster and a multitude of its environmental and socio-economic implications associated with the mismanagement of irrigation in water-deficient Central Asia.

The 'Aral region', as a term, normally means the drainage basin of the Aral Sea. Of this territory, 83% is located within the former Soviet Union and is shared by the five former Soviet republics, Kazakstan, Kyrgyzstan, Tajikistan, Turkmenistan and Uzbekistan. The remaining part of the basin encompasses sections of northern Afghanistan and north-eastern Iran. Political divisions in the region are shown in Figure 10.1. Two mighty rivers flow to the Aral Sea: the Amudarya, which travels for 2540 km from the slopes of the Hindu Kush mountains through the Pamirs and across the Karakum desert, and the Syrdarya, which originates in the Tien Shan mountains and stretches for 3019 km to the Aral delta.

Another commonly used term important for this case study, is 'Priaralye', or Circum-Aral region which includes the territory directly adjacent to the Aral Sea and the deltaic parts of the Amudarya and

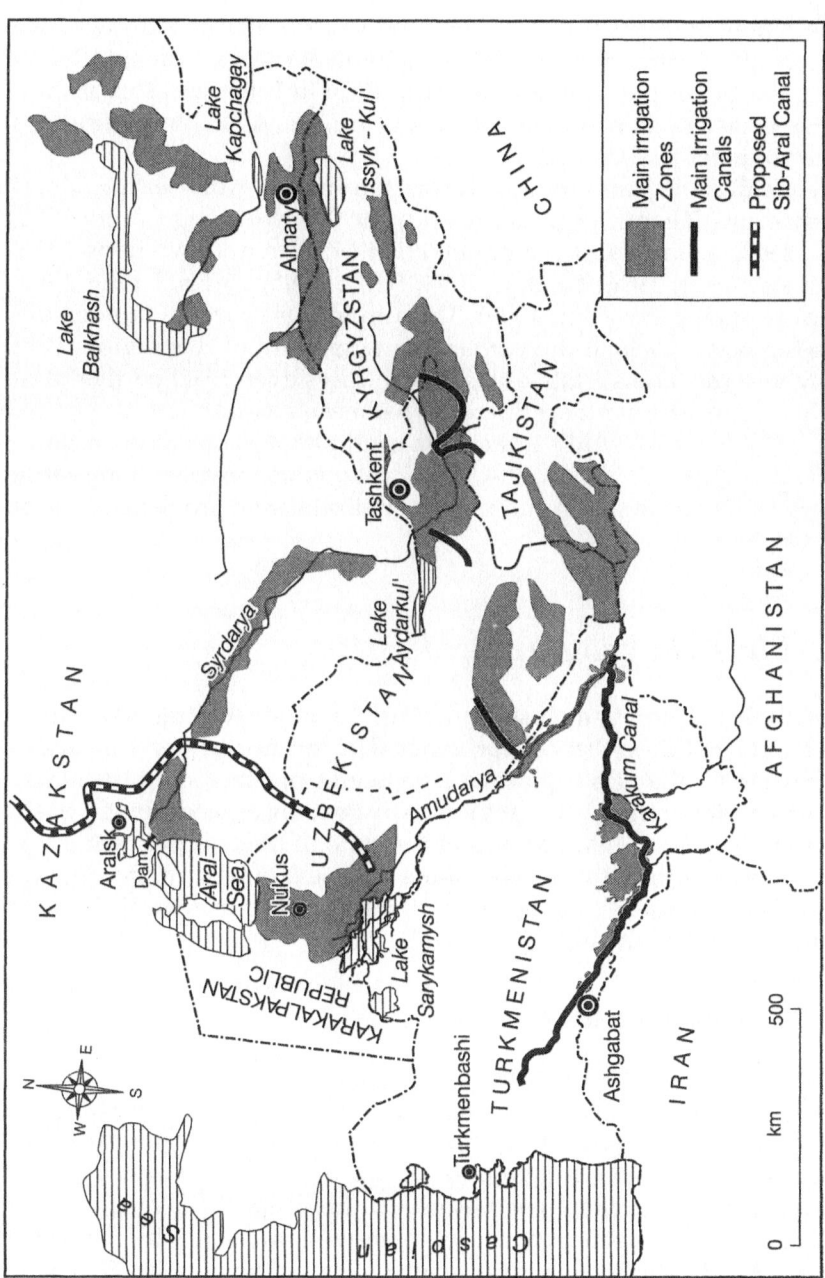

Figure 10.1 Political divisions in the Aral Sea region and the main areas of irrigated lands

Syrdarya, with a total area of 546,000 sq km. The total population of the Aral basin is approximately 40 million people including the five million people in Priaralye.

The exponential increase in cotton production which caused excessive withdrawals of water from the Amudarya and Syrdarya ultimately led to the desiccation of the Aral and desertification in Priaralye. This anthropogenic disaster has had major socio-economic and health implications for the population of the region.

A considerable amount of information on the Aral Sea problems already exists (Glantz 1999; Glazovskiy 1990; Kobori and Glantz 1998; Kuksa 1994; Letolle and Mainguet 1993; Micklin and Williams 1996; Ptichnikov 1991, 1996; Rafikov 1994; Saiko 1995, 1998b; UNEP 1992; Zaletayev *et al*. 1989; Zonn 1999). The aim in this chapter is to demonstrate the way in which the physical environment of this region went deeply into the 'crisis' state and in some cases even reached the 'catastrophic' stage after a critical threshold had been passed.

The Aral Sea case study is a classic example of a 'creeping environmental phenomena', as described by Glantz, which involve 'long-term, low-grade, incremental but cumulative environmental problems' (Glantz 1999, vii).

10.2 Physical geographical setting

The Aral Sea region is a closed drainage basin located mainly within desert and semi-desert landscape zones. The location of Central Asia in the heartland of Eurasia explains a high degree of climatic continentality and aridity inherent to this region. Application of a 'drainage basin approach' is essential for analysis of the state of the environment in the direct vicinity of the Aral Sea and of the sea itself. In its turn, the desiccation of the Aral Sea has had a feedback effect on the ecological situation in Priaralye and on the basin in general.

10.2.1 Geographical location and relief

The Aral region incorporates the drainage basins of two major rivers, the Amudarya and Syrdarya; two minor ones, the Tedjen and the Murghab; the basins of small rivers which flow from glaciers on the western slopes of Tien Shan and Kopet Dag mountains; and also drainless desert territories, lying between the river basins. In the west, the Aral region is bordered by the Caspian Sea; in the south-west and south by the Kopet Dag and Hindu Kush mountains, which serve as the orographic barrier between the typical subtropical climate of Iran and Afghanistan and the temperate continental climate of Central Asia. In the south-east, the high mountain ranges of Tien Shan and the Pamirs stretch deep into the extensive Turan lowland and play an important role in the formation of physical environment. Glaciers on the mountain tops feed the upper reaches of the Amudarya and Syrdarya, which ultimately reach the Aral Sea. In the north, the boundary of the drainage basin is adjacent to a few closed drainless river basins in the immense Kazak hummocky lowland, while in the east it borders the basin of the beautiful Balkhash lake. The Turan lowland traversing the region from north to south is the major plain of the Aral Sea basin. Approximately 80% of the territory has plain and 20% mountainous topography.

Two major deserts are located here – the Karakum, bordering the Caspian Sea in Turkmenistan with an area of 375,000 sq km, and the Kyzylkum, which stretches between the Amudarya and Syrdarya. The specific geographical location of this region determines the climatic peculiarities and diversity of its relief and creates unique and contrasting features of physical environment. According to their varying substrata, various types of deserts can be distinguished within the Aral drainage basin: sandy, rocky, gypsum, clayey (locally called 'takyrs'), loess, saline (called 'solonchak'). For the most part, the zonal type of landscapes found here are deserts. In the north they border with the semi-desert landscapes of Central Kazakstan.

10.2.2 Climate and hydrology

The **climate** of Central Asia features high degrees of aridity and continentality. The latter is expressed by great contrasts in temperature and precipitation between seasons. Only two main seasons are distinguished: dry, lasting from mid-May to mid-October, and wet, lasting for the rest of the year. The value of total solar radiation reaches 140 kcal/sq cm at the Aral Sea latitudes and 160 in the south. The annual radiation balance is greatest compared to other landscape zones of the former Soviet Union and amounts to 40–55 kcal/sq cm (Table 1.2). Bioclimatic potential of the region is very high and is sufficient for growth of almost all kinds of agricultural crops. However, the growing of many of them requires the use of irrigation.

Climatically, the Aral Sea basin can be subdivided into two parts. In the northern and central sections, climate is moderate to extreme continental and is related to the temperate climatic belt, while in the southern part it is subtropical. An extremely high degree of 'continentality' in the central part of the basin is reflected in a great temperature difference between the absolute minimum and maximum temperatures. This range is probably the most spectacular in the world and exceeds 80 degrees C in the hottest Karakum desert. There is also a great contrast between diurnal surface temperatures.

The climate of the central part of the basin is characterized by very hot and long summers with mean July temperatures reaching +28 to +34 degrees C. Summer maximums can reach as high as +50 degrees Celsius. Hot dry winds called 'sukhoveys' are common in summer. Winters are cold but normally short, with mean January temperatures ranging from –15 to –13 degrees C in the north and to 0 to +2 degrees C in the south. Normally, snowfall is limited and no permanent snow cover is formed in the central and southern parts of the basin. Ice cover on the Aral Sea normally stays for 4–5 months while in particularly cold winters the sea is completely frozen. Major hydrological changes, which will be discussed later, have occurred since the 1960s.

Droughts are a normal feature of the climatic pattern. Rainfall occurs mostly in spring, often in the form of torrential storms, and is highly erratic and variable both in time and volume. In summer, precipitation is almost negligible, particularly over the plains. Maximum rainfall occurs normally in March–April. Annual precipitation in the central part of the basin amounts to only 80–120 mm, brought by the atmospheric masses from the Atlantic Ocean; in the piedmont areas this value reaches 250 mm per year. At the same time, potential evaporation is almost 20 times the amount of available precipitation and reaches 1400–2300 mm. This combination reflects the region's high degree of 'aridity' and generates 'soil drought', where the top layer of soil desiccates and plant transpiration greatly increases. Strong winds which generate dust storms are quite common, particularly in the south-east of the Karakum desert.

The **hydrological** network is poorly developed. Insufficient precipitation, the high infiltration capacity of the soils, and high air temperatures explain the lack of surface runoff. Most local rivers are intermittent and often end in the sands of deserts, forming dry river beds and deltas. The most important feature of the hydrographic system is that the whole of Central Asia is a drainless (closed) basin or, in other words, there is no outlet to the ocean. The Aral Sea basin is dominated by the sea itself and two major transit rivers, the Amudarya and the Syrdarya, which cross the two main deserts of Central Asia. These rivers play an important role in recharging ground water aquifers through subterranean flow. In 1960 the total annual surface flow amounted to 118 cubic km, from which Amudarya's discharge made up 73 cubic km and Syrdarya's was 37 cubic km. Before 1960 the regime of the sea was relatively stable with the annual evaporation from the surface amounting to 65 cubic km being compensated by the perennial rivers' inflow of 56 cubic km and atmospheric precipitation equivalent to about 9 cubic km. The level of the Aral Sea fluctuated around 53 m above sea level, and its salinity was only 9.6–10.3 mg/l which gave grounds for considering it a lake rather than a sea.

There also exist a significant potential of fresh and brackish ground water resources amounting to about 100 cubic km per year, with 45% of this being ground waters not connected with the surface flow. The present withdrawal of ground waters comprises 14 cubic km annually (Zonn 1992). Long-term irrigation and application of excessive watering norms have created a system of artificial 'lakes' in depressions filled with mineralized or saline drainage water. The largest of these is Lake Sarykamysh south-west of the Aral Sea (Figure 10.1).

10.2.3 Vegetation and soil cover

Both soil and vegetation cover of the region is very diverse, patchy and contrasting due to the reasons mentioned above. In the northern part of the Aral Sea basin a **semi-desert** type of vegetation is represented by an alternation or patchy distribution of gramineous steppe and desert plant communities. Species composition is poorer than in the steppes and the sparse plant cover comprises mostly perennials such as *Stipa orientalis* and ephemeral plants. Light chestnut soils with slightly increased salinity have developed in semi-desert landscapes. Saline soils called 'solonchaks' are normally covered by halophyte plants (e.g. *Atriplex cana* and *Anabasis salsa*).

Despite a common perception of **deserts** as being barren sandy expanses, in spring, fixed sandy areas can have a relatively dense plant cover. A typical feature of desert landscapes is the presence of 'ephemeral' vegetation with a short life cycle which dries up when summer drought starts and remains 'dormant' in the form of seeds, bulbs and roots until there is enough moisture to support their growth. Ephemerals are annual plants which can generate seeds over 3–4 weeks from the beginning of the growing period while ephemeroids are perennial grasses. Their root systems are spread close to the surface to collect the maximum available moisture or they go deep into the ground to tap ground water aquifers.

Vegetation composition in each desert depends largely on its substratum. Thus, psammophyte vegetation is characteristic of sandy deserts; gypsophyte plants grow on rocky and gravelly deserts; and halophytes grow on solonchaks. Wormwood (*Artemisia spp.*) and saltwort (*Salsola spp.*) are typical of loess and clayey surfaces, while ephemeral plants are common in all types of deserts.

In the deserts of the north of the basin, sparse sagebrush and thistle communities are most common. Vegetation cover of the Karakum sand desert is dominated by xerophytic shrubs and semi-shrubs (Plate 18). A common species is sand acacia (*Ammodendron conollyi*) up to 3 m high with silvery leaves. Typical shrubs of the Karakums are white and black saxaul (*Haloxylon persicum* and *H. aphyllum*), the latter being a hardwood species with green branches devoid of leaves and very deep roots tapping ground water aquifers. Both shrubs are highly valued by the local inhabitants for their multiple uses. Also widely spread is *Salsola richteri*

Plate 18 A typical landscape of the Karakum sand desert

and ephemeral plants such as *Carex physodes* and *Bromus rectorum*, along with halophyte (salt-resistant) shrubs and grasses along with groups of sparse woody plants from the *Chenopodiacae* family. An interesting feature of the Karakums is that a reasonable amount of grazing is beneficial for this type of desert, as it improves the soil surface structure protecting it from the formation of moss cover which depresses grass vegetation (Gvozdetskiy and Mikhailov 1987).

Plant biodiversity in deserts is comparable to that of the forested tundra zone and amounts to 200–300 species per 100 sq km, while the value of phytomass is similar to subarctic tundra landscapes with only 10–30 tonnes per ha (Table 1.2). Mean annual biological production is also extremely low at 2–5 tonnes per ha. Duration of primary succession is, in contrast, much shorter and equals 100–300 years compared to over a thousand years in the tundra. All the above features explain the low rehabilitative ability of desert landscapes and their high vulnerability to external disturbances.

Natural vegetation in the **river deltas** include reeds and in the floodplain areas the so-called *tugai*, dense riparian woodlands with shrubs and meadows composed of poplar, willow, tamarisk, sallow-thorn and other species. It should be noted that desert vegetation forms the basis for large-scale rangeland pastoralism, which was and still is one of the main occupations of the Turkmens and Kazaks.

Despite the harsh climatic conditions, animal life is diverse and includes rodents and reptiles as well as larger mammals such as camels and antelopes, etc. Lack of moisture, heat and hot winds are the main limiting factors for their life and activity. During a long history of evolution these creatures have developed various adaptations, which enable them to survive and make maximum use of the scarce resources of deserts.

Soil cover of desert landscapes is extremely diverse. Most common types of soils are grey and brown desert soils called 'serozems'. Soil formation is limited by the lack of leaching which has a pulse regime in these moisture-deficient landscapes. Lack of vegetation cover limits the development of an organic layer which is generally shallow and poor in humus content. High evaporation causes soil salinization as the predominant upward movement of soil moisture brings soluble salts closer to the surface. Formation of 'solonchak' soils with an excess of salts and 'solonets' with an excess of sodium is a common process. 'Takyrs', or clay polygonal soils, are also quite widely spread. Because of the ancient history of irrigation in this region, a specific type of cultured irrigated 'oasis' soils has developed in the flood plains of the Amudarya and Syrdarya.

The soil cover in these arid environments is very vulnerable to intensive use and external disturbance. Thin topsoil can easily be eroded and blown away by the typically strong desert winds if it is not protected by vegetation cover. Therefore soils are highly susceptible to desertification processes which include accelerated water and wind erosion, deterioration of physical properties e.g. soil compaction, decline in organic matter and ultimately soil fertility, salinization and waterlogging.

10.2.4 Overall vulnerability of landscape

Deserts are arguably the most vulnerable to human impact compared to other types of landscapes. A very high degree of aridity, inherent to this part of Eurasia, means that the main limitation of this territory is a water deficit caused by low precipitation and limited surface runoff. In addition, precipitation is highly spatially and temporally variable, and often occurs in convectional form which creates major natural erosion hazards. At the same time, vegetation cover is discontinuous and biological production restrained by all the above-mentioned factors.

Great temperature variations, insufficient moisture along with lack of vegetation and increased natural salinity of soils are the main features which determine the high vulnerability of natural landscapes in arid environments and their extreme sensitivity to anthropogenic influences. The high degree of aridity indigenous to these fragile environments also indicates that agricultural practices require the use of irrigation in farming and provision of water for livestock.

10.3 Historical perspective and economic development

The Aral Sea basin is a region of ancient farming culture. During the peak of irrigation development in antiquity (4th century BC to 2nd century AD), the total irrigated area in the lower Amudarya and Syrdarya exceeded the contemporary area by several times. However, since that time, wars, natural disasters and desertification have turned much of the ancient irrigated lands into a state of disrepair.

While European Russia of the pre-First World War period had already entered a stage of very rapid industrial development, Central Asia continued to remain predominantly rural. At that time, it had relatively diversified agriculture and was self-sustainable in terms of food production. Cultivation of grain, mainly rice, was very important, while cotton of rather poor quality was grown only on land left over from the main staple crops. Animal husbandry was the main economy in many parts of the region, and in terms of livestock numbers and meat consumption per capita, Tsarist Asiatic Russia was the leading region in the Russian Empire. Central Asia exported much of its produce, particularly 'karakul' sheepskin and fruits to European Russia and foreign countries.

The core of the cotton processing industry was historically centred around Moscow even during the first half of the 19th century. The total number of cotton-processing factories was 1017. The plants of Russia annually processed over three times more cotton than was produced within the Empire, mainly in Turkestan. In 1889 it amounted to 278,460 tonnes of which 78,329 tonnes came from Turkestan (Brokgauz and Efron 1995).

The principal objectives of Tsarist Russian agricultural policies in Central Asia were two-fold: to ameliorate the hydrological regime in the deltaic regions and, more importantly, to replace imports from America by cultivation of more cotton by the local population. As Massal'skiy (1892, 158) noted, 'The extensive delta of the Amudarya is being flooded annually, which is often disastrous for the population ... Sometimes, for example in 1878, this flooding occurred on such a grand scale, that it destroyed dams and flooded cultivated lands to an enormous degree ... There is no doubt, that ameliorative works, which would divert the excessive water to the currently non-irrigated areas, would contribute to the expansion of cultivated lands and, consequently, of those which can be planted with cotton.'

In 1884 only 810 acres of American-brand cotton was sown in the whole of Central Asia, while by 1890 its area increased to 159,000 acres. Irrigation development for cotton production was initiated under the Tsarist government at a sometimes amazing pace. One of the main centres supplying cotton to Russia became the Ferghana oblast. Between 1885 and 1915 the area sown to cotton in this region rose from 14% to 44% (Allworth 1994). Despite this exceptional example, elsewhere the situation remained less altered. Before the First World War the area under cotton in Central Asia amounted to 438,131 hectares (Rossiya... 1995b).

During the Soviet period extensive development of irrigation was for the most part also associated with cotton growing and to a lesser extent with rice production. The Bolsheviks actually followed the pattern pursued by the Tsarist regime. Reconstruction and expansion of the irrigation network was stimulated by the Bolsheviks after the end of the Civil War in Central Asia in the 1920s. Estimates show that the area under cotton in Central Asia more than doubled between 1913 and 1940. The further acceleration of cotton production was stimulated by the Soviet government's post-war policy of achieving self-sufficiency in cotton for the USSR as a whole and for the East European socialist countries. This was the time when the foundation for the environmental catastrophe was laid. A large-scale irrigation campaign to attain 'cotton independence' was started by a 1950 decree on the transition to a new mechanized irrigation system.

By 1987 the irrigation system in Central Asia included 52,800 km of main canals, 390 reservoirs, and 230 dams and other hydrotechnical works. One of the campaign's major hydrotechnical giants, the hand-dug Karakum Canal, stretches for over 1150 km through the deserts of Turkmenistan, and 'no other single Soviet measure has found such a wide degree of support and appreciation among Turkmens' (Murat 1975). As Orlovskiy noted, 'Water supply to the old irrigated areas increased from about 40% to 85%. It allowed the introduction of crop rotation and the creation of a stable basis not only for cotton growing but for cattle breeding, vegetable growing and horticulture whose water demands had not been fully satisfied in low-water years' (Orlovskiy 1999, 225). The expansion of irrigated lands was stimulated by the major investments in agricultural and industrial development in Central Asia during this period.

The environmental reference book published by the Commonwealth of Independent States (CIS) in 1996, stated '... large-scale land reclamation (irrigation) works, development of industry and other economic activities have contributed to the transformation of these economically backward regions of the Aral Sea basin into the major agro-industrial regions. Output of agricultural produce ... increased from 1950 to mid-1990s by almost four times' (Okruzhayushchaya... 1996, 171). However, these positive developments were followed by a major ecological disaster.

In 1950 the area under irrigation in Central Asia accounted for 5.4 million ha. The overall expansion of irrigation was particularly rapid between 1976 and 1988 when the irrigated land area in all republics of Central Asia reached its peak of 9.4 million ha, an increase of 1.7 times for the region as a whole (Narodnoye... 1977, 1982, 1986, 1989); this figure includes approximately 7.6 million ha within the Aral Sea basin. Expansion was most striking in Turkmenistan, where the irrigated area almost trebled in size. The location of irrigated lands is shown in Figure 10.1. A cotton monoculture gradually became dominant. The share of cotton in crop rotation systems reached 75–78% while in some regions and on farms it reached 100%. At the same time, on the Syrdarya, in the Kzyl-Orda region of Kazakstan, a monoculture of an even more water-consuming crop, rice, was introduced.

The dynamics of the expansion of cotton can also be illustrated by the value of its production output. Table 10.1 shows that the annual output of cotton in Central Asia in 1980, at the peak of its production, exceeded the 1913 level by 13.7 times. Compared with the pre-war levels of 1940, cotton production by 1980 had increased by 4.5 times and it had almost doubled in comparison with the average value for 1961–1965.

The expansion of cotton production was also stimulated by a system of rewards and incentives for the leaders of Central Asian republics and chairmen of collective and state farms. Harvesting of this so-called 'white gold' remained largely manual and involved the work not only of farmers, mostly women, but also students, employees of various institutes and enterprises and even schoolchildren, as can be seen in Plate 19. It is interesting that the peak of cotton production in Central Asia was reached in 1980, after which time it gradually decreased despite an overall increase in irrigated land. This was clearly a reflection of the overall decline in productivity due to the loss of soil fertility, water and soil salinization and the cultivation of unproductive marginal lands.

The dissolution of the former USSR had a whole set of geopolitical and socio-economic implications for each of the new independent countries of the Aral Sea region (Saiko 1995). As Olcott noted, 'In Central Asia the collapse of the Soviet Union was more than just a blow to their part of an interdependent, all-USSR economy. All five republics faced serious economic problems well before the Soviet Union's economic crisis of the late 1980s. But independence also brought an abrupt end to subsidies from Moscow ...' (Olcott 1996, 9).

Table 10.1 Cotton production in Central Asia in 1997 and in 1980 compared to 1913, 1940 and 1961–1965

Republic	Cotton production output (000 tonnes)		Cotton production in 1980 compared to: (number of times)		
	1997	1980	1913	1940	average for 1961–1965
Kazakstan	198	358	23.9	4.0	1.6
Kyrgyzstan	62	206	7.4	2.2	1.3
Tajikistan	358	1011	31.6	5.9	1.9
Turkmenistan	632	1258	18.2	6.0	2.8
Uzbekistan	3700	6245	12.1	3.8	1.9
Total for Central Asia	4950	9078	13.7	4.5	1.8

Sources of data: Narodnoye... (1977, 1982); Rossiyskiy... (1997)

Plate 19 Schoolchildren of Uzbekistan harvesting the 'white gold'

Table 10.2 Per capita water withdrawal in Central Asian republics and Russia, 1985–1994 (cubic m)

	1985	1990	1991	1992	1993	1994
Kazakstan	2473	2111	2077	1936	1911	1841
Kyrgyzstan	2333	2494	2521	2509	2467	2458
Tajikistan	2836	2583	2530	2305	2333	2365
Turkmenistan	7505	6193	7139	6482	6513	5933
Uzbekistan	3899	3380	3350	3338	3271	ND
Russia	745	717	725	672	641	587

Sources of data: Okruzhayushchaya... (1996); Rossiyskiy... (1997)

Each of the five republics started the post-socialist period attempting to develop their own economic policies and regain sovereignty and independence. It should be emphasized that none of them had existed as a sovereign state in its present borders ever before. Each had a different 'starting capital', as mineral and fuel resources in the region are distributed quite unevenly between the republics, but most of them had a common problem – deficient shared water resources and an increasing demand for them. As Table 10.1 shows, by 1997 cotton production in all countries had decreased. It was accompanied by slight reduction in water withdrawals in most countries (Table 10.2). However, this decrease was not associated with the needed agricultural restructuring and transition to the use of more water-saving technologies but with a general production decline and reduction in crop productivity.

10.4 Ecological consequences and recent changes

The exponential expansion of irrigated lands and mismanagement of irrigation have created major environmental repercussions on the local, regional and even global scale, involving the desiccation of the Aral Sea, desertification in Priaralye and throughout the basin. Desertification in the region is expressed through such processes as salinization and waterlogging of irrigated lands; accelerated soil erosion and loss of fertility; deflation and dust storms; vegetation degeneration and destruction; and loss of biodiversity.

10.4.1 Desiccation of the Aral Sea

As a result of irrigation expansion, diversion of water from the Amudarya and Syrdarya rivers, which support the existence of the Aral Sea, has dramatically increased. Water withdrawal in all Central Asian republics has greatly exceeded the average values for other regions of the former Soviet Union. For example, in 1985 per capita water withdrawal in Kyrgyzstan was approximately 2300 cubic m; in Kazakstan 2500; Tajikistan 2800; Uzbekistan 3900; and Turkmenistan 7500 compared to the Russian republican mean of 745 cubic m (Table 10.2). This can be further contrasted to about 450 cubic m per capita in Israel. A continuous growth in water consumption during the Soviet period was associated with major losses: for example, from the unlined Karakum canal losses on infiltration initially reached 70–99% (Orlovskiy 1999) from the total losses value estimated at 18% of 13.5 cubic km of the water annually diverted from the Amudarya.

Our calculations show that in 1985 the overall water losses in transit in the Central Asian region amounted to 17.8%. Both in Turkmenistan (not accounting for losses from the Karakum Canal) and Kyrgyzstan they exceeded 21%; in Uzbekistan they reached 18.9%; and in Tajikistan and Kazakstan, 15.2% and 13.6% respectively. By the time of dissolution of the former Soviet Union they had further increased in most countries except Kyrgyzstan, and in 1991 reached 28.9% in Turkmenistan and 20.8% in Uzbekistan with a regional total of 20.2% (Okruzhayushchaya... 1996). It should be emphasized that since 1993 for Uzbekistan and 1994 for Turkmenistan, data on water withdrawal, consumption and losses does not appear in the intergovernmental statistical yearbooks (Sodruzhestvo... 1997, 1998).

In addition huge losses were caused by on-farm irrigation as the most water-wasteful practice, furrow irrigation, was widely used. Large-scale collective and state farms created in the 1930s during the notorious 'collectivization campaign' in the Soviet Union were even encouraged to use more water. The reason for this was the fact that quotas for water withdrawal, which were annually allocated by 'Minvodkhoz', the All-Union Water Management Ministry in Moscow, were linked to the size of lands planned for irrigation in the following year.

Estimates show that from 1960 to 1990 about 1000 cubic km of the Amudarya and Syrdarya waters did not reach the Aral Sea (Kuznetsov 1993). Desiccation of the sea ensued and the rates of its drying up accelerated from year to year. The volume of water reaching the sea reduced from 56 cubic km characteristic for the period from 1911 to 1960, to 3.5 cubic km for 1981–1987. In addition during the latter period it fell to zero three times. A catastrophic decline in sea level resulted. Figure 10.2 shows the retreat of the sea shoreline from 1960 to 1995. By 1987 the level of the sea declined by 13 m and in 1989–1990 the sea separated into two parts, Large Aral and Small Aral; major changes have been recorded in the parameters of the inflowing rivers and the sea itself.

As illustrated by Table 10.3, between 1960 and 1995 the total volume of discharge of both rivers into the Aral Sea decreased by 37–48 cubic km; the area of the sea more than halved; its volume decreased by over two

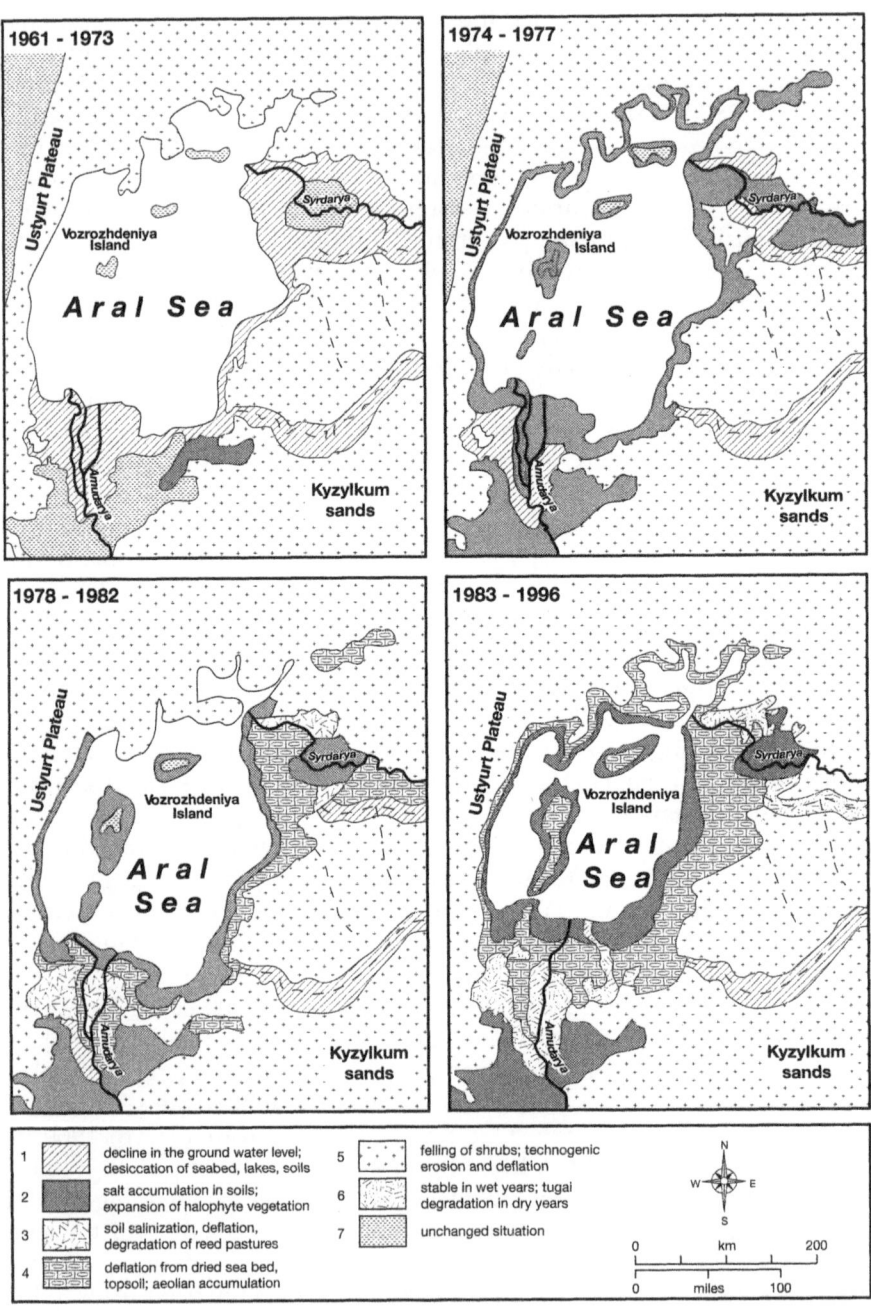

Figure 10.2 Dynamics of desertification processes in the Circum-Aral region, 1960–1996 (after Rafikov 1999 with amendments)

Table 10.3 Environmental changes in the Aral Sea basin, 1960-1995

Parameter	1960	1995	Total magnitude of change
Area of the Aral Sea, sq km	66,900	32,000 (3,000)*	-34,900
Volume of the Aral Sea, cubic km	1,064	310 (23)*	-753
Level of the Large Aral Sea, m	53.4	36.6	-17.4
Length of shoreline, km	4,430	3,950	-480
Salinity of the Aral Sea, g/l	11-14	34	+20-23
Volume of Amudarya & Syrdarya discharge into the sea, cubic km	52.9 – 56.0	5-19	-47.9 – 37.0
Inflow of fertile silts into the Amudarya, million tonnes per year	12	0.4	-11.6
Inflow of dissolved salts into the Amudarya, million tonnes per year	2.8	5.3	+2.5
Salinity of ground water in irrigated lands, g/l	1-3	10-12	+8-10
Commercial fish catch, 000 tonnes per year	30-40	nil	30-40

Source: after Zonn (1995a)
* data in brackets for Small Aral Sea

thirds. At the same time, the sea level in Large Aral has dropped by 17.4 metres. The length of shoreline has been reduced by 480 km and in some places the sea coast receded by 40–100 km, leaving former fishing villages in the heart of the newly formed anthropogenic desert. The annual rates of sea water mineralization during 1930–1960 amounted to 0.10–0.15 grams per litre while during 1960–1990 they reached 0.8–0.9 grams per litre. By 1995 salinity of the Aral Sea trebled, thus exceeding by one-third the value of salinity in the Black Sea and reaching the level of salinity of open sea waters such as the North Sea. This was inevitably followed by a dramatic loss of commercial fisheries.

During a relatively short historical time span, i.e. 1960–1996, human actions have disrupted the structure of the natural hydrographic network, supplementing it with an artificially created water system which has changed the territory's hydrological regime. More recently, since gaining independence, these wasteful practices have remained actually unchanged (O'Hara and Hannan 1999). As a result, the regime and pattern of functioning of the natural ecosystems have been radically transformed, ensuing in desertification.

One of the most important implications of the shrinkage of the Aral Sea was the contraction of the flooded area and shortening of flooding duration in the deltaic parts of the Amudarya and Syrdarya. It should be noted that desiccation of the sea has affected the deltaic plant communities composed of reeds and 'tugai' vegetation which served as a precious habitat for wildlife; it has resulted in continuous persistent desertification in the deltaic territories and ultimately in the whole of Priaralye. Evolution of desertification here is directly linked to the water supply of ecosystems or to the volume of the two rivers' discharge within their deltas and the amount of flow into the sea.

10.4.2 Processes of desertification in Priaralye

Before the 1960s desertification in the region adjacent to the sea was already occurring. Individual processes of desertification were already recorded in the deltaic areas of the region during the first half of the 20th century; however, the reason for these was natural climatic fluctuations. The first massive evidence of the beginning of anthropogenic desertification was recorded in the 1960s, when the ever-expanding irrigation needs reduced the amount of water reaching the deltaic areas of the Amudarya and Syrdarya (Saiko and Zonn 1994).

As Figure 10.2 shows, from 1961 to 1995 the nature of desertification processes has been constantly changing. During the period from 1961 to 1973 **desiccation** of the sea, lakes, meadows and soils along with an overall **decline in the ground water level** were recorded. In their turn, other components of natural landscapes were affected, vegetation in the first place. Shallowing of deep lakes contributed to the development of hydrophyte plants, while on some plains grasses started to replace reed vegetation. To the east of the Amudarya delta, processes of salt accumulation in soils and some expansion of halophyte vegetation have started. An additional cause of desertification in the areas adjacent to the sea has been the excessive felling of shrubs and overgrazing. The processes here took form of technogenic erosion and deflation in the vast territories to the west, north and south-east of the sea.

Over the next four years, from 1974 to 1977, the predominant desertification process was **salinization**, including intensive mineralization of surface and ground waters and salt accumulation in soils. As a result, **expansion of halophyte vegetation** followed. Desiccation of the sea continued, leaving exposed dry saline areas along the former coastline of the sea and islands, and was followed by an increase in aridity. This was aggravated by a number of dry years during this period. It appears that the impetus for the catastrophic development of desertification was the combination of excessive water withdrawals and a series of dry years between 1974 and 1977. Aridization, in its turn, contributed to desiccation of many minor delta arms, a decline in the level of ground waters and an increase in their salinity levels. Natural vegetation acquired more xeromorphic features, particularly on the elevated surfaces. As a result of these changes areas under swamps, tugai and reed vegetation reduced and desiccation of soils followed. As Vostokova (1999) noted, between 1974 and 1979 ground water rise and the increase in salinity of soils were accompanied by an ecological disaster for biota.

The period from 1978 to 1982 (Figure 10.2) experienced a marked intensification of desertification processes throughout the region. Salt accumulation remained one of the dominant processes and was supplemented by deflation and degradation of reed pastures. According to Rafikov (1994), accumulation of salts in the upper three metres of soil in the Amudarya delta reached from 1200 to 6000 tonnes per hectare. Vegetation destruction during this period reached 20–40% on sandy loam soils (Rafikov 1999). Extensive areas with solonchak soils have been formed; tugai and reed vegetation have been replaced by saltwort and tamarisk. Vostokova recorded that hydromorphic landscapes were replaced by halomorphic and automorphic (Vostokova 1999).

Desiccation of the sea has been greatly accelerated, and massive development of the processes of **deflation** from the dried sea bed and topsoil of adjacent territories have started. These are common along with processes of **aeolian accumulation** throughout the deltaic areas. Processes of erosion have been particularly intensive in elevated parts of both deltas, where the level of ground waters has fallen to the depth of 5–7 m from the soil surface. As Ptichnikov (1991) noted, the hydrological role of rivers from the late 1970s has changed from a feeding to a draining one, while deltaic depressions have become salt accumulators for the whole of the Aral Sea drainage basin.

One of the factors which contributed to the development of aeolian processes was related to the nature of mechanical substrate, sand or sandy loam in most areas. Another factor was a marked decrease in the proportion of continuous vegetation cover. By the early 1980s percentage cover was only 10–40% thus greatly facilitating development of aeolian processes. In addition, as was recorded by Rafikov (1999, 1994), acceleration of the rates of aeolian processes in the Amudarya delta was also attributable to the character of vegetation successions. When in the course of desertification the swamp and lacustrine communities dried out, reeds were replaced by more xerophytic vegetation including mixed grasses and saltwort associations.

The pattern of desertification processes during the latest period from 1983 to 1995 was characterized by a dramatic reduction of the rivers' discharge into the sea in the late 1980s: a total extinction of inflow in case of the Syrdarya, and a fall to a negligible amount of discharge in case of the Amudarya. Figure 10.2 shows that the Aral Sea area has been further reduced in size and processes of salinization have spread to the former sea bed, giving way to **activation of aeolian processes**.

Starting in the late 1970s, **salt and sand storms** originated in the dried sea bed. They had an effective radius of up to several hundred km but sometimes they reached territories located over a thousand kilometres away from the Aral. Annually from 15 to 75 million tonnes of dust and salt were blown out and carried away from the former sea bed. The total mass of sediments in one of the major storms reached 1.68 million tonnes (Murzayev 1991). Deposition of airborne salt particles in the late 1970s reached one tonne per hectare annually. As a result of salt fallout, productivity of crops, rice in particular, has declined, for example in Uzbekistan (EIU 1996b). The salinity level of rainfall, which was naturally quite high at 30–100 mg/l, has also increased to 120–150 mg/l (Zonn 1992).

Novikova considered that 'the most "dangerous" change in the region is [associated with] the salt and dust that is carried from the exposed seabed' (Novikova 1999, 123). She estimated that the 'normal' annual content of salts in the atmosphere near the earth is 10,000–20,000 tonnes over an area with a radius of 300 to 500 km. Landscapes in Priaralye have been therefore subjected to exceptionally heavy salt pressures. In addition, a reduction in the vegetation cover has created a situation where previously fixed barchans have started to move at a rate of 900–1200 m per year (Murzayev 1991). During the period between 1980 and 1989 the formation of automorphic landscapes was completed and, in the opinion of Vostokova, an *ecological catastrophe* for biota had occurred (Vostokova 1999).

10.4.3 *Environmental changes in the Aral Sea drainage basin*

Various processes of land degradation evident throughout the basin have been the result of mismanagement of irrigation, cultivation of marginal lands, excessive felling of trees and shrubs and overgrazing. The two most widely spread physical processes within irrigated lands have been

waterlogging or decline in natural drainage conditions and water and soil mineralization (Table 10.3). The former process developed quite early in the course of irrigation expansion, because construction of the drainage system, which normally should precede creation of irrigation canals, was not provided for from the beginning of the hydraulic works. Major land degradation changes have occurred throughout irrigated areas. In addition, the ground water regime was adversely affected by water infiltration through the canals' unlined beds and banks. This problem was most acute within the impact zone of the major canals such as the Karakum canal. Immediately after its construction the level of ground water rose by 25–28 m at a distance of 50 m from the canal (Orlovskiy 1999). In addition the rate of ground water movement decreased three-fold over an 11-year period between 1965 and 1976. More recent problems associated with the operation of this immense project have been discussed by O'Hara and Hannan (Hannan and O'Hara 1998; O'Hara and Hannan 1997).

The first signs of **secondary salinization of soils** in the above region were already recorded in 1968. High air temperatures and evaporation bring soil moisture plus salt up to the soil surface. By the beginning of the 1990s, three to five waterings per year were needed to desalinate the upper one metre of soil (Glazovskiy 1990). Soil salinization processes became widespread throughout the region. A study has revealed that currently about 44% of irrigated land is severely salinized in Uzbekistan alone (Lubin 1994). In Kyrgyzstan over one-third of the irrigated lands are salinized to a moderate or high degree and about one-third of Kazakstan's total irrigated area needs rehabilitation. Farmers are shifting from growing rice and cotton to fodder, because no other crops can be grown on the degraded lands and 110,000 ha of formerly irrigated lands have been withdrawn from agricultural use due to salinization in Kazakstan.

Salinization of soils and water have affected the quality of drinking water in the region, particularly in areas closer to the sea. For example, one survey has revealed that in Karakalpakstan 68% of respondents considered their drinking water salty in taste (Sociological... 1995). The cotton monoculture has required the application of increased amounts of fertilizers and defoliants to facilitate cotton harvesting. An excessive use of toxic chemicals has also led to extensive **chemical pollution** of surface and ground waters and soil contamination. In 1975 the average annual fertilizer use in Uzbekistan was 238 kg per ha; Turkmenistan 241 kg; Tajikistan 220 kg and Kyrgyzstan 146.5 kg.

Due to an overall production decline during the period of transition, the levels of fertilizer application have decreased in all countries. Nevertheless, in 1993 they still amounted to 150 kg per ha of cropland in Uzbekistan; 97 in Turkmenistan; 81 in Tajikistan and 20 in Kyrgyzstan. These values can be compared to 29 kg per ha applied in the Russian Federation (World... 1996). In addition, from 1980 to 1996 the volume of discharge of polluted waters into rivers and canals increased by 2.6 times in Uzbekistan, 1.8 in Kazakstan and 1.6 in Turkmenistan. Furthermore, the share of toxic chemicals in the total pollution load in 1996 reached 42% in Uzbekistan and 32% in Turkmenistan.

One additional implication associated with the dissolution of the USSR was an increase in the animal pressure on rangelands and their further overgrazing and desertification. This was due to the absence of previous restrictions on the size and maintenance of private herds, which have sometimes become the only source of support for rural dwellers.

10.4.4 Overall ecological situation

The irreversibility of the ecological situation related to the **Large Aral Sea** permits us to classify it as an *ecological catastrophe*. There is a theoretical possibility of ecological improvement in the adjacent parts of **Priaralye** and some positive change has already occurred. Thus the situation there can still be considered as an *ecological crisis*. For the rest of the drainage basin it is *critical* with a *crisis* situation e.g. along the Karakum canal, Ferghana valley and in some other parts.

The magnitude of this anthropogenic disaster is large scale as it affects in some way the 'health' of ecosystems over an area of almost 2 million sq km. As well as the desiccation of the Aral Sea, transformation of its aquatic system has greatly reduced its biodiversity and created an anthropogenic desert on the former dried sea bed. Additionally, it has involved a decline in the ground water level and desertification of the adjacent territories including the drying up of deltaic areas, salinization of soils, acceleration of aeolian processes as well as the generation of salt and sand storms.

Furthermore, the ecological crisis in Priaralye has involved a local climate change which has been made evident in a reduction of the growing season by approximately two weeks. This has sometimes been critical and has brought about a shift to growing less valuable crops. The latter process has also been attributed to extensive soil salinization problems which ultimately decreased the soil fertility and consequently crop productivity thus affecting the agricultural economy.

10.5 Socio-economic and health issues

Generally, people living near the Aral Sea think of themselves as victims of the politics of the former Soviet government. They understand that the sea has been sacrificed for the sake of the cotton industry, needed for the FSU as a whole, but that they have to keep paying its cost with their lives and well-being. Most of the region was already predominantly agricultural and poor, but the drying up of the sea and desertification of the deltaic territory have further impoverished the resource base of the local population. Environmental disaster has been accompanied by such grave implications as the decline in family incomes, deterioration of health and has resulted in a basic change in the historical mode of life of the people living near the sea.

10.5.1 Social repercussions

The transition to the market economy and the associated economic crisis in all new independent countries have added to the impact of the Aral Sea catastrophe by further degrading the local **standard of living** which is currently well below the poverty line. Calculations of the US dollar-equivalent for the average monthly earnings in Turkmenistan from the third quarter of 1994 to the second quarter of 1996 show that they ranged between US$18.2 and 42.6. In Uzbekistan they were slightly higher and varied between US$38.8 and 68.6 (EIU 1996c, 1997). An important point is that, for example, in Karakalpakstan 78.5% to 93% of the average monthly income is spent on food (Sociological... 1995). Additionally, their standard of living was affected by a rise in consumer prices. Estimates show that during 1991–1995 the annual average percentage of growth was about 900% in Turkmenistan, over 600% in Tajikistan and Kazakstan and almost 500% in Uzbekistan (EIU 1996c).

Emigration from the affected region was particularly high during the late 1980s when ecological problems became most urgent. In Karakalpakstan, the local population migrated to cities such as Kungrad, Nukus and Chimbai and outside the republic (Karimsakov 1994). Recently it has slowed down, mainly because people simply cannot afford to move. One survey has revealed that one-third of potential migrants who wish to move, do so because of financial difficulties and one third because of ecological problems. The ecological factor accounted for 79% of respondents in the Khorezm region (Sociological... 1995). The main limiting factors are the cost of housing and travel which have rocketed during the post-Soviet period.

In addition the new independence was accompanied by a large-scale emigration of the highly-skilled labour force of European origin following an increasing wave of anti-Russian nationalism in all new states. Social tensions in the region have also been noted by Lubin: 'It is often difficult to fathom the degree to which the social fabric of the former Soviet Union has been rent in the chaos, dislocation and frustration caused by the break-up of the USSR ... Tensions have certainly surfaced between Russians and Central Asians. By the end of 1992, well over 100,000 Russians, many of whom were valued specialists, had already left these new Central Asian states... On the whole, the social dislocations in Central Asia today are massive. Many people no longer know where they fit ...' (Lubin 1994, 267–9).

This trend was evident throughout the post-socialist period and, most recently, in 1998 over 210,000 people emigrated from the states of Central Asia to Russia. This figure included over 160,000 people from Kazakstan, more than 33,000 from Uzbekistan, and over 8000 from Turkmenistan (Chislennost'... 1999). Some of them officially stated the adverse ecological situation as the main reason for migration along with inter-ethnic tensions.

10.5.2 Economic implications

One aspect of economic losses has been due to the widespread soil salinization processes. A study has revealed that by the mid-1990s about 44% of irrigated land was severely salinized in Uzbekistan alone (Lubin 1994). In Kyrgyzstan over one-third of the irrigated lands are salinized to a moderate or high degree and 110,000 ha of formerly irrigated lands have been withdrawn from agricultural use due to salinization in Kazakstan.

A serious economic implication of the desiccation of the Aral Sea arose in the loss of commercial fisheries (Table 10.3). In the city of Muynak in the 1960s canning factories used to can 45 types of fish caught in the Aral Sea. In 1958 the fish catch reached its peak of 24,400 tonnes (Karimsakov 1994). Before 1972 about 41.1% of the food industry in the Muynak region was related to fishery products. The city of Aralsk on the north-east border is located now at a distance of over 100 km from the sea (Plate 20). The sea produced 30 to 40 tonnes of fish and supplied tens of thousands

Plate 20 Will the sea come back to Aralsk?

of people with jobs in the fishing and canning industries. Currently only 2000 workers are employed in the canning industry in Muynak. In both towns canning factories keep working, though at greatly reduced volumes, and process fish delivered from distant locations.

Another important economic issue associated with cotton expansion, only fully evaluated after the dissolution of the USSR, was a basic **change in the structure of agricultural lands** which has occurred during the Soviet period. Before the Bolshevik revolution the structure of sown lands was basically different from the pattern prevailing later. As shown in Figure 10.3, in 1913 in Central Asia, not including Kazakstan, 75.1% of the total sown area was under grain, 14.9% under technical crops, 6.6% fodder crops and 3.4% under potatoes and vegetables. In Kazakstan the share of grain crops was even greater reaching 93.3% (McCauley 1976; Rossiya... 1995b).

By the start of the 'cotton independence policy' the share of grain crops had diminished to 41.3%, while the proportion of technical crops in the region increased to 37.3%. In the irrigated lands of Uzbekistan, Turkmenistan and Tajikistan by 1976 the share of technical crops even exceeded 60% of the total sown area. At the peak of cotton production in 1981, grain crops accounted for less than a third of the total area of sown lands, while technical crops reached 40.6%. By the end of the Soviet period the structure had slightly changed: the proportion of grain crops had fallen even lower to only 26% while the share of technical crops had slightly decreased to 38.1% (Figure 10.3). The only exception was Kazakstan which, along with the Ukraine, became a granary of the USSR after a notorious campaign of 'virgin land development' pursued in the 1950s; this had resulted in another environmental disaster comparable to the 'dust-bowl' of the Great Plains in the USA. Despite a consequent and current decline in grain productivity, Kazakstan produces a sufficient amount of wheat.

An important aspect of the above change in the structure of sown lands was that it was also expressed through a decline in per capita grain production (Table 10.4). In 1913 the values of per capita grain production were as follows: in Kazakstan 386.3 kilograms per capita; Kyrgyzstan 504.6; Uzbekistan 236.5; Tajikistan 195.4 and Turkmenistan 152.6 kg. In Kazakstan, which has become a net exporter of grain, the value of per capita production further increased reaching 1752 kg in 1992. At the same time, Table 10.4 demonstrates that in all other republics per capita grain production has *never reached* the levels of 1913. The values of per capita grain production were particularly low during the period between 1971 and 1975 (Narodnoye... 1977, 1982, 1986, 1989). At the same time, over 90% of all Soviet cotton was produced in Central Asia! In this way, the initially diversified and self-sustained agriculture in Central Asia has given way to a distorted cotton-orientated economy which undermined the ability of each republic to feed their population after gaining independence (Saiko 1998).

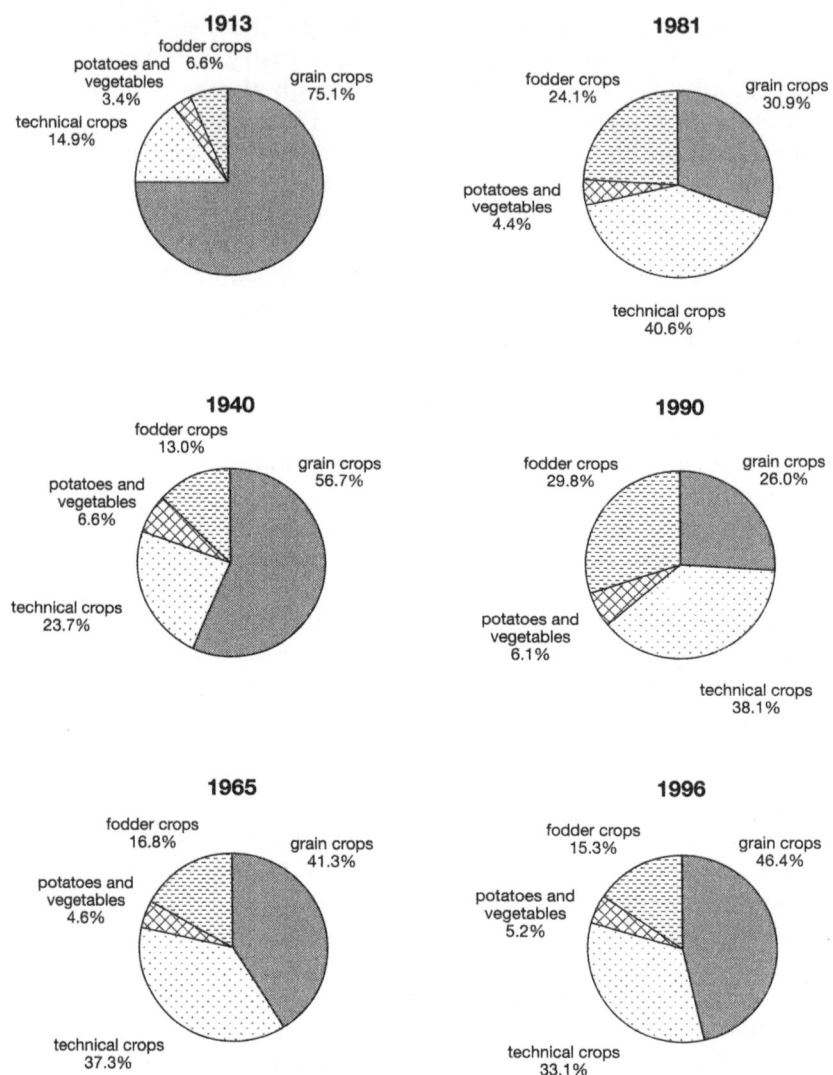

Figure 10.3 Change in the structure of sown agricultural lands in Central Asia excluding Kazakstan, 1913–1996 (sources of data: Narodnoye... 1987, 1989; Rossiya... 1995b; Rossiyskiy... 1996, 1998)

During the post-Soviet period there has been an insufficient shift in direction towards a more sustainable structure of agricultural production as is shown in Figure 10.3. This illustrates that the share of grain crops in the sown lands of Central Asia apart from Kazakstan increased from 26% in 1990 to 46.4% in 1996. However, this increase was achieved not at the

Table 10.4 Per capita grain production in Central Asian republics, 1913–1997 (kg)

	1913	1940	1965	1975	1985	1990	1995	1997
Kazakstan	386	409	1205	1593	1434	1702	570	774
Kyrgyzstan	505	385	252	346	357	342	221	376
Tajikistan	195	213	94	63	68	57	42	91
Turkmenistan	153	95	42	63	98	122	249	131*
Uzbekistan	237	92	75	81	81	93	142	162

Source of data: Narodnoye... (1977, 1982); Strany... (1992); Rossiyskiy... (1997)
* data for 1996

expense of a decrease in areas under cotton, which shrunk by only 5%, but mostly due to a reduction of lands under fodder crops, potatoes and vegetables, which is a worrying change. Furthermore, cotton still accounts for the greater share in the total agricultural output of major cotton-producers.

Expansion of land under grain crops was particularly substantial in Turkmenistan and Uzbekistan. The area under cotton in the former state was reduced by 5% in 1991, while that under wheat increased by 63% (EIU 1996b). However, cultivation of marginal salinized lands and successive droughts led to disastrous grain crops in 1995 and 1996. It should be noted that this shift to sustainability in Turkmenistan has been carried out along with realization of a major plan to extend the length of the gigantic Karakum canal for 300 km and to withdraw another six cubic km of water from the Amudarya (Ilamanov 1993).

10.5.3 Human health effects

Water pollution and salinization along with airborne pollutants have adversely affected the health of people inhabiting the Aral Sea region and those in Priaralye most severely. These problems have affected at least 70% of the local population. Contamination of drinking water due to the excessive use of chemicals has often been cited as the main reason for health problems (Downing 1995; Feshbach and Friendly 1992; Ishida *et al.* 1995). Along with ecological factors the lack of proper water treatment systems and appalling sanitary conditions in the region are responsible for people's poor health. One study in the Karakalpakstan republic and Khorezm oblast of Uzbekistan has recorded that only 5% of respondents had a personal bath or shower; 34% of people did not have soap for washing and 68 of every 100 water sources did not meet sanitary requirements (Sociological... 1995).

It was also found that in the Karakalpakstan republic, which ecologists call 'Asiatic Chernobyl' (*Zelyoniy Mir* 15, 1999), frequencies of viral hepatitis and tuberculosis increased by 50% from 1985 to 1991, and skin diseases are twice as frequent as average in Uzbekistan (Sociological... 1995). Another substantive report made in 1993 on the survey of medical and ecological aspects of the Aral Sea disaster has shown that in 1990 the death rates from viral hepatitis in the Karakalpakstan republic were three times higher and from typhoid fever almost five times greater than the average for the USSR (Sergiev *et al.* 1993). Many health studies have revealed that 60–80% of women in Priaralye have anaemia and frequent miscarriages. The demographic situation in the region is also becoming desperate.

One of the most comprehensive indices of health status is infant mortality. In 1990–1995 the average infant mortality in Uzbekistan was 41 persons per 1000 live births; Kazakstan 30; Turkmenistan 57; Kyrgyzstan 35 and Tajikistan 48, compared with seven in the UK and nine in the USA (World... 1996). In some regions of the Aral Sea, for example, in the Bozatau district infant mortality rates were even higher and exceeded 110 persons per 1000 (UNEP 1992). A local study revealed that the average life expectancy in villages in the Aral Sea region is about 38 years (Lubin 1994). The existing extensive socio-economic and health implications enable us to classify the contemporary situation in Priaralye as an *environmental* rather than just an *ecological catastrophe*.

10.6 Prospects for the future

It is currently obvious that rehabilitation of the Aral Sea to its initial level of 53 m is practically impossible. It has been estimated that, even to support its current level of 36–37 m, would require providing an inflow of water into the sea amounting to at least 30–35 cubic km. These figures can be compared to the planned value of discharge at 16–18 cubic km by 2000 and 22–25 cubic km by 2010.

During the Soviet period, the revival of the Aral Sea was associated with the notorious project of diverting the water from Siberia via the so-called Sib-Aral canal shown in Figure 10.1. However, it was actually never intended to refill the sea but was aimed to support further expansion of irrigated lands in Priaralye, as is evident from the map. It was planned that this unlined canal would stretch for over 2600 km across southern Siberia and the arid regions of Kazakstan. Clearly, water losses from evaporation and infiltration would have been comparable to those from the Karakum canal. This project was abandoned in 1986, paradoxically, for economic rather than ecological reasons.

Recently this breathtaking project has reappeared as Central Asian countries equate Russia with the former Soviet government and assume that it should take all the responsibility for the ecological disaster.

However, the Russian Federation is just as burdened with its own environmental problems inherited from the Soviet period and is experiencing major financial difficulties. No other major external source of potential water exists for this moisture-deficient region, but although technically such a project is feasible, it seems very doubtful that it will ever be implemented. The new market economy conditions also suggest that it will definitely not happen before the republics sharing the Aral Sea basin became rich enough to be able to pay the full price for Siberian water. Again, even in this case, the additional water will not be used for recharging the sea but to support further agricultural development. Therefore, it is difficult to expect that the overall ecological situation in the Aral Sea basin will radically improve in the near future or in the medium term.

Some small-scale positive changes are, however, already occurring. These are associated with the construction of a 14 km earth dike across the mouth of the Syrdarya which was carried out in 1997 under the initiative of the local government in Aralsk. Protected from the polluted body of the Large Aral, the Small Aral started to recover and its former shoreline is slowly creeping closer to the ships 'cemetery' (Plate 20). The water level has already increased by about three metres and by June 1999 approximately 9 million cubic m have been reclaimed.

The ecological situation in the Syrdarya delta has improved and some birds including swans and gulls have already reappeared. As more water has reached the delta, the former sea has started to recede, coming closer to Aralsk. This improvement has been achieved at the expense of communities further south who have ceased to receive any water from the Syrdarya. The dam has already been broken a few times and the financing needed to upkeep it and pay wages to the workers is running out. So people in Aralsk are racing against time and can only live in hope of a better future.

However, the future of the Large Aral Sea appears to be less hopeful. Based on a combination of indices, it is possible to make some forecasts for the future. Considering the current trends in water management implemented by each of the five countries sharing the sea's basin, it is obvious that the process of desiccation of the Large Aral Sea is *irreversible*. The area of the dried former sea bed amounted in 1994 to over 35,000 sq km. More remote dried parts of this zone are currently undergoing processes of wind erosion and barchan (sand dune) formation. With the continuing desiccation of the sea these will be threatening distant areas further from the sea and will affect the productivity of crops at a greater scale.

The deltaic areas of the Amudarya and Syrdarya are affected by processes of desertification up to a distance of 500 km from the former sea shore. The formerly fertile delta lands have turned into a fruitless desert; the productivity of agricultural crops has already declined here. It appears that no major positive changes can be expected in this area in the foreseeable future. The area of deserts around the former Aral Sea bed outside the deltaic areas will experience a further decline in ground water level. Under the current conditions of increasing aridization, the aeolian processes here will be further accelerated.

At the present time, the possibilities for any major improvement are complicated by the existing political, economic and social factors. At the national level this is due to the difficulties associated with the very painful process of transition to market economies. The values of gross national product have declined in most countries of the region. In 1996 the total gross domestic product (GDP) of all Central Asian countries accounted for only 16.8% of the Russian Federation value. The values of per GDP per capita in each of the countries are shown in Figure 10.4. It is clear that throughout the region there has been a steady decline in the value of GDP per capita, except for Uzbekistan, where it had been relatively stable during the last few years. Estimates show that from 1989 to 1996 GDP per capita has decreased in Kazakstan by 1.7 times; Kyrgyzstan by 1.9; Tajikistan by 2.7 and Turkmenistan by 1.8. While in 1989 the difference between per capita value for the Russian Federation and Kazakstan was only 34%, by 1996 it had already reached 60%.

Water is the main limiting factor for economic development in the Aral Sea region. During the Soviet period all water allocation was decided centrally by the government in Moscow and the Ministry of Water Management and Land Reclamation of the USSR. It is obvious that to

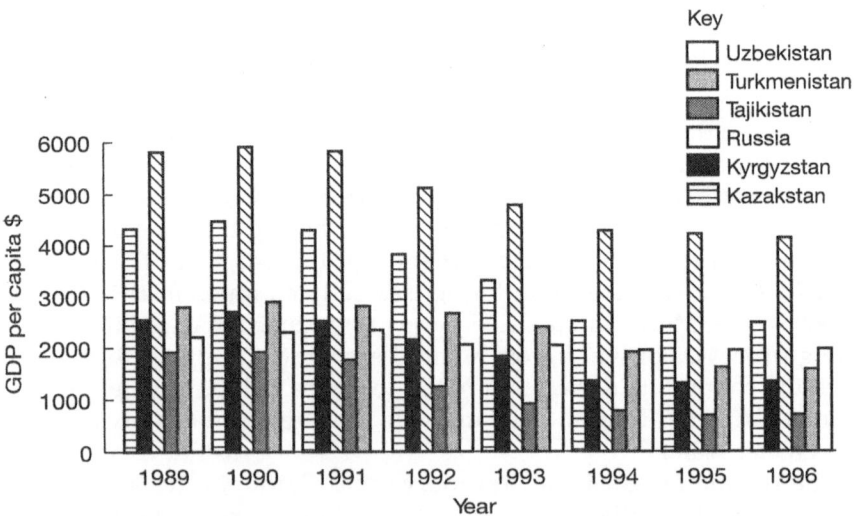

Figure 10.4 Per capita gross domestic product in the countries of Central Asia and Russia, 1989–1996 (in purchasing power parities) (from Saiko 1998)

solve the Aral Sea crisis it is essential, firstly, to reach a joint agreement on the general regional strategy of water use, management and pricing. However, this seems to be an impossible task and water consumption is becoming an increasingly potential cause for dispute. Ethnic conflicts and political instability have occurred in many parts of the Central Asian region, sometimes occurring, directly or indirectly, over water issues (Saiko 1995). Recently conflicts have been recorded between Uzbekistan, which needs water for irrigation in the summer period, and Kyrgyzstan, which 'owns' the upper reaches of Amudarya and uses water all year round for hydroelectric energy generation.

The potential for conflict will increase further with the competition for water particularly between the states located upstream and downstream of the main rivers which consume water of different quality. It should be noted that water withdrawal in Uzbekistan and Turkmenistan increased between 1990 and 1993 (Okruzhayushchaya... 1996) while after 1994 no official data was submitted to the interstate Statistical Committee by both countries. The official data for Kazakstan revealed that between 1991 and 1994, despite an overall reduction in water withdrawal by 9.2% in the main water-consuming Kzyl-Orda oblast, its water losses in transport more than doubled over the same period (Okhrana... 1996).

Recently, Turkmenistan extended the Karakum canal that now stretches for 1370 km to irrigate new farmlands and pastures in the western part of the desert. Despite a number of measures taken to improve the operation and maintenance of this canal over recent years, it remains in crisis (Hannan and O'Hara 1998). The needed infrastructure for effective inter-state water management exists in the region, but international co-operation between all five states is essential for any real changes. At the same time, agricultural and land reform in most countries of the region has been limited, and only a few planned goals were achieved on the way of restructuring and self-sustainability (e.g. Lerman et al. 1996; O'Hara and Hannan 1997).

During the last two decades the Aral Sea region has become the focus of international attention and dozens of projects have been started or implemented with the help of such international organizations as the World Bank, UNDP and the UNEP. In the view of Micklin (1998), the main failings in the efficiency of international projects carried out within the Aral Sea basin were due to their lack of co-ordination, low pace of programme implementation and creation of a donor-orientated mentality in the region.

The local people believe that the Aral Sea crisis has created a living for many aid workers and Central Asian bureaucrats. They also suspect that Westerners use the Aral Sea crisis merely as an excuse to come to the region and stay there, but that they are actually attracted by its potential wealth. On the whole, the local people feel that there are too many conferences and that much of the money spent on these projects is wasted. The most popular local joke is that if every foreigner who visited this region brought with him one bucket of water, the Aral Sea would be full again.

It is apparent that in the case of the Aral Sea and Priaralye, socio-economic and health implications have been so dramatic, along with ecological changes, that both have reached the stage of the *environmental catastrophe*. As regards the future of the former, a forecast made by the satellite data analysis shows that the sea could disappear some time in the 21st century (Sakata 1998).

The outlook for the future is bleak. Some experts believe that to achieve any major improvement in the socio-economic and environmental sphere, action should be taken only after this potentially oil- and gas-rich region earns enough from the exploitation of these deposits. However, this will be too late for the local population and the sea. As Downing noted, 'For the population of Turkmenistan, and of Central Asia as a whole, the problem is grim and inescapable' (Downing 1995, 399). In addition, 'turning the Aral basin into the "oil and gas Klondike" will conclusively bury all life in the region, while Karakalpak people as an ethnic group will only exist in the written archives within 10–20 years' (*Zelyoniy Mir* 15, 1999, 7).

Conclusion

... a tremendous future is wide open for mankind, if only we realised it and would not use our minds and labour for self-destruction.

(Vernadskiy 1969, 327)

The Cold War images

The 20th century will probably remain in the human memory as a period of gigantic human and social experiments. The attempted and failed construction of an egalitarian society on over one-sixth of the world's surface has dramatically changed both the minds of people and their natural environment, leaving many scars and sometimes even ruins throughout its domain. Before socialism ultimately collapsed, the magnitude of damage was concealed by an elaborate system of ideological and censorship controls in the former Soviet Union and its socialist partners. An objective evaluation was also difficult to achieve due to political reasons.

Although in many industrially developed socialist bloc countries factual data about the state of the environment actually existed, it was politically embarrassing to admit the real extent of environmental problems, which 'by socialist definition' could only be inherent to the capitalist system. Much of these data was classified and concealed from the population inhabiting the threatened regions. As a result, only since the advent of 'glasnost' (publicity) during the Gorbachev era in the mid and late 1980s have circumstances improved sufficiently to allow the revelation of environmental disasters in many countries of the region.

At the same time, Manser noted that 'the image of black billowing clouds and dilapidated buildings was one of the last propaganda coups of the cold war. Some parts of both the Western media and the local opposition overstated the extent of the ecological disaster in order to condemn and if possible bury communism and central planning' (Manser

1993, 18). The case of the former German Democratic Republic serves as the best example of this 'misunderstanding'.

According to Boehmer-Christiansen, 'Prior to 1990 East Germany appeared to be an environmental contradiction. Those relying on West German media stories and government documents perceived East Germany as an ecological disaster zone where air, water and soil were so terribly polluted that human health was impaired and communism stood condemned. To others, less interested in proving the superiority of the "market economy" and possessing more historical knowledge, large parts of the GDR seemed pristine compared to the environmental disasters of the West .... "Disaster areas" were found less in rural areas than in crumbling inner cities and industrial regions, from which rising outputs were demanded with declining investment' (Boehmer-Christiansen 1998, 67–68).

The last decade of the 20th century proved to be a period of the most radical and dynamic transformations in the political, economic and social spheres but, most importantly, in the mentalities of hundreds of millions of people. These swift changes were associated with a continuous process of painful reassessment of the true virtues and faults of the past regimes and histories based on an 'avalanche' of new information previously unavailable. This was the time of unprecedented large-scale human confusion and disappointment in relation not only to the past values but also to the newly emerging 'market' ideals. It appears there is still not enough appreciation in the West that the goals of communism were *genuinely shared* by the majority of the population in the former socialist (and not 'communist' as communism has never been constructed there!) countries. Many positive aspects of the socialist system remain ignored by Western scholars and experts.

Unfortunately some political factors that affected an impartial evaluation of the situation did not cease to operate after the collapse of socialism. They were operating in the former GDR even after the re-unification of Germany. As pointed out by the above author, '... the exaggeration of environmental catastrophes include the assertion that a slightly lower life expectancy was caused by pollution (rather than diet, available health care, etc.) and the claim that the entire East German housing stock would need rebuilding because of asbestos problems ... The positive environmental achievements of the former GDR, such as its household waste recycling system and small, slow cars, became completely overshadowed by the domination of negative images ...' (Boehmer-Christiansen 1998, 69).

In addition, as has been shown by the author of this book in an earlier publication, the generally accepted idea of a one-way environmental impact of Central and Eastern European countries on West Europe as a case of transboundary pollution is a myth. Due to the predominantly westerly direction of winds in Europe, Russia, for example, receives more pollutants than it 'exports' to Western countries (Saiko 1998a). However, it is still impossible to draw a precise and comprehensive overall picture of the current state of the environment in the post-socialist world because of insufficiency or inadequacy of data but also due to political reasons.

In the Russian Federation, environmental concerns remain influenced by the electoral interests of different levels of power circles or 'green movements'. In the Ukraine, they are sometimes veiled by the interest of the state to attract maximum foreign aid and investments. In the Central Asian region, environmental matters are being assigned a new level of secrecy and, in some republics, the most active environmentalists serve sentences in jail. Although we are still nowhere near the point where an objective comprehensive assessment of the state of the environment in post-socialist Eurasia can be made, it appears that an attempt in this direction is useful. This is why the author of this book has tried, in the course of the above case studies, to avoid drawing the customary 'black' picture of the ecological situation and instead has pointed out the achievements in environmental protection over recent years.

Physical causes of the slow progress in environmental improvement

Although the major problem regions have been highlighted above, it does not mean that other areas remained unscathed. On the contrary, many other regions in the former socialist countries have had a whole range of environmental problems with varying degrees of severity. To quote Alcamo once more: 'The damage is as diverse as the landscape' (Alcamo 1992, 1). While, for the most part, the adversity of this situation has been due to the 'inheritance' of the socialist system, some new challenges in the more recent period have been associated with the nature of the transition period and the values of the market economy.

Investigation of various case studies of environmental deterioration and destruction examined in the course of this book have given the author an opportunity to draw some comparative conclusions about the contemporary state of the environment in the former socialist world. The magnitude and nature of the inherited environmental problems were extremely variable in each of the former socialist countries, but the overall situation could with good reason be called a crisis. It appears that the environmental crises were primarily the consequence of three major factors: the nature of the economy, the influence of the politico-legal and social environment, and the sensitivity of the specific landscapes.

An important point is that the former two factors can operate within different natural environments, featured by varying degrees of vulnerability to human impacts. In some of them, such as the Arctic tundra landscape, a relatively 'modest' economic development has already resulted in some major or irreversible changes, which will remain in the environment for decades, even if the cause of the disturbance is removed. In other landscapes, which are naturally more resistant to human influence, such as the deciduous forest zone, a similar scale of impact could be overcome by the natural rehabilitative capacity of the environment, if the cause of the impact was dealt with or the necessary financial means for restoration available. Therefore, to be able to examine the potential consequences of various types of economic development and magnitudes of impact, it is essential to study the natural features of a specific environment.

It has also been pointed out by Nefedova and Treivish that 'It was difficult to combat the ecological crisis because the economic crisis was not recognised and no structural reformation took place. With the transition to the market economy, the production decline has automatically relieved the acuteness of the ecological problems' (Nefedova and Treivish 1994, 6). However, as was shown in several of the case studies in this book, the latter effect was not observed in all the affected areas and in many cases such partial improvement was outweighed by other forms of environmental deterioration. In 1997 it was recorded that the rates of reduction in air pollution had slowed down and an increase in the volume of atmospheric emissions was evident in many regions of the Russian Federation (Gosudarstvenniy... 1998a). This fact supports the point made by the author, that the nature of *'transitional environmental improvement'* is rather ephemeral. Most certainly in the foreseeable future the adverse ecological conditions will further become aggravated in many parts of the former Soviet Union. Any real success was only evidenced in cases when both the means and the will were available to reverse the situation, e.g. in the case of the Black Triangle.

The overall situation during the transition period has somewhat improved in Central European countries and, partially, in some other areas. At the same time, it has, at the best, stabilized or even worsened in the former Soviet Union. The actual improvement during the transition period was in many cases far less than expected even when compared to the degree to which the pollution load or anthropogenic pressure had declined as was shown e.g. in the case study of the Volga basin (Chapter 8). The pace of progress in environmental enhancement has been much slower than expected, and new challenges have appeared, adding to the effects of old ones. This was arguably caused by the following reasons:

(a) the threshold of irreversibility in the natural landscapes was exceeded, e.g. in the case of the desiccation of the Aral Sea;

(b) the assimilative and/or rehabilitation capacity of natural ecosystems was exceeded long before the reduction started, e.g. near Monchegorsk;

(c) the assimilative and/or rehabilitative capacity of natural ecosystems is so low, e.g. in the tundra zone of Yamal, that any apparent improvement will be seen only in the long term if the impact substantially abates or completely ceases;

(d) the actual amount of pollutants reaching the environment was greater than recorded by official data, e.g. in case of the Volga basin, where substantial amounts of unaccountable pollutants enter the river system with surface runoff or in the cases of oil spills in the Tyumen oblast when the reported data was falsified;

(e) along with a reduction in one source of pollution or impact, a new, more powerful source appears or increases in magnitude, e.g. in the case of the Moscow region where air pollution from transport became a much greater threat than the receding amount of pollutants from stationary sources;

(f) while the amount of pollutants decreased with the decline in industry, enterprises tried to save money on environmental protection such as would be spent on installation of new advanced treatment facilities or operation of the existing ones e.g. in many parts of the Volga basin;

(g) the overall reduction in the pollution load or impact was achieved at the expense of less environmentally dangerous branches of economy, e.g. machinery construction, high-tech engineering, aviation and defence industries, while more hazardous resource extracting and processing industries are increasing their impact e.g. oil and gas production in West Siberia and mining and metallurgy in Norilsk;

(h) the current economic difficulties and potential social implications prevent enterprises from shifting to less environmentally hazardous productions, e.g. in the case of the Baikalsk pulp and paper mill;

(i) the share of obsolete and antiquated equipment including treatment facilities, enterprises and technologies which exert the crucial adverse impact on the environment is increasing in most states of the former Soviet Union.

In addition, the current system of economic incentives for environmental protection is still very weak in many countries of the former socialist bloc with the exception of some Central European states like Poland, which is being already motivated by the objective to conform to the standards of the European Union. At the same time, in most countries of the former Soviet Union environmental issues, along with other long-term needs and issues such as demographic problems, remain outside the main state priorities.

It also appears that in the countries of the former Soviet Union environmental enhancement will also require them to elaborate and implement a whole system of state and market economic mechanisms as well as a set of effective punitive measures. At present an overwhelming law-evasion attitude penetrates all levels of society, reflecting an overall disappointment with the many revolutionary transformations of the post-Soviet period.

The existing practice of 'pollutant pays' principle appears to have only a limited effect as the impact on natural landscapes involves many other influences which are not so readily assessed by quantitative indices but are no less important for the 'health' of the environment. The latter term is being more and more frequently used by experts in Russian ecological literature. The current system of economic incentives which includes payment for natural resources and payment for damage from ecological destruction not sufficient to provide an efficient regulation of the 'resource use – environmental impact – waste generation' system. The contemporary level of payments for natural resources or for the impact on the environment is clearly insufficient to stimulate the introduction of advanced ecologically friendly technologies by enterprises.

Some of the above-mentioned reasons for slow progress could be overcome with the aid of adequate financial investments directed to environmental enhancement when the economic situation substantially improves and when environmental priorities gain a sufficient rating. However, it appears that in the short-term future, revival of economic activity will be associated not with the reduction of anthropogenic impact but with its further acceleration.

In addition, as is now widely appreciated, an important outcome of this extensive and intensive environmental damage has been its detrimental impact on people's health, which has caused some authors to name it an 'ecocide' (Feshbach and Friendly 1992). Although the Russian Federation is comparably one of the world's best provided with doctors and still has a free health care system (Sallnow and Saiko 1996), only 14% of children in Russia are practically healthy; half have various anomalies in their state of health and 35% are chronically ill (Gosudastvenniy... 1995). This is also typical for other countries of the former Soviet Union.

While in Western countries it is often almost impossible to prove that an adverse ecological situation is responsible for ill health (Matthews and Saiko 1994), the environmental situation in some parts of the former socialist world has become so devastating that a whole set of 'industrial' or 'ecologically related' illnesses had affected whole towns or regions. Furthermore it appears that, at present, the load of accumulated environmental problems plays an increasing role in affecting the demographic issues in many states of the former Soviet Union. A seemingly more 'favourable' situation in Central Asian countries is only due to their typically high birth rates despite an apparently poor quality of medical care.

All the above-stated medical and demographic facts clearly reflect the actual unfavourable sanitary-hygienic situation in the post-socialist countries. The deep transformation of human habitat due to anthropogenic impact has had an unfavourable feedback effect on people's health. For example, in Russia approximately 60 million people live in conditions of permanently exceeded maximum permissible levels of pollutants in the atmosphere and over half of the population drink water of poor quality (Gosudarstvenniy... 1995).

Human causes of the lack of progress in environmental improvement

It is still not sufficiently realized that many causes of the current situation and of the obvious lack of actual progress in environmental enhancement are also rooted in the mentalities of people. Paradoxical as it may sound, the availability of many pristine and relatively untouched areas throughout the former socialist states could have served, for some of them, as one of the causes for the lack of concern and protective action in the past. In the world's largest country, Russia, the existence of the 'limitless' expanses of 'virgin forests' in Siberia has been probably engraved in the mentality of people for centuries since the time serfs headed there in search of freedom. This availability of 'empty lands' beyond the Ural mountains clearly inspired Nikita Khrushchev to launch his notoriously failed 'Virgin Lands development' campaign rather than invest in intensive farming methods within the European USSR.

'Free common land that belongs to everyone' was another dubiously fated concept that stimulated environmental abuse and neglect, particularly in the most populated parts of the country. There is also the religious belief of Muslims that water is a free gift from Allah. In the Central Asian states this belief appears to be a stumbling block against local people accepting the idea of water pricing.

Another common concept dominant in the mentality of the predominantly rural population of Russia is best reflected in the words of the Russian poet Nikolai Nekrasov, 'The landlord will come, and HE will decide it all ...'. The latter typical attitude was used to its maximum by the socialist system that made an individual into a small insignificant 'screw' in the overall important 'mechanism of social progress'. The state readily took on the responsibility for everything and, more importantly, the power to make all decisions. The lack of personal responsibility for environmental degradation still remains one of the main causes of the lack of progress in its control. An apparent difference between the grassroots environmental activism in the West and the former 'East' originates from the above concept. It also explains the lack of local belief that anything can be changed for the better.

In the case of the Central Asian situation, the formerly feudal khanates which existed in the pre-Soviet period were governed by the principles of absolute obedience of peasants to their ruler. It appears that the following Soviet totalitarian period only confirmed these attitudes in the minds of local people who at present, generally, lack any initiative, for example in attempting to apply water-saving practices. A more recent example of this over-reliance on somebody else is a widespread belief in the region that international organizations will pay for the notorious gigantic 'Sibaral' project, which was supposed to bring water from Siberia to Central Asia.

Linked with the previous notion was an extremely low value that the totalitarian form of society *de facto* attached to an individual human life and human issues in general. This was arguably the most important feature of the socialist system in terms of its indirect environmental implications. All individuals were expected to sacrifice their personal interests 'for the sake of society at large' and for the benefit of future generations who would live under communism. Therefore, their own living conditions, health and the environment were only of limited importance. This kind of attitude would be totally inconceivable for a Western individual, for whom private property and their own personal interests were always strictly guarded by the law and society.

How could the environment, which for the most part was supposed to be the source of natural resources, be valued highly when the cost of human life was so negligibly small? The system which so easily sacrificed lives of millions of the Soviet people during the Stalinist period, which moved whole nations from their motherlands to Siberia for 'cooperation with an enemy' after the Civil and the Second World Wars, never attributed much value to an individual's needs and aspirations.

It was assumed that the totally controlled system could easily overcome any possible negative 'side-effects' of the highly prioritized industrial development that was essential for the construction of communism. 'Moral persuasion' was used as a mechanism for the enforcement of environmental protection measures and, as Fullenbach noted, a frequent argument presented by the socialist system was that 'pollution is a structural problem in capitalist societies, therefore in principle insoluble, while the environmental problem in a socialist system is simply a question of motivation, and thus easily capable of solution via political decisions' (Fullenbach 1981, 106).

The regional ability to cope with problems

In terms of the gravity of the current ecological situation and the ability to cope with the related problems, four groups of countries can be distinguished. Arguably, the worst affected appears to be the **Central Asian region** where the acuteness of the inherited ecological deterioration is compounded by an inherently very high natural vulnerability of landscapes to human impact and by the increasing pressure on the very component responsible for this vulnerability, i.e. water. The adverse large-scale health implications that are evident in Priaral'ye complicate the matter. This combination, compounded by an obvious lack of regional and local financial abilities in the region to reverse the current trends, does not leave much hope in the foreseeable future for a substantial improvement of the situation.

Apparently this is an *environmental catastrophe region* that needs urgent international assistance and advice on ecologically sustainable agricultural and water management policies. The situation here is further undermined by large-scale emigration over the last decade of highly skilled engineering and academic staff of non-Central Asiatic ethnic origin who are being 'squeezed out' by nationalist attitudes and policies.

The second group includes the **Ukraine and Belarus**, the countries that are affected by the Chernobyl accident along with many other ecological problems. The current large-scale health threats within the contaminated areas are complicated by longer-term 'invisible' genetic effects that will remain in the environment for millennia. The Ukraine, Europe's largest country apart from Russia, will, in the long run, be in the position to look at its other devastated landscapes e.g. in the Donetsk region and to attempt to rehabilitate them. Environmental issues in this republic are recognized by the state and are the inherent part of government foreign policies.

However, the economic situation in this country is still rather desperate. At the present time, the Ukraine stands third last in the list of 22 former socialist countries reviving after the 'transitional' economic collapse and it will require a lot of international assistance (*The Economist* 20 November 1999). Although Belarus ranks higher in the list, in some ways it is in an even worse position with its predominantly agrarian and forestry potential. In addition, it is worst hit by the Chernobyl accident and some experts compare the economic losses and damage with those caused by the Second World War.

The **Russian Federation** stands separately for many clear reasons. The range of ecological conditions is the greatest, from pristine landscapes to anthropogenic deserts. The scope of various environmental problems is enormous along with the magnitude of their effects on the natural landscapes. These include the devastated reindeer pastures in the tundra; extended tracts of dead forest stands in the fragile landscapes of the Arctic; highly polluted water courses throughout the Volga basin; formation of a desert in the steppe zone; catastrophic decline in sturgeon stock in the Volga-Caspian region, etc. At the other extreme, Russia arguably has the largest wilderness area on the planet.

With its greatest diversity of landscapes and ecological problems, potentially it is the richest of all post-socialist states and theoretically could independently solve its own problems. In addition it has a powerful human potential in terms of its intellectual elite, advances of fundamental science and developed and applied technologies. A good albeit limited example of a local success story is the recent rehabilitation of over 300,000 ha of completely degraded pastures in the Black Lands of Kalmykia. Potentially effective rehabilitation programmes have been already designed for the vast Volga and Baikal regions but their implementation is halted due to the lack of finances.

The paradox of this country is that it is extremely rich and poor at the same time. It is extremely rich in its resource potential and in actual revenues received currently by private individuals and placed in their foreign accounts. It is so rich that, after many centuries of stealing its wealth by various groups of people and individuals, a lot of it is still 'left over', as Russians like to joke. The problem with Russia is that the appeal of this huge and sweet 'cake' for politicians and businessmen is enormous. Therefore, they are much more preoccupied with issues related to their next elections or short-term profit than with any actual developments in favour of the common people and environment.

At the same time, contemporary Russia is poor, primarily because the majority of the population of Russia is impoverished, both as judged by Western standards and internal prices. Indeed, there are several objective reasons that go beyond this fact. One is that it is overburdened by the past debts of the whole USSR that it took on in the aftermath of its dissolution. Another cause is rooted in the ways and means of restructuring, particularly its voucher privatization campaign. The latter was carried out during a period of rampant inflation and at a frenzied pace. This condition allowed interested parties and individuals to buyout vouchers from the population literally, for virtually nothing.

Ultimately, the 'symbolic' costs of shares in many enterprises enabled a swift transition of valuable national assets, such as the aluminium or nickel industries, into the hands of a few extremely rich oligarchs. Thus they have effectively escaped state controls including those relating to environmental matters. Both industries mentioned above are among the worst polluters. At the same time, they are very powerful as they provide much of the revenue for regional governments e.g. in the Krasnoyarsk kray of Siberia. Furthermore, bureaucratic corruption and the widespread practice of bribing poorly paid government officials has enabled the guilty parties to avoid paying the fees for violation of environmental legislation.

In addition, the present realities of the 'stalled transition' in Russia (*Moscow Times* 16 November 1999) or of any other forms of economic restructuring in the countries of the FSU, mean that the impoverished population are almost exclusively preoccupied with desperate efforts to make ends meet. In such a situation any other concerns, including environmental matters and sometimes even health issues, seem of much less importance. It was noted recently that this is the case at the grass-roots level even in the ecological disaster areas (*Zelyoniy Mir* 23, 1999). Therefore, socio-economic revival appears to be the main prerequisite for any significant environmental improvement.

As Danilov-Danilyan stated, 'The acute adversity of the state of the environment in the most densely populated and industrially developed regions of Russia is brought about not so much by the scale of the impact on nature in the past as by the criminal ignorance of the emerging ecological problems and also by the present catastrophic lack of financial means for their solution or, at least facilitation' (*Zelyoniy Mir* 10–11,1999, 6). To aggravate the situation, a recent demilitarization campaign has removed additional budget allocations that could have been potentially used for environmental goals.

The current situation is further undermined by the failure of the political elite to realize the importance of science in the socio-economic revival of the country. Heavily underfinanced academic institutions are only keeping afloat at present due to the international assistance coming in the form of joint projects. With their meagre salaries, scientists cannot afford to buy the basic statistical data sources that currently cost almost world prices. Recently one scientist from Moscow State University commented on the latter circumstance noting that such statistical yearbooks can now only be purchased by Western scholars. He complained that this was probably the reason why the *Post-Soviet Geography and Economics* journal has almost ceased to publish articles by Russian authors.

In addition, Russia is probably the only European country in which environmental issues were recently almost officially 'downgraded' by the transformation in 1996 of the former Ministry for the Protection of the Environment and Natural Resources into the State Committee for Environmental Protection. At the same time this 'restructuring experiment' has effectively 'divorced' natural resources from their conservation by their reallocation to a separate Ministry of Natural Resources, which is clearly more concerned with their utilization than protection.

Further adding to the lack of interest of politicians to environmental matters, come the mass media sources that seem to be only involved with political campaigns of the parties that finance them. All the above factors contribute to the formation of a vicious circle in Russian society which explains the amazingly low progress in environmental improvement against the background of the country's immense potential.

The fourth group includes the former socialist **Central European** countries. These countries are affected by a multitude of environmental problems. Many of the affected areas are still in a very poor condition. This region also faces the new challenges associated with the transition to the market economy. However, Central Europe is nevertheless one of the few 'success stories' of the post-socialist period. The reason for this is that in many cases environmental enhancement has been achieved in the course of and due to the overall socio-economic transformation and restructuring.

Of course, a lot more investment and effort are needed to sustain these achievements and for substantial and larger-scale rehabilitation of the deteriorated environment, but it is clear that the situation is getting under control and the direction of action has been chosen correctly. The prospect of joining the EU will probably remain the major stimulus for countries of Central Europe to harmonize their environmental legislation with EU requirements. However, reliance on Western European and international assistance cannot form the main environmental strategy for the new democracies.

Back to the future?

At the turning point between the 'era of the ecological barbarianism' (*Zelyoniy Mir* 23, 1999) when 'human dominance over nature was regarded as the strength of socialism' (Juhasz *et al*. 1993, 227) and the era of the democratic societies, it was noted that 'As Central and East European societies move into a new era of their histories, they drag along with them the environmental costs of their recent past' (Alcamo 1992, 1). A decade later, at the start of the new millennium, we can only hope that such attitude is not automatically transferred into the future.

It is apparent that economic difficulties will remain the main obstacle for the progress in ecological sphere for years to come. However, the environmental problems, in their turn, could make the implementation of economic reforms even more difficult. Therefore, the time has clearly come for the 'East' and 'West' to stop thinking in terms of 'environmental guilt' but to proceed to effective co-operation measures in protecting our 'common' environment. This can be successful only on the basis of the 'equal partnership' concept. A new type of thinking is also needed for the integration of environmental enhancement into the national programmes of sustainable socio-economic development.

Ideally, the situation requires a long-term sustainable development strategy, designed both to ameliorate past damage and prevent further deterioration. Arguably, this strategy would incorporate a set of environmental objectives related to (a) urgent action to minimize further adverse health effects in the ecological disaster areas; (b) improvement of the environmental situation in the region as a whole; (c) provision of nation-wide monitoring networks; (d) protection of the pristine nature areas; and (e) development of effective economic mechanisms to avoid further damage to the environment.

Unfortunately the initial collapse of the economies and further economic problems suffered by all the countries of the region, combined with the low priority given to environmental issues during this period, became serious impediments to the evolution of such an integrated strategy, making it difficult to achieve significant progress towards many of the above goals. In many countries of post-socialist Eurasia urgent measures are still needed to avert the current adverse tendencies in the state of the environment. An integral approach should be used to the complex of ecological, economic and social problems. At the same time extensive further research should be focused on the vulnerability of different types of natural ecosystems and landscapes and their critical loads. In the future, anthropogenic impacts should be planned taking into account not only the available natural resource base but also the sensitivity of local ecosystems and their ability to withstand human influence and thus to rehabilitate.

It should finally be emphasized that the notion of a 'crisis' is dynamic and not totally pessimistic. According to Lyuri (1997), the ecological crisis implies not only decline and degradation but represents a stage in the development of the use of resources at which humans strive to achieve material wealth as quickly as possible. He stated that this apparently risky phase, which is associated with depletion of natural resources and environmental transformation, can result in a defeat and a catastrophe. 'But it can also end up in a successful achievement of the set goal and restoration of the ecological balance on a higher resource-technological level' (Lyuri 1997, 169). The latter obviously will depend on a radical change in the human attitude to nature and the environment.

In view of the above, it should be emphasized that many human beliefs and concepts mentioned above are still strongly engraved in people's minds and memories. They are responsible for the overall passive attitude to environmental values at grass-roots level that has been prevalent in the past. It appears that in the future, the task that will be most difficult and take the longest time will be to change this attitude in order to achieve any real progress.

Thus, the turning point in environmental recovery and enhancement will hopefully occur in this new millennium when humans begin to perceive themselves as an inherent part of the biosphere, their home. They will then stop looking indifferently at the drying seas and at rivers overflowing with oil, in which only aluminium cans can multiply successfully.

References

Agranat, G.A. (1992) Vozmozhnosti i real'nosti osvoyeniya Severa: global 'nyie uroki [Possibilities and realities of the Northern development]. *Itogi Nauki i Tekhniki. Teoretichesliye i Obshchiye Voprosy Geografii*, 10

Agranat, G.A. (1998) The Russian North at a dangerous crossroads. *Polar Geography*, **22**, 4, 268–282

Alcamo, J. (ed.) (1992) *Coping with Crisis in Eastern Europe's Environment.* Parthenon Publishing Group, Carnforth

Aleksandrova, V.D. (1980) *The Arctic and Antarctic: Their Division into Geobotanical Areas.* Cambridge University Press, Cambridge

Allworth, E. (ed.) (1994) *Central Asia: 130 Years of Russian Dominance: a Historical Overview.* Duke University Press, Durham

Altshuler, I.I., Golubchikov, Y.N. and **Mnatsakanian, R.A.** (1992) Glasnost, perestroika and eco-sovietology, in Stewart, J.M. (ed.) *The Soviet Environment: Problems, Policies and Politics.* Cambridge University Press, Cambridge, 197–212

Aplin, G., Mitchell, P., Cleigh, H., Pitman, A. and **Rich, D.** (1996) *Global Environmental Crises: an Australian Perspective.* Oxford University Press, Oxford

Armand, D.L. (1975) *Nauka o Landshafte* [Scientific basis of the landscape studies]. 'Mysl', Moscow

Atlas of Chernobyl Exclusion Zone (1996). 'Kartografiya', Kiev

Atlas. Okruzhayushchaya Sreda i Zdorov'ye Naseleniya Rossii [Environmental and health atlas of Russia] (1995), Feshbach, M. (ed.). PAIMS, Moscow

Atlas Radioaktivnogo Zagriazneniya Evropeiskoi Chasti Rossii, Belorussii i Ukrainy [Atlas of radioactive contamination of the European part of Russia, Belarus and the Ukraine] (1998). Federal'naya Sluzhba Geodezii i Kartografii Rossii, Moscow

Babaev, A.G. (1997) Ecological, social and economic problems of the Turkmen Caspian zone, in Glantz, M.H. and Zonn, I.S. (eds) (1997) *Scientific, Environmental, and Political Issues in the Circum-Caspian Region.* Kluwer Academic Publishers, Dordrecht, 97–103

Baikal. Atlas (1993), Galaziy, G.I. (ed.). Sluzhba Geodezii i Kartografii Rossii, Moscow

Baikal'skiy Region. Byulleten' [Baikal Region Bulletin] (1998), 1

Bananova, V.A. (1993) *Antropogennoye opustynivaniye aridnyh territoriy Kalmykii* [Anthropogenic desertification of arid territories in Kalmykia]. Synopsis of thesis for the Doctor's Degree of Geographical Sciences, Ashgabat

Bazilevich, N.I., Grebenshchikov, O.S. and **Tishkov, A.A.** (1986) *Geograficheskiye Zakonomernosti Struktury i Funktsionirovaniya Ekosistem* [Geographical regularities of the structure and functioning of ecosystems]. Moscow

Belopukhova, E.B., Proshkin, V.I. and **Feldman, G.M.** (1976) Vliyaniye rastitel'nyh nadpochvennyh pokrovov na temperaturniy rezhim gruntov na primere Zapadnoi Sibiri [Impact of vegetation cover on the temperature regime of the ground by the example of West Siberia], in *Inzhenerno-Geologicheskiye i Geokriologicheskiye Issledovaniya v Zapadnoi Sibiri* [Engineering, geological and geocryological studies in West Siberia]. Akademiya Nauk SSSR, Moscow, 48–63

Bingham, S. (1997) Report and Recommendations for an Application by the Kalmyk Republic to the Global Environmental Facility, based on a mission to Kalmykia 21 July to 1 August 1997 (unpublished)

Blais, J.M., Duff K.E., Laing T.E. and **Smol, J.P.** (1999) Regional contamination in lakes from the Noril'sk region in Siberia. *Water, Air and Soil Pollution,* **110,** 389–404

Bliznyuk, A.I. (1995) Rol' khozyaistvennogo osvoyeniya territorii v izmenenii chislennosti kalmytskoi populyatsii saigaka [The role of economic development in the change in Kalmyk saiga population, in Zonn, I.S. and Neronov, V.M. (eds) *Biota i Prirodnaya Sreda Kalmykii* [Biota and environment of Kalmykia]. TOO 'Korkis', Moscow-Elista, 222–244

Boehmer-Christiansen, S. (1998) Environment-friendly deindustrialisation: impacts of unification on East Germany, in Tickle, A. and Welsh, I. (eds) *Environment and Society in Eastern Europe,* Addison Wesley Longman, New York, 67–96

Bradshaw, M.J. (1995) The Russian North in transition: general introduction. *Post-Soviet Geography,* **36,** 4, 195–203

Bradshaw, M.J. (ed.) (1997) *Geography and Transition in the Post-Soviet Republics.* John Wiley & Sons, Chichester

Bridges, O. and **Bridges J.** (1996) *Losing Hope: the Environment and Health in Russia.* Avebury, Aldershot

Brokgauz, F.A. and **Efron, I.A.** (eds) (1995) *Rossiya.* Lenizdat, St Petersburg

Brueggemann, E. and **Rolle, W.** (1998) Changes of some components of precipitation in East Germany after the unification. *Water, Air, and Soil Pollution,* **107,** 1–23

Brueggemann, E. and **Spindler, G.** (1999) Wet and dry deposition of sulphur at the site Melpitz in East Germany. *Water, Air, and Soil Pollution*, **109**, 81–99

Budyko, M.I. (1977) *Global'naya Ekologiya* [Global ecology]. 'Mysl', Moscow

Burrough, P.A., Perk Van Der, M., Howard, B.G., Prister, B.S., Sansone, U. and **Voitsekhovitch, O.V.** (1999) Environmental mobility of radeocaesium in the Pripyat catchment, Ukraine/Belarus. *Water, Air, and Soil Pollution*, **110**, 35–55

Carter, F.W. and **Turnock, D.** (eds) (1993, 1996) *Environmental Problems in Eastern Europe*. Routledge, London

Centre for International Projects (1995) *National Report of the Russian Federation on the Consequences of Climatic Change in the Caspian Region*. Draft. Centre for International projects. Moscow (unpublished)

Chaturvedi, S. (1996) *The Polar Regions. A Political Geography*. John Wiley & Sons, Chichester

Chernobyl: Sobytiya i Uroki [Chernobyl: events and lessons] (1989). Izdatel'stvo Politicheskoi Literatury, Moscow

Chernov, Y.I. (1985) *The Living Tundra*. Cambridge University Press, Cambridge

Chislennost' i Migratsiya Naseleniya Rossiiskoi Federatsii v 1998 Godu [Population numbers and migration statistics of the Russia Federation in 1998] (1999). (Statistical bulletin), Goskomstat Rossii, Moscow

Chislennost' Naseleniya Rossiyskoi Federatsii na 1 Yanvarya 1998 Goda [Population numbers statistics of the Russian Federation by 1 January 1998] (1998). Goskomstat Rossii, Moscow

Clark, A.N. (1998) *The Penguin Dictionary of Geography* 2nd edn. Penguin Books, Longman, London

Cole, J.P. (1984) *Geography of the Soviet Union*. Butterworths, London

Colls, J. (1997) *Air Pollution: an Introduction*. E & FN Spon, London

Cullen, R. (1999) The Caspian Sea. *National Geographic*, **195**, 5, 2–35

Danilov-Danilyan, V.I., Gorshkov, V.G., Arskiy, Yu.M. and **Losev, K.S.** (1994) *Okruzhayushchaya Sreda mezhdu Proshlym i Budushchim: Mir i Rossiya* [Environment between the past and future: the world and Russia]. VINITI, Moscow

Danilov-Danilyan, V.I. and **Kotlyakov, V.M.** (eds) (1993) *Rossiya v Ekologicheskom Krizise* [Russia in environmental crisis]. Federal'nyi Ekologicheskiy Fond, Moscow

DeBardeleben, J. and **Hannigan, J.** (eds) (1994) *Environmental Security and Quality after Communism in Eastern Europe and the Soviet Successor States*. Westview Press, Boulder

Doklad o Katastroficheskom Ekologicheskom Polozhenii v Volzhskom Regione [Report on the catastrophic ecological situation in the Volga region] (1989). Minvodkhoz SSSR, Moscow (unpublished)

Downing, D. (1995) Health consequences of the Pre-Aral disaster. *Journal of Nitrition and Environmental Medicine*, **5**, 391–399

Dregne, H., Kassas, M. and **Rozanov, B.** (1991) A new assessment of the world status of desertification. *Desertification Control Bulletin,* **20,** 6–18

Eastern Europe and the Commonwealth of Independent States (1994). Regional Surveys of the World, Europa Publications Limited, London

Economika Sodruzhestva Nezavisimykh Gosudarstv [Economy of the Commonwealth of Independent States] (1993). Finstatinform, Moscow

EIU Country Profile 1996–97 (1996a) *Czech Republic.* The Economist Intelligence Unit Ltd, London

EIU Country Profile, 1996–97 (1996b) *Kazakstan.* The Economist Intelligence Unit Ltd, London

EIU Country Profile, 1996–97 (1996c) *Kyrgyz Republic, Tajikistan, Turkmenistan and Uzbekistan.* The Economist Intelligence Unit Ltd, London

EIU Country Report, 1st quarter 1996 (1996d) *Albania, Belarus, Bulgaria, Czech Republic, Estonia, Hungary, Latvia, Lithuania, Moldova, Poland, Romania, Slovakia, Slovenia, Ukraine.* The Economist Intelligence Unit Ltd, London

EIU Country Report, 1st quarter 1996 (1996e) *Azerbaijan, Kazakhstan and Turkmenistan.* The Economist Intelligence Unit Ltd, London

EIU Country Report, 2nd quarter 1997 (1997) *Kazakstan, Kyrgyz Republic, Tajikistan, Turkmenistan, Uzbekistan.* The Economist Intelligence Unit Ltd, London

Ekologicheskaya Karta Rossii. Sostoyaniye Okruzhayushchei Prirodnoi Sredy [The ecological map of Russia: the state of the natural environment] (1999). Scale 1:8,000,000. PKO 'Kartografiya', Moscow

The Europa World Yearbook (1999) 40th edition, Volumes 1, 2. Europa Publishing Ltd, London

The Europa World Yearbook (1995) 36th edition, Volumes 1, 2. Europa Publishing Ltd, London

Europe's Environment: the Dobris Assessment (1995a). European Environment Agency, Copenhagen

Europe's Environment: Statistical Compendium for the Dobris Assessment (1995b). Statistical Office of the European Communities, Luxembourg

Europe's Environment: the Second Assessment (1998). European Environment Agency, OOPEC, Luxembourg

Feshbach, M. and **Friendly, A.** (1992) *Ecocide in the USSR.* BasicBooks, HarperCollins Publishers Inc., New York

Finlayson, C.M., Chuikov, Y.S., Prentice, R.C and **Fischer, W.** (eds) (1993) *Biogeography of the Lower Volga, Russia: an overview.* IWRB Spec. Publ. 28, Slimbridge

Fisher, D. (1992) *Paradise Deferred: Environmental Policymaking in Central and Eastern Europe.* Royal Institute of International Affairs, London

Fiziko-Geograficheskoye Raionirovaniye SSSR [Physical-geographical zonation of the USSR] (1967) Map of scale 1: 10,000,000. GUGK, Moscow

Fodor, I. (1994) Characteristics of environmental problems in Eastern-Central Europe, in Fodor, I. and Walker, G.P. (eds) *Environmental Policy and Practice in Eastern and Western Europe*. Centre for Regional Studies, Pecs, 33–41

Fondahl, G.A. (1995) The status of indigenous peoples in the Russian North. *Post-Soviet Geography*, **36**, 4, 215–224

Fullenbach, J. (1981) *European Environmental Policy: East and West*. Butterworths, London

Furman, A.E. and **Livanova, G.S.** (1978) *Krugovoroty i Progress v Razvitii Material'nyh System* [Cycles and progress in the evolution of material systems]. Izdatel'stvo Moskovskogo Universiteta, Moscow

Glantz, M.H. (1994) Creeping environmental phenomena in the Aral Sea basin. *NATO Advanced Research Workshop on Critical Scientific Issues of the Aral Sea Basin: State of Knowledge and Future Research Needs*, 2–4 May 1994, Tashkent

Glantz, M.H. (ed.) (1999) *Creeping Environmental Phenomena in the Aral Sea Basin*. Cambridge University Press, Cambridge

Glantz, M.H., Rubinstein, A.Z. and **Zonn, I.S.** (1993). Tragedy in the Aral Sea basin. Looking back to plan ahead? *Global Environmental Change*, **6**, June, 174–198

Glantz, M.H. and **Zonn, I.S.** (eds) (1997) *Scientific, Environmental, and Political Issues in the Circum-Caspian Region*. Kluwer Academic Publishers, Dordrecht

Glazovskiy, N.F. (1990) *Aralskiy Krizis. Prichiny Vozniknoveniya i Puti Vykhoda*. [The Aral crisis. Causes of emergence and ways out]. Nauka, Moscow

Glazovskiy, N.F., Koronkevich, N.I., Kochurov, B.I., Krenke, A.N. and **Sdasyuk, G.V.** (1991) Kriticheskiye ekologicheskiye raiony: geograficheskiye podkhody i printsipy izucheniya [Critical ecological regions: geographical approaches and principles of study]. *Izvestiya Geograficheskogo Obshchestva*, **123**, 1, 9–17

Golubchikov, Yu.N. (1992) Kryogennoye pochvoobrazovaniye [Cryogenic soil formation], in Solomatin, V.I. (ed.) *Geoekologiya Severa* [Geoecology of the North]. Izdatel'stvo Moskovskogo Universiteta, Moscow, 21–25

Golubchikov, Yu.N. (1996) *Geografiya Gornyh i Polyarnih Stran* [Geography of the mountainous and polar domains]. Izdatelstvo Moskovskogo Universiteta, Moscow

Golubev, G.N. (1997) Closed areas and the case of the Caspian Sea basin, in Glantz, M.H. and Zonn, I.S. (eds) *Scientific, Environmental, and Political Issues in the Circum-Caspian Region*. Kluwer Academic Publishers, Dordrecht, 11–16

Golytsyn, G.S. (1997) Overview: history and causes of Caspian Sea level change, in Glantz, M.H. and Zonn, I.S. (eds) *Scientific, Environmental, and Political Issues in the Circum-Caspian Region*. Kluwer Academic Publishers, Dordrecht, 17–25

Gorshkov, S.P. (1998) *Kontseptual'nyie Osnovy Geoekologii* [Conceptual bases of geoecology]. Izdatel'stvo Smolenskogo Gumanitarnogo Universiteta, Smolensk

Gosudarstvenniy Doklad 'O Sostoyanii Okruzhayushchei Prirodnoi Sredy Rossiyskoi Federatsii v 1991 Godu [State Report 'On the state of the environment in the Russian Federation in 1991'] (1992). Ministerstvo Ekologii i Prirodnyh Resursov Rossiyskoi Federatsii, Moscow

Gosudarstvenniy Doklad 'O Sostoyanii Okruzhayushchei Prirodnoi Sredy Rossiyskoi Federatsii v 1992 Godu [State Report 'On the state of the environment in the Russian Federation in 1992'] (1993). *Evraziya*, **9**, 17, 2–106

Gosudarstvenniy Doklad 'O Sostoyanii Okruzhayushchei Prirodnoi Sredy Rossiyskoi Federatsii v 1993 Godu [State Report 'On the state of the environment in the Russian Federation in 1993'] (1994). Ministerstvo Okhrany Okruzhayushchei Sredy i Prirodnyh Resursov Rossiyskoi Federatsii, and Centre for International Projects, Moscow

Gosudarstvenniy Doklad 'O Sostoyanii Okruzhayushchei Prirodnoi Sredy Rossiyskoi Federatsii v 1994 Godu [State Report 'On the state of the environment in the Russian Federation in 1994'] (1995). Ministerstvo Ekologii i Prirodnykh Resursov RF, Centre for International Projects, Moscow

Gosudarstvenniy Doklad 'O Sostoyanii Okruzhayushchei Prirodnoi Sredy Rossiyskoi Federatsii v 1995 Godu [State Report 'On the state of the environment in the Russian Federation in 1995'] (1996). Ministerstvo Okhrany Okruzhayushchei Sredy i Prirodnykh Resursov RF, Moscow

Gosudarstvenniy Doklad 'O Sostoyanii Okruzhayushchei Prirodnoi Sredy Rossiiskoi Federatsii v 1996 Godu [State Report 'On the state of the environment in the Russian Federation in 1996'] (1997). Gosudarstvenniy Komitet Rossiiskoi Federatsii po Okhrane Okruzhayushchei Sredy, Moscow

Gosudarstvenniy Doklad 'O Sostoyanii Okruzhayushchei Prirodnoi Sredy Rossiiskoi Federatsii v 1997 Godu [State Report 'On the state of the environment in the Russian Federation in 1997'] (1998a). *Zelyoniy Mir*, **25** (289), 1–31

Gosudarstvenniy Doklad 'O Sostoyanii Okruzhayushchei Prirodnoi Sredy Rossiiskoi Federatsii v 1997 Godu [as above] (1998b). *Zelyoniy Mir*, **26** (290), 1–31

Gosudarstvenniy Doklad 'O Sostoyanii Okruzhayushchei Prirodnoi Sredy Rossiiskoi Federatsii v 1997 Godu [as above] (1998c). *Zelyoniy Mir*, **27** (291), 1–31

Goudie, A. (ed.) (1997) *The Encyclopedic Dictionary of Physical Geography* 2nd edn. Blackwell Publishers Ltd, Oxford

Grachev, M.A., Kumarev, V.P., Mamaev, L.V., Zorin, V.L., Baranova, L.V., Denikina, N.N., Belikov, S.I. and **Petrov, E.A.** (1989) Distemper virus in Baikal seals. *Nature*, **338**, 209–210

Grigoriev, A.A. and **Budyko, M.I.** (1956) O periodicheskom zakone geograficheskoi zonal'nosti [On the periodical law of the geographical zonality]. *DAN SSSR*, 110, 1

Gukalova, I. (1998) Socio-economic consequences of the Chernobyl catastrophe and regional development in Ukraine, in Graute, U. (ed.) *Sustainable Development for Central and Eastern Europe: Spatial Development in the European Context*. Springer-Verlag, Berlin-Heidelberg

Gvozdetskiy, N.A. and **Mikhailov, N.I.** (1987) *Fizicheskaya Geografiya SSSR. Aziatskaya Chast.* [Physical geography of the USSR. The Asiatic part]. Vysshaya Shkola, Moscow

Halkos, G.E. (1996) Implementing optimal sulphur abatement strategies in Europe. *Water, Air and Soil Pollution*, **87**, 329–344

Hannan, T. and **O'Hara, S.** (1998) Managing Turkmenistan's Kara Kum Canal: problems and prospects. *Post-Soviet Geography and Economics*, **39**, 4, 225–235

Haynes, V. and **Bojcun, M.** (1988) *The Chernobyl Disaster*. Hogarth, London

Herzman, C. (1995) *Environment and Health in Central and Eastern Europe*. The World Bank, Washington, D.C.

Ilamanov, A. (1993) Garagumsky kanal: proshloye i nastoyashcheye [The Karakum canal: past and present]. *Turkmenskaya Iskra*, 16 September 1993

Ishida, N., Tsuimura, S., Kubota, H. and **Izumi, K.** (1995) *Environmental Problems in the Area of Syr-Darya and Aral Sea*. Paper presented at the Symposium on Aral Sea and the Surrounding region, convened at Lake Biwa Research Institute, Otsu, Japan, 29 March 1995

IUCN (1990) *Environmental Status Reports: 1988–1989, Volume One: Czechoslovakia, Hungary, Poland*. International Union for Conservation of Nature and Natural Resources, Geneva

Ivanov, Y.A., Lewyckyj, N., Levchuk, S.E., Prister, B.S., Firsakova, S.K., Arkhipov, N.P., Arkhipov, A.N., Kruglov S.V., Alexakhin R.M., Sandalls J. and **Askbrant S.** (1997) Migration of 137 Cs and 90Sr from Chernobyl Fallout in Ukrainian, Belarussian and Russian Soils. *Journal of Environmental Radioactivity*, **35**, 1, 1–21

Jones, A. (1995) *The New Germany. A Human Geography*. John Wiley and Sons, Chichester

Juhasz, J., Vari, A. and **Tolgyyesi, J.** (1993) Environmental conflict and political change: public perception on low-level radioactive waste management in Hungary, in Vari, A. and Tamas, P. (eds) *Environment and Democratic Transition: Policy and Politics in Eastern Europe*. Kluwer, Dordrecht, 227–248

Kamyshev, A.P. (1999) *Metody i Tekhnologii Monitoringa Prirodno-Tekhnicheskih Sistem Severa Zapadnoi Sibiri* [Methods and technologies of monitoring natural-technical systems of the North of West Siberia]. BNIPIGAZODOBYCHA, Moscow

Karimsakov, O. (1994) *The voice from the Aral*. RO 'Karakalpakvodkhoz' (unpublished)

Kasimov, N.S. (ed.) (1997) *Geoekologicheskiye izmeneniya pri kolebaniyakh urovnya Kaspiyskogo morya* [Geoecological changes from fluctuations of the Caspian sea's level]. Geograficheskiy Fakul'tet MGU, Moscow

Katushov, K. (1998) *Kalmykia v Geoprostranstve Rossii*. APP 'Jangar', Elista

Kazakstan. Economic Trends. Quarterly Issue, April-June 1999 (1999). Tacis, European Commission, Brussels

Kennedy, P. (1994) *Preparing for the Twenty First Century*. Fontana Press, London

Khesina, A.Ya., Kolyadich, M.N., Krivosheyeva, L.V., Sokolskaya, N.N., Shcherbak, N.P. and **Levinsky, S.S.** (1996) Assessment of pollution of the Moscow city air with carcinogenic PAH and N-nitrosamines. *Experimental Oncology*, **18**, 14–18

Kiessling, K.L. (1998) Conference on the Aral Sea – women, children and environment. *Ambio*, **27**, 7, 560–564

Klarer, J. and **Francis, P.** (1997) Regional overview, in Klarer, J. and Moldan, B. (eds) (1997) *The Environmental Challenge for Central European Economies in Transition*. John Wiley and Sons, Chichester, 1–66

Klarer, J. and **Moldan, B.** (eds) (1997) *The Environmental Challenge for Central European Economies in Transition*. John Wiley and Sons, Chichester

Klarer, J., Sitnicki S. and **Zlinszky J.** (1994) Strategic environmental issues in Central and Eastern Europe, in Fodor, I. and Walker, G.P. (eds) *Environmental Policy and Practice in Eastern and Western Europe*. Centre for Regional Studies, Pecs, 73–97

Kobori, I. and **Glantz, M.H.** (ed.) (1998) *Central Asian Water Crisis: Caspian, Aral and Dead Seas*. United Nations University Press, Tokyo

Kotlyakov, V.M. and **Agranat, G.A.** (1995) Zapoliarye: Priroda, Ekonomika, Prioritety [Polar regions: nature, economy, priorities]. *Vestnik RAN, seriya Geographiya*, **65**, 3, 214–220

Kovalevskaya, L. (1995) [Chernobyl classified: consequences of Chernobyl]. Abris, Kiev

Krasovskaya, T.M. (ed.) (1998) *Kol'skiy Poluostrov* [The Kola peninsula]. Smolenskiy Gumanitarniy Universitet, Moscow-Smolensk

Krasovskaya, T.M. and **Yevseev, A.V.** (1990) *Ratsional'noye Prirodopol'zovaniye na Kol'skom Poluostrove* [Rational nature utilization on the Kola peninsula]. Izdatelstvo Moskovskogo Universiteta, Moscow

Kremenetsky, C., Vaschalova, T., Goriachkin, S., Cherbinsky, A. and **Sulerzhitsliy, L.** (1997) Holocene pollen stratigraphy and bog development in the western part of the Kola peninsula, Russia. *BOREAS*, **26**, 91–102

Kryuchkov, V.V. (1987) *Sever na Grani Tysyacheletiy* [The North on the edge of milennia]. 'Mysl', Moscow

Kudelsky, A.V., Smith, J.T., Ovsiannikova, S.V. and **Hilton, J.** (1996) Mobility of Chernobyl-derived Cs-137 in a peatbog system within the catchment of the Pripyat River, Belarus. *The Science of the Total Environment*, **188**, 101–113

Kuksa, V.I. (1994) *Yuzhnyie Morya (Aral'skoye, Kaspiiskoye, Azovskoye im Chernoye) pod Vozdeistviyem Antropogennogo Stressa* [Southern seas (Aral, Caspian, Azov and Black) under the impact of anthropogenic stress]. Gidrometeoizdat, St Petersburg

Kuznetsov, N. (1993) Luchshe byt pessimistom chem optimistom. *Zelyoniy Mir* **12**, 14

Lawrence, E., Jackson, A.R.W. and **Jackson, J.M.** (1998) *Dictionary of Environmental Science*. Addison Wesley Longman Ltd, Harlow

Lebedev, A.T., Poliakova, O.O., Karakhanova, N.K., Petrosyan, V.S. and **Renzoni, A.** (1998) The contamination of birds with organic pollutants in the Lake Baikal region. *The Science of the Total Environment*, **212**, 153–162

Lerman, Z., Garcia-Garcia, G. and **Wichelns, D.** (1996) Land and water policies in Uzbekistan, *Post-Soviet Geography and Economics*, **37**, 3, 145–174

Letolle, R. and **Mainguet, M.** (1993) Aral. Springer-Verlag, Berlin-Paris

Levy, M.A. (1994) East-West environmental politics after 1989: the case of air pollution, in Keohane, R.O., Nye, J.S. and Hoffman, S. (eds) *After the Cold War. International Institutions and State Strategies in Europe, 1989–1991*. Harvard University Press, Harvard, 310–339

Likhacheva, E.A. and **Smirnova, E.B.** (1994) Ekologicheskiye problemy Moskvy za 150 let [Ecological problems of Moscow for the last 150 years]. RFFI/ Institut Geografii RAN, Moscow

Lofstedt, R.E. and **Sjostedt, J.** (1995) Environmental aid to Eastern Europe: problems and possible solutions. *Ambio*, **24**, 6, 366–370

Lubin, N. (1994) Central Asia: issues and challenges for United States policy, in Banuazizi, A. and Weiner, M. (eds) *The New Geopolitics of Central Asia and its Borderlands*. I.B.Taurus, London, 259–272

Luzin, G.P., Pretes, M. and **V. Vasiliev** (1994) The Kola Peninsula: geography, history and resources. *Arctic*, **47**, 1, 1–15

L'vov, G.N. (ed.) (1998) *Ekologiya Moskvy: Resheniya, Problemy, Perspektivy* [Ecology of Moscow: decisions, problems, prospects]. Meriya. Pravitel'stvo Moskvy, Moscow

Lychagina, N.Yu., Kasimov, N.S. and **Lychagin, M.Yu.** (1998) *Biogeokhimiya Makrofitov Delty Volgi* [Biogeochemistry of macrophytes in the Volga's delta]. Geograficheskiy Fakul'tet MGU, Moscow

Lyuri, D.I. (1997) *Razvitiye Resursopol'zovaniya i Ekologicheskiye Krizisy ili Zachem nam Nuzhny Ekologicheskiye Krizisy* [Evolution of resource use and ecological crises, or why do we need ecological crises]. Institut Geografii RAN, Izdatel'stvo 'Del'ta', Moscow

Makarova, O.A., Andreev, G.N., Pokhilko, A.A., Filippova, L.N. and

Shklyarevich, F.N. (1997) *Rastitel'niy i Zhivotniy Mir Murmanskoi Oblasti* [Vegetation and wildlife of the Murmansk province]. MOIP-KRO, Murmansk

Makarova, T.D. (1988) *Posledstviya Vybrosov Vrednyh Veshchestv v Atmospheru dlya Okruzhaiyushchei Prirodnoi Sredy Krainego Severa* [Consequences of pollutant emissions into the atmosphere for the environment of the Extreme North]. Avtoreferat diss. na soisk. uch. st. kand. geogr. nauk, Moscow

Mankovska, B. (1997) Variations in sulphur and nitrogen foliar concentration of deciduous and conifers vegetation in Slovakia. *Water, Air, and Soil Pollution*, **96**, 329–345

Manser, R. (1993) *Squandered Dividend*. Earthscan Publications Ltd, London

Markert, B., Herpin, U., Berlekamp, J., Oehlmann, J., Grodzinska, K., Mankovska, B., Suchara, I., Siewers, U., Weckert, V. and **Lieth, H.** (1996) A comparison of heavy metal deposition in selected Eastern European countries using the moss monitoring method, with special emphasis on the 'Black Triangle'. *The Science of the Total Environment*, **193**, 85–100

Marples, D.B. (1995) Belarus' ten years after Chernobyl. *Post-Soviet Geography*, **36**, 6, 323–350

Marples, D.R. (1996) *Belarus. From Soviet Rule to Nuclear Catastrophe*. Macmillan Press Ltd, Houndmills, Basingstoke, Hampshire

Massal'skiy, V.I. (1892) *Khlopkovoye delo v Srednei Azii* [Cotton production in Central Asia]. V. Kirshbaum, St Peterburg

Matthews, J.A. and **Saiko T.A.** (1994) Environmental policies and public participation during transition periods: a comparison between Britain and Russia, in Fodor, I. and Walker, G.P. (eds) *Environmental Policy and Practice in Eastern and Western Europe*. Centre for Regional Studies, Pecs, 223–234

Matzner, E. and **Murach, D.** (1995) Soil changes induced by air pollution deposition and their implications for forests in Central Europe. *Water, Air and Soil Pollution*, **85**, 63–76

McCauley, M. (1976) *Khrushchev and the Development of Soviet Agriculture. The Virgin Land Programme 1953–1964*. Holmes & Meier Publishers, NY

Medvedev, Zh.A. (1990) *The Legacy of Chernobyl*. Basil Blackwell, Padstow

Medvedev, Zh.A. (1992) The global impact of the Chernobyl accident five years after, in Stewart, J.M. (ed.) *The Soviet Environment: Problems, Policies and Politics*. Cambridge University Press, Cambridge, 174–196

Medvedev, Zh.A. (1994) The Chernobyl Legacy. *The Moscow Times*, 6 May

Mekhtiev, A.Sh. and **Gul, A.K.** (1997) Ecological problems of the Caspian Sea and perspectives on possible solutions, in Glantz, M.H. and Zonn, I.S. (eds) *Scientific, Environmental, and Political Issues in the Circum-Caspian Region*. Kluwer Academic Publishers, Dordrecht, 79–103

Mellor, R.E.H. (1975) *Eastern Europe: a Geography of the Comecon Countries.* Macmillan, London

Micklin, P.P. (1992) Water management in Soviet Central Asia: problems and prospects, in Stewart, J.M. (ed.) *The Soviet Environment: Problems, Policies and Politics.* Cambridge University Press, Cambridge, 88–114

Micklin, P. (1998) International and regional responses to the Aral Crisis: an overview of efforts and accomplishments. *Post-Soviet Geography and Economics,* **39**, 7, 399–416

Micklin, P.P. and **Williams, W.D.** (eds) (1996) *The Aral Sea Basin,* NATO ASI Series, Partnership sub-series 2: Environment, Vol. 12. Springer-Verlag, New York

Milanova, E.V. and **Kushlin, A.V.** (eds) (1991) *Report on Methodology of Compiling the Maps of the Present Status of Landscapes.* Centre for International Projects, Moscow

Milanova, E.V. and **Ryabchikov, A.M.** (1987) *Ispolzovaniye Prirodnyh Resursov I Okhrana Prirody* [The use of natural resources and the environmental protection]. Vysshaya Shkola Publishers, Moscow

Mil'kov, F.N. and **Gvozdetskiy, N.A.** (1986) *Fizicheskaya Geografiya SSSR. Obshchiy Obzor. Evropeiskaya Chast SSSR. Kavkaz* [Physical geography of the USSR. Overview. European part of the USSR. Caucasus]. Vysshaya Shkola, Moscow

Mnatsakanian, R.A. (1992) *Environmental Legacy of the Former Soviet Republics.* Centre for Human Ecology, University of Edinburgh

Moldan, B. (1997) Czech Republic, in Klarer, J. and Moldan, B. (eds) *The Environmental Challenge for Central European Economies in Transition.* John Wiley and Sons, Chichester, 107–129

Moscow City Ecological Profile (1999) UNEP/HABITAT, Moscow

Moskvovedeniye: Ekologiya Moskovskogo Regiona [Moscow studies: ecology of the Moscow region] (1996a). Ekopros, Moscow

Moskvovedeniye: Geografiya Moskvy i Moskovskoi oblasti [Moscow studies: geography of Moscow and the Moscow region] (1996b). Ekopros, Moscow

Moskovskiy Gorodskoi Ekologicheskiy Profil' [Moscow urban ecological profile] (1998). Komitet po telekommunikatsiyam i sredstvam massovoi informatsii Pravitel'stva Moskvy, Moscow

Mote, V. (1992) BAM after the fanfare: the unbearable ecumene, in Stewart, J.M. (ed.) *The Soviet Environment: Problems, Policies and Politics.* Cambridge University Press, Cambridge, 40–56

Muir, R. (1998) Landscape: a wasted legacy. *Area,* **30**, 3, 263–271

Murat, A.B. (1975) Turkmenistan and the Turkmen, in Katz, Z. (ed.) *Handbook of Major Soviet Nationalities.* Macmillan, New York, 262–282

Murmansk Region. Environmental Pollution Impact, 1990–1995. (1996) Map 1: 1,000,000. KSC RAS INEP, MSU Department of Geography, Moscow

Murmanskaya Oblast v Tsifrah [The Murmansk region in figures] (1996). Murmanskiy Oblastnoi Komitet Gosudarstvennoi Statistiki, Murmansk

Murmanskaya Oblast v Tsifrah [The Murmansk region in figures] (1997). Murmanskiy Oblastnoi Komitet Gosudarstvennoi Statistiki, Murmansk

Murzayev, E. (1991) Obzor issledovaniy Aral'skogo morya i Priaral'ya [A brief review of the study of the Aral Sea and its region]. *Izvestiya AN SSSR, Seriya Geografiya*, **4**, 22–35

Myach, L.T. (1996) Himicheskoye zagryazneniye [Chemical pollution], in Yablokov, A.V. (ed.) *Rossiyskaya Arktika: na Poroge Katastrofy.* Tsentr Ekologicheskoi Politiki Rossii, Moscow, 18–24

Myalo, E.G. and **Volodina, I.A.** (1995) Pochvenniye banki semyan i ikh rol' v funktsionirovanii ekosistem Chornyh Zemel' [Soil seed banks and their role in ecosystem functioning in the Black Lands], in Zonn, I.S. and Neronov, V.M. (eds) *Biota i Prirodnaya Sreda Kalmykii* [Biota and environment of Kalmykia]. TOO 'Korkis', Moscow-Elista, 93–105

Narodnoye Khozyaistvo SSSR, 1922–1982 [National economy of the USSR in 1922–1988] (1982). Goskomstat SSSR, Finansy i statistika, Moscow

Narodnoye Khozyaistvo SSSR v 1985 Godu [National economy of the USSR in 1985] (1986). Goskomstat SSSR, Finansy i statistika, Moscow

Narodnoye Khozyaistvo RSFSR za 70 let [The economy of RSFSR for 70 years. Statistical yearbook] (1987). Goskomstat SSSR, Moscow

Narodnoye Khozyaistvo SSSR v 1987 Godu [National economy of the USSR in 1987] (1988). Goskomstat SSSR, Finansy i Statistika, Moscow

Narodnoye Khozyaistvo SSSR v 1988 Godu [National economy of the USSR in 1988] (1989). Goskomstat SSSR, Finansy i statistika, Moscow

Narodnoye Khozyaistvo SSSR za 60 let [National economy of the USSR for 60 years] (1977). Goskomstat SSSR, Statistika, Moscow

Natsional'niy Doklad Rossiyskoi Federatsii 'Posledstviya Klimaticheskih Izmeneniy v Prikaspiyskom Regione' [National Report of the Russian Federation 'Consequences of the climatic change in the Caspian Sea Region'] (1995). Centre for International Projects, Moscow (unpublished)

Nefedova, T. (1994) Industrial development and the environment in Central and Eastern Europe. *European Urban and Regional Studies*, **1**, 2, 168–171

Nefedova, T. and **Treivish, A.** (1994) *Raiony Rossii mi Drugikh Evropeiskih Stran s Perekhodnoi Ekonomikoi v nachale 90h* [Regions of Russia and other European countries with economies in transition at the beginning of the 1990s]. Institut Geografii Rossiiskoi Akademii Nauk, Moscow

Novikova, N.M. (1999) Priaralye ecosystems and creeping environmental changes in the Aral Sea, in Glantz, M.H. (ed.) *Creeping Environmental Problems and Sustainable Development in the Aral Sea Basin.* Cambridge University Press, Cambridge, 100–127

Nowicki, M. (1997) Poland, in Klarer, J. and Moldan, B. (eds) *The Environmental Challenge for Central European Economies in Transition.* John Wiley and Sons, Chichester, 193–227

OECD (1991) *Energy Statistics and Balances of non-OECD Countries 1988–1989.* Paris

O'Hara, S.L. and **Hannan, T.** (1997) Agriculture and land reform in Turkmenistan since independence. *Post-Soviet Geography and Economics*, **38**, 7, 430–444

O'Hara, S.L. and **Hannan, T.** (1999) Irrigation and water management in Turkmenistan: past systems, present problems and future scenarios. *Europe-Asia Studies*, **51**, 1, 21–41

Okhrana Okruzhayushchei Sredy i Ratsional'noyeIspol'zovaniye Prirodnyh Resursov v Kazahstane (1996). Goskomstat Respubliki Kazakstan, Almaty

Okhrana Okruzhayushchei Sredy v Rossii. Statisticheskiy Sbornik. [Protection of the environment in Russia. Statistical collection] (1998). Goskomstat Rossii, Moscow

Okruzhayushchaya Sreda v Sodruzhestve Nezavisimykh Gosudarstv [Environment in the Commonwealth of Independent States. Statistical Reference Book] (1996). Statkomitet SNG, Moscow

Olcott, M.B. (1996) *Central Asia's New States. Independence, Foreign Policy, and Regional Security.* US Institute of Peace Press, Washington

Orlovskiy, N.S. (1999) Creeping environmental changes in the Karakum Canal's zone of impact, in Glantz, M.H. (ed.) *Creeping Environmental Problems and Sustainable Development in the Aral Sea Basin.* Cambridge University Press, Cambridge, 225–244

Osherenko, G. (1995) Indigenous political and property rights and economic/environmental reform in Northwest Siberia. *Post-Soviet Geography* **36**, 4, 225–237

Ozturk, M., Ozdemir, F. and **Yucel, E.** (1997) An overview of the environmental issues in the Black Sea Region. Paper presented at the NATO Advanced Research Workshop on *Scientific, Environmental and Political Issues in the Circum-Caspian Region*, 13–16 May 1996, Moscow

Peterson, D.J. (1993) *Troubled Lands. The Legacy of Soviet Environmental Destruction.* Westview Press, Oxford

Perel'man, A.I. (1975) *Geohimiya Landshafta* [Geochemistry of landscape]. 'Vysshaya Shkola', Moscow

Poelzer, G. (1995) Devolution, constitutional development, and the Russian North. *Post-Soviet Geography*, **36**, 4, 204–214

Popov, S.N. (1987) Tendentsii i perspektivy razvitiya promyshlennosti v raionakh Severa [Trends and perspectives of industrial development in the regions of the North], in *Problemy Sovershenstvovaniya Territorial'noi Struktury Ekonomiki RSFSR* [Problems of optimization of the territorial structure of the RFSSR economy]. VINITI, Moscow

Poyasnitel'naya Zapiska k Nauchno-spravochnoi Ekologo-Geograficheskoi Karte Rossiiskoi Federatsii Masshtaba 1: 4,000,000 [Explanatory note to scientific and information ecological geographical map of the Russian Federation of scale 1: 4M] (1996). Roskartografiya i Geograficheskiy Fakul'tet MGU, Moscow

Problemy Okhrany Ozera Baikal mi Prirodopol'zovaniya v Baikalskom Regione v 1993 Godu [Problems of Lake Baikal's protection and the use of natural resources in the Baikal region in 1993] (1994). Gidrometeoizdat, Moscow

Problemy Okhrany Ozera Baikal mi Prirodopol'zovaniya v Baikalskom Regione v 1994 Godu [Problems of Lake Baikal's protection and the use of natural resources in the Baikal region in 1994 (1995). Meteorologiya i Gidrologiya, Moscow

Problemy Okhrany Ozera Baikal mi Prirodopol'zovaniya v Baikalskom Regione v 1995 Godu [Problems of Lake Baikal's protection and the use of natural resources in the Baikal region in 1995] (1996). Meteorologiya i Gidrologiya, Moscow

Problemy Okhrany Ozera Baikal mi Prirodopol'zovaniya v Baikalskom Regione v 1996 Godu [Problems of Lake Baikal's protection and the use of natural resources in the Baikal region in 1996] (1997). Tsentr Mezhdunarodnyh Proyektov, Moscow

Prokhorov, B.B. (1996) Okruzhayushchaya sreda i zdoroviye cheloveka v Arktike [The environment and human health in the Arctic], in Yablokov, A.V. (ed.) *Rossiyskaya Arktika: na Poroge Katastrofy*. Tsentr Ekologicheskoi Politiki Rossii, Moscow, 118–123

Protasov, V.F. and **Molchanov, A.V.** (1995) *Ekologiya, Zdorov'ye i Prirodopol'zovaniye v Rossii* [Ecology, health and the use of natural resources in Russia]. Finansy i Statistika, Moscow

Pryde, P.R. (1997) The post-Soviet environment, in Bradshaw, M.J. (ed.) *Geography and Transition in the Post-Soviet Republics.* John Wiley & Sons, Chichester, 131–144

Pryde, P.R. (1991) *Environmental Management in the Soviet Union.* Cambridge University Press, Cambridge

Ptichnikov, A.V. (1991) Fiziko-geograficheskiye posledstviya antropogennogo opustynivaniya v Priaraliye [Physical-geographical implications of anthropogenic desertification in Priaraliye], in *Aral'skiy Krizis: Istoricheskaya i Geograficheskaya Retrospektiva* [The Aral crisis: historical and geographical retrospective]. AN SSSR, Moscow, 28–47

Ptichnikov, A.V. (1996) Environmental and landscape changes in the Aral Sea region as detected from remote sensing, in Micklin, P.P. and Williams, W.D. (eds) *The Aral Sea Basin.* NATO ASI Series, Partnership Sub-Series, 2. Environment, Vol. 12, NATO, Washington

Rafikov, A.A. (1999) Desertification in the Aral Sea region, in Glantz, M.H. (ed.) *Creeping Environmental Problems and Sustainable Development in the Aral Sea Basin.* Cambridge University Press, Cambridge

Rafikov, A.A. (1994) Osnovnyie stadii razvitiya antropogennogo opustynivaniya v Yuzhnom Priaraliye [The main phases of development of anthropogenic desertification in Southern Priaraliye]. *Problemy Osvoyeniya Pustyn'*, **3**, 17–25

Reznichenko, G. (1982) *Aral'skaya katastrofa.* 'Novosti', Moscow

Reznikov, N.I. (1995) Opyt bor'by s opustynivaniyem v Zapadnom Prikaspii [Experience of desertification control in the Western Circum-Caspian region], in Zonn, I.S. and Neronov, V.M. (eds) *Biota i Prirodnaya Sreda Kalmykii* [Biota and environment of Kalmykia]. TOO 'Korkis', Moscow-Elista, 84–92

Roginko, A.Yu. (1992) Environmental issues in the Soviet Arctic and the fate of Northern natives, in Stewart, J.M. (ed.) (1992) *The Soviet Environment: Policies and Politics.* Cambridge University Press, Cambridge, 213–222

Romanova, E.P. (1997) *Sovremenniye Landshafty Evropy* [Contemporary landscapes of Europe]. Izdate'stvo MGU, Moscow

Rosgidromet (1998) Obzor zagryazneniya okruzhayushchei prirodnoi sredy v Rossiiskoi Federatsii za 1997 god [A review of the pollution of the environment in the Russian Federation in 1997]. *Zelyoniy Mir*, **20** (284), 1–15

Rossiya v Tsifrakh, 1995 [Russia in Figures, 1995] (1995a). Goskomstat Rossiiskoi Federatsii, Moscow

Rossiya v Tsifrakh, 1999 [Russia in Figures, 1999] (1999). Goskomstat Rossii, Moscow

Rossiya 1913 [Russia in 1913. Statistical Reference Book] (1995b). Blitz, St Peterburg

Rossiyskiy Statisticheskiy Ezhegodnik 1994 [Russian Statistical Yearbook] (1994). Goskomstat Rossii, Moscow

Rossiyskiy Statisticheskiy Ezhegodnik 1995 [Russian Statistical Yearbook] (1995). Goskomstat Rossii, Moscow

Rossiyskiy Statisticheskiy Ezhegodnik 1996 [Russian Statistical Yearbook] (1996). Goskomstat Rossii, Moscow

Rossiyskiy Statisticheskiy Ezhegodnik 1997 [Russian Statistical Yearbook] (1997). Goskomstat Rossii, Moscow

Rossiyskiy Statisticheskiy Ezhegodnik 1998 [Russian Statistical Yearbook] (1998). Goskomstat Rossi, Moscow

Rossiyskiy Statisticheskiy Ezhegodnik 1999 [Russian Statistical Yearbook] (1999). Goskomstat Rossii, Moscow

Rozanov, B.G. (1990) Assessment of global desertification, in *Desertification Revisited*. United Nations Environment Programme, Nairobi, 45–122

Russia and Eurasia Facts and Figures Annual (1996), **21**. Academic International Press, Gulf Breeze

Sagers, M.J. (1994) Oil spill in Russian Arctic. *Polar Geography*, **18**, 2, 95–102

Saiko, T.A. (1995) Implications of the disintegration of the former Soviet Union for desertification control. *Environmental Monitoring and Assessment*, **37**, 289–302

Saiko, T.A. (1997) Environmental problems of the Caspian Sea Region and the conflict of national priorities, in Glantz, M.H. and Zonn, I.S.

(eds) *Scientific, Environmental, amd Political Issues in the Circum-Caspian Region*. Kluwer Academic Publishers, Dordrecht, 41–52

Saiko, T.A. (1998a) Environmental challenges in the new democracies, in Pinder, D. (ed.) *The New Europe. Economy, Society and Environment*. John Wiley and Sons, Chichester, 381–399

Saiko, T.A. (1998b) Geographical and socio-economic dimensions of the Aral Sea crisis and their impact on the potential for community action. *Journal of Arid Environments*, **39**, 225–238

Saiko, T.A. (1998c) Kol'skiy poluostrov [The Kola peninsula], in Krasovskaya, T.M. (ed.) *Kol'skiy Poluostrov*. Smolenskiy Gumanitarniy Universitet, Moscow-Smolensk, 12–19

Saiko, T.A. and Zonn, I.S. (1994) Deserting a dying sea. *Geographical Magazine*, **66**, 7, 12–15

Saiko, T.A. and Zonn, I.S. (1997) Europe's first desert, in Glantz, M.H. and Zonn, I.S. (eds) *Scientific, Environmental, and Political Issues in the Circum-Caspian Region*. Kluwer Academic Publishers, Dordrecht, 141–144

Sakata, T. (1998) Satellite image maps of the Aral Sea and Central Asia, in Kobori, I. and Glantz, M.H. *Central Eurasian Water Crisis: Caspian, Aral and Dead Seas*. United Nations University Press, Tokyo, 75–77

Sakwa, R. (1993) *Russian Politics and Society*. Routledge, London

Sallnow, J.A. (1998) Murmanskaya oblast [The Murmansk province], in Krasovskaya, T.M. (ed.) *Kolskiy Poluostrov*. Smolenskiy Gumanitarniy Universitet, Moscow-Smolensk, 20–29

Sallnow, J. and Saiko, T. (1996) Russia. Country Fact Files. Macdonald Young Books, Hove

Sandjiev, B.C. (1998) Respublika Kalmykia [Republic of Kalmykia]. Kalmytskiy Institut Gumanitarnyh i Prikladnykh Issledovaniy, 50 pp (unpublished)

Savchenko, V.K. (1995) *The Ecology of the Chernobyl Catastrophe: Scientific Outlines of an International Programme of Collaborative Research*. UNESCO and the Parthenon Publishing Group Ltd, Paris

Selskoye Khozyaistvo Respubliki Kazakhstan [Agriculture of Kazakstan. Statistical Reference Book] (1997). Komitet po Statistike i Analizu, Almaty

Selskoye Khozyaistvo Rossii [Agriculture of Russia. Statistical Reference Book] (1995). Goskomstat Rossii, Moscow

Sergiev, V.P., Beer, S.A., Elpiner L.I. and Vinogradov, V.G. (eds) (1993) *Medical and Ecological Aspects of the Aral Sea Crisis*. VINITI, Moscow

Shahgedanova, M., Burt, T.P. and Davies, T.D. (1999) Carbon monoxide and nitrogen oxides pollution in Moscow. *Water, Air, and Soil Pollution*. **112**, 107–131

Shaw, D.J.B. (1999) *Russia in the Modern World*. Blackwell Publishers, Oxford

Shaw, D.J.B. and **Oldfield, J.** (1998) The natural environment of the CIS in the transition from Communism. *Post-Soviet Geography and Economics*, **39**, 3, 164–177

Shcherbak, Yu.N. (1991) Chernobyl kak novoye yavleniye v istorii tsivilizatsii [Chernobyl as a new phenomenon in the history of civilisation], in *Evrochernobyl-2*, Proceedings of the International Conference, Kiev, 1991. Informtsentr Vsemirnogo Soveta Mira, Moscow-Helsinki

Shlikhunov, V.M. (1993) The coastal sea level rise: problems and solutions. *Report at the World Coastal Conference*, the Hague, Netherlands, 1–5 November 1993

Simagin, Yu.A. (1997) *Sovremenniy Etap Suburbanizatsii v Moskovskom Stolichnom Regione* [Present stage of suburbanization in the Moscow capital region]. NITS 'Geovektor', Moscow

Skrivan, P., Rusek, J., Fottova, D., Burian, L. and **Minarik, L.** (1995) Factors affecting the content of heavy metals in bulk atmospheric precipitation, throughfall and stemflow in Central Bohemia, Czech Republic. *Water, Air and Soil Pollution*, **85**, 841–846

Sochava, V.B. (1974) Geotopologiya kak razdel ucheniya o geosistemah [Geotopology as a part of the study of geosystems], in *Topologicheskiye Aspekty Ucheniya o Geosistemah* [Topological aspects of geosystem studies]. AN, Novosibirsk

Sociological 'Expert' Center (1995). Assessment of population requirements in water, sanitation and public health in the Aral region of Uzbekistan. Synopsis or Report. Tashkent-Nukus, 26 pp

Sodruzhestvo Nezavisimykh Gosudarstv in 1996, (1997) Statistical Yearbook of CIS, Mezhgosudarstvenniy Statisticheskiy Komitet SNG, Moscow

Sodruzhestvo Nezavisimykh Gosudarstv in 1997, (1998a) Statistical Abstract of CIS, Mezhgosudarstvenniy Statisticheskiy Komitet SNG, Moscow

Sodruzhestvo Nezavisimykh Gosudarstv in 1997, (1998b) Statistical Yearbook of CIS, Mezhgosudarstvenniy Statisticheskiy Komitet SNG, Moscow

Solomatin, V.I. (ed.) (1992) *Geoekologiya Severa* [Geoecology of the North]. Izdatel'stvo Moskovskogo Universiteta, Moscow

Sokhraneniye Biologicheskogo Raznoobraziya v Rossii. Pervyi Natsional'nyi doklad Rossiiskoi Federatsii [Conservation of biological diversity in Russia. The First National Report of the RF] (1997). Gos. Komitet RF po okhrane okruzhayushchei sredy, Project 'Conservation of biodiversity' of GEF, Moscow

Sostoyaniye Okruzhayushchei Sredy v Turkmenistane [State of the environment in Turkmenistan] (1999). State Report, Ministerstvo Prirodopol'zovaniya i Okhrany Okrushayushchei Sredy Turkmenistana, United Nations Development Programme, Ashgabat

Sostoyaniye Prirodnoi Sredy v SSSR v 1988 Godu [State of the environment in the USSR in 1988] (1990). Goskompriroda SSSR, Lesnaya Promyshlennost', Moscow

Statisticheskiy spravochnik SNG 1996 [Statistical Yearbook of CIS] (1996). Mezhgosudarstvenniy Statisticheskiy Komitet SNG, Moscow

Stewart, J.M. (1992a) Air and water problems beyond the Urals, in Stewart, J.M. (ed.) *The Soviet Environment: Problems, Policies and Politics.* Cambridge University Press, Cambridge, 223–237

Stewart, J.M. (ed.) (1992b) *The Soviet Environment: Problems, Policies and Politics.* Cambridge University Press, Cambridge

Strany – Chleny SNG [Countries – members of CIS. Statistical Reference Book] (1992). Mezhgosudarstvenniy Statisticheskiy Komitet SNG, Moscow

Sultangazin, U.M. and **Tsukatani, T.** (1995) Modelling of the Kazakhstan economy and environment. Discussion Paper No. 416, *International Forum on Aral, Caspian and Dead Seas* held on 27–29 March 1995, Tokyo and Ohtsu, Japan

Svitoch, A.A. (1997) *Extremal'niy Pod'yom Urovnya Kaspiiskogo Morya i Geoekologicheskaya Katastrofa v Primorskih Gorodah Dagestana* [Extreme rise of the Caspian Sea's level and the geoecological catastrophe in the adjacent towns of Dagestan]. Geograficheskiy Fakul'tet MGU, Moscow

Symons, L. (1990) *Soviet Union: A Systematic Geography.* Hodder and Stoughton

Tishkov, A.A. (1993) *Sovremennye Problemy Biogeografii* [Contemporary problems of biogeography]. Rossiyskiy Otkrytyi Universitet, Moscow

Tickle, A. and **Welsh, I.** (ed.) (1998) *Environment and Society in Eastern Europe.* Addison Wesley Longman Ltd, Harlow

Thompson, J. (1991) East Europe's dark dawn. *National Geographic,* **179**, 6, 36–69

Trofimov, I.A. (1995) Prirodnye kormovyye ugod'ya Chornyh zemel' [Natural fodder resources of the Black lands], in Zonn, I.S. and Neronov, V.M. (eds) *Biota i Prirodnaya Sreda Kalmykii* [Biota and environment of Kalmykia]. TOO 'Korkis', Moscow-Elista, 53–83

United Nations Conference on Environment and Development (UNCED) (1992). Rio de Janeiro, Brazil

United Nations Environment Programme (UNEP) (1992) *Diagnostic Study for the Preparation of an Action Plan for the Rehabilitation of the Aral Sea.* UNEP, Nairobi

Ust'yevaya Oblast' Volgi: Gidrologo-Morfologicheskiye Protsessy, Rezhim Zagryazniayushchih Veshchestv i Vliyaniye Kolebaniy Urovnya Kaspiiskogo Morya [The Volga's mouth area: hydrological-morphological processes, regime of contaminants and influence of the Caspian Sea level's changes] (1998). GEOS, Moscow

Vari, A. and **Tamas, P.** (eds) (1993) *Environment and Democratic Transition: Policy and Politics in Eastern Europe.* Kluwer, Dordrecht

Vashchalova, T.V. (1986) Istoriia izmeneniya prirody Khibin v golocene, in Miagkov, S.M. (ed.) *Prirodnyie Usloviia Khibinskogo Uchebnogo Poligona.* Izdatelstvo Moskovskogo Universiteta, Moscow, 22–25

Vernadskiy, V.T. (1969) *Himicheskoye Stroyeniye Biosfery Zemli i eyo Okruzheniye* [Chemical composition of the Earth's biosphere and its surroundings]. Nauka, Moscow

Vil'check, G.E. (1996) Degradatsiya rastitel'nogo pokrova i rastitelnyh resursov [Degradation of vegetation cover and resources], in Yablokov, A.V. (ed.) *Rossiyskaya Arktika: na Poroge Katastrofy*. Tsentr Ekologicheskoi Politiki Rossii, Moscow, 87–92

Vil'check, G.E., Krasovskaya, T.M., Tsyban, A.V. and **Chelyukanov, V.V.** (1996a) The environment in the Russian Arctic: status report. *Polar Geography*, **20**, 1, 20–43

Vil'chek, G.E., Serebryannyy, L.R. and **Tishkov, A.A.** (1996b) A geographic perspective on sustainable development in the Russian Arctic. *Polar Geography*, **20**, 4, 249–266

Vinogradov, B.V. (1993) Sovremennaya dinamika I ekologicheskoye prognozirovaniye prirodnyh usloviy Kalmykii [Contemporary dynamics and ecological forecasting of natural environment in Kalmykia]. *Problemy Osvoyeniya Pustyn'*, **1**, 29–37

Vinogradov, B.V., Sorokin, A.D. and **Fedotov, P.B.** (1995) Kartografirovaniye klimaticheskoi aridnosti territorii Kalmykii [Mapping of the climatic aridity of Kalmykia], *in Biota i Prirodnaya Sreda Kalmykii*. TOO 'Korkis', Moscow-Elista, 253–258

Vinogradov, S.V. (1997) Toward regional cooperation in the Caspian: a legal perspective, in Glantz, M.H. and Zonn, I.S. (eds) *Scientific, Environmental, and Political Issues in the Circum-Caspian Region*. Kluwer Academic Publishers, Dordrecht, 53–66

Vorob'yev, V.V. (1988) Current problems of Baikal, in *Geographical Problems in Siberia*. Institut Geografii SO AN SSSR, 3–21

Voropayev, G. (1997) The problem of the Caspian Sea level forecast and its control for the purpose of management optimization, in Glantz, M.H. and Zonn, I.S. (eds) *Scientific, Environmental, and Political Issues in the Circum-Caspian Region*. Kluwer Academic Publishers, Dordrecht, 105–117

Voskresenskiy, K.S. (1992) Sovremennye tempy denudatsii ravnin kriolotozony [Contemporary rates of denudation in the kryolithozone], in Solomatin, V.I. (ed.) *Geoekologiya Severa* [Geoecology of the North]. Izdatelstvo Moskovskogo Universiteta, Moscow, 83–94

Vostokova, E.A. (1999) Ecological disaster linked to landscape composition changes in the Aral Sea region, in Glantz, M.H. (ed.) *Creeping Environmental Problems and Sustainable Development in the Aral Sea Basin*. Cambridge University Press, Cambridge, 26–46

Vozrozhdeniye Volgi – shag k spaseniyu Rossii [Revival of the Volga – a step to Russia's salvation] (1996). Ekologiya, Moscow-Nizhniy Novgorod

Vozrozhdeniye Volgi – shag k spaseniyu Rossii. Kniga 2: Sub'ekty Federatsii i goroda basseina [Revival of the Volga – a step to Russia's salvation. Book 2: Federation subjects and towns of the basin] (1997). Ekologiya, Moscow

Wolchik, S. (1991) *Czechoslovakia in Transition*. Pinter Publishers, London

Wolfson, Z. (1992) The massive degradation of ecosystems in the USSR, in Stewart. J.M. (1992) *The Soviet Environment: Problems, Policies and Politics*. Cambridge University Press, Cambridge, 57–63

World Bank (1993) Humanitarian aid to the republics of Central Asia. Report. World Bank, Washington

Wolfson, Z. (1994) *The Geography of Survival. Ecology in the post-Soviet Era*. M.E. Sharpe, London

World Resources 1996–97 (1996) Oxford University Press, Oxford

Yablokov, A.V. (ed.) (1996) *Rossiyskaya Arktika: na Poroge Katastrofy* [The Russian Arctic: on the threshold of a catastrophe]. Tsentr Ekologicheskoi Politiki Rossii, Moscow

Yanshin, A.L. and **Melua, A.I.** (1991) Uroki Ekologicheskih Proschetov [Lessons from ecological failures]. 'Mysl', Moscow

Yaroshinskaya, A. (1994) *Chernobyl: the Forbidden Truth*. Jon Carpenter, Oxford

Yaroshinskaya, A (1992) *Chernobyl. Sovershenno Sekretno* [Chernobyl. Top Secret]. Drugiye Berega, Moscow

Yatsukhno, V. and **Kozlovskaya, L.** (1998) The ecological impact of the Chernobyl catastrophe on sustainable development in Belarus, in Graute, U. (ed.) *Sustainable Development for Central and Eastern Europe: Spatial Development in the European Context*. Springer-Verlag, Berlin-Heidelberg

Yevseev, A.V. (1998) Ekologicheskaya obstanovka [Ecological situation], in Krasovskaya, T.M. (ed.) *Kol'skiy Poluostrov* [The Kola peninsula]. Smolenskiy Gumanitarniy Universitet, Moscow-Smolensk, 119–133

Yevseev, A.V. (1996) Zagriazneniye nazemnyh ekosistem [Pollution of terrestrial ecosystems], in Yablokov, A.V. (ed.) *Rossiyskaya Arktika: na Poroge Katastrofy* [The Russian Arctic: on the threshold of a catastrophe]. Tsentr Ekologicheskoi Politiki Rossii, Moscow, 47–63

Yevseev, A.V. and **Krasovskaya, T.M.** (1996) *Ekologo-Geograficheskiye Osobennosti Prirodnoi Sredy Raionov Krainego Severa Rossii* [Ecological and geographical features of the natural environment of the Extreme North of Russia]. Smolenskiy Gumanitarniy Universitet, Smolensk

Yevseev, A.V. and **Krasovskaya, T.M.** (1998), Regions of adverse environmental impact in the Russian Arctic and Subarctic. *Polar Geography*, **22**, 2, 136–142

Zaidfudim, P. (1998) Na kraiyu zemli [On the edge of the Earth]. *Nezavisimaya Gazeta. NG-Regiony*, **21**, 24, 8 December, 9–11

Zaletayev, V.S. (1989) *Ekologicheski Destabilizirovannaya Sreda* [Ecologically destabilised environment]. Nauka, Moscow

Zonn, I. S. (1992) Environmental stress and the search for sustainable development in arid and semi-arid Central Asia: the case of the Aral Sea basin. *Report at the International Conference on Climatic Impacts and Sustainable Development*, Fortaleza, Brazil, 27 January–3 February 1992

Zonn, I.S. (1993) Aral'skaya problema v svete novoi geopolitiki [The Aral's problem in the light of new geopolitics]. *Problemy Osvoyeniya Pustyn'*, 3, 9–17

Zonn, I.S. (1994) A creeping environmental phenomenon in Russia: Desertification in Kalmykia, in Glantz, M. (ed.) *Workshop Report on Creeping Environmental Phenomena*. National Center for Atmospheric Research, Boulder, Colorado, 169–174

Zonn, I.S. (1995a) Desertification issues in the Aral Sea basin. Discussion Paper. *International Forum on Aral, Caspian and Dead Seas* held on 27–29 March 1995, Tokyo and Ohtsu, Japan

Zonn, I.S. (1995b) Desertification in Russia: problems and solutions (an example in the Republic of Kalmykia – Khalm-Tangch). *Environmental Monitoring and Assessment*, 37, 347–363

Zonn, I.S. (1997a) Assessment of the state of the Caspian Sea, in Glantz, M.H. and Zonn, I.S. (eds) *Scientific, Environmental, and Political Issues in the Circum-Caspian Region*. Kluwer Academic Publishers, Dordrecht, 27–39

Zonn, I.S. (1997b) *Kaspiyskiy Memorandum* [The Caspian memorandum]. TOO 'Korkis', Moscow

Zonn, I.S. (1999) *Kaspiy: Illyuzii i Real'nost'* [The Caspian: myths and realities]. TOO 'Korkis', Moscow

Zonn, I.S. and **Neronov, V.M.** (eds) (1995) *Biota i Prirodnaya Sreda Kalmykii* [Biota and environment of Kalmykia]. TOO 'Korkis', Moscow-Elista

Zonn, S.V. (1995c) Opustynivaniye prirodnyh resursov agrarnogo proizvodstva Kalmykii za posledniye 70 let i mery bor'by s nim [Desertification of natural resources in agricultural production of Kalmykia and methods of its control], in Zonn, I.S. and Neronov, V.M. (eds) *Biota i Prirodnaya Sreda Kalmykii*. TOO 'Korkis', Moscow-Elista, 19–52

Zonn, S.V. (1986) 'Black Lands' of Kalmykia, in Glantz, M.H. (ed.) *Arid Land Development and the Combat against Desertification: an Integrated Approach*. Centre for International Projects, Moscow, 124–127

Index

Numbers in **bold** indicate a figure, table or plate.

accidents, oil and gas industry 54–55
acid rain, distribution in Central Europe
    149, 150
aeolian processes 259–260
air pollution
    ammonia/ammonium **104**, 106
    Arctic North 49–53
    Baikal, Lake 86
    Black Triangle 147–148
    as cause of soil pollution 208
    copper 49–51
    Kalmykia 178
    metallurgical industries 49–52
    meteorological conditions 102
    mining activities 52
    Moscow region 104–107
    Murmansk oblast 51
    oil and gas industries 53
    Volga basin 204–205
    vulnerability of deciduous trees 59
airborne pollutants, from various Russian
        cities **50**
Alcamo, J. 146, 275, 284
alders 38, 122
Allworth, E. 250
altitudinal zonality 19, 39
Altschuler, I.I. *et al* (1992) 1, 6, 26
ameritum–241 138
Ammofos chemical complex 202
ammonia/ammonium, air pollution **104**, 106
Amudarya river 242, 244, 245, 246, 255, 269
anaemia 63
Angara river 72, 90, 91
Angarsk **50**, 86
animal pressure on rangelands 262
annual biological production 17, **18**
annual plants 163
annual precipitation **18**
anthropegenic disturbances, definition 22
anthropogenic desert 4
anthropogenic modifications 19
Apatity combine **45**, 52
Aplin, G. *et al* (1996) 137
apple trees 191
Aral Sea 25
    dessication 4, 254–258
        irrigation water losses in transit 255
    Large Aral 255
    Small Aral 255
Aral Sea basin 242–272

climate 245–246
cotton production 244, 250–252, 267
creeping environmental phenomena 244
deserts 245
geographical location and relief 244–245
hydrological network 246–247
political divisions **243**
population 244
rivers flowing into 242
soil cover 249
sukhovey winds 246
use of term 242
vegetation 247–249
see also Priaralye
Aral Sea region
    agricultural land, change in crops sown
        265, **266**, 267
    animal life 249
    cotton industry 250–252, 267
    desertification process 255, **256**
    drinking water quality 261, 267
    economic implications 264–267
    emigration 263–264
    environmental changes **257**, 260–262
    ethnic conflicts over water issues 271
    fertilizer use 261
    fisheries, loss of 264–265
    future prospects 268–272
    grain crops 265
    health effects 267–268
    historical perspective and economic
        development 250–254
    infant mortality 268
    international aid 271–272
    irrigation 250–252
    overall ecological situation 262
    per capita GDP **270**
    social repercussions 262–264
    soil salinization 261, 264
    standard of living 263
    vulnerability of landscape 249
    waterlogging 261
Aralsk 264, 269
Arctic desert 30
Arctic haze 35
Arctic North 31–65, 275–276
    air pollution 49–53
        extent 49
        Kola peninsula **50**, 51, 52
        transboundary 53

Arctic North *continued*
  apatite-nepheline ores 44
  Arctic haze 35
  climate 34–35
  cobalt 44, 48, 49–51
  copper 44, 48, 49–51
  decline in industrial production 48
  degradation of natural vegetation and
      pastures 58–60
  economic development and ecological change,
      summary 1929 to date **46–47**, 48
  emigration 63
  future prospects 64–65
  gas 43, 44
  gas industry 53
  geographical location and relief 33–34
  gold 43, 44, 48
  health issues 63–64
  heavy metals 56
  historical perspective and economic
      development 43–48
  hydrology 35–36
  inflation 63
  iron ores 44
  metallurgical industries 49–52
  mineral deposits 43, 44
  mineral fertilizers 44
  mining activities 52
  native people 63, 64
  natural resources 43–44
  nickel 43, 44, 48, 49–51
  oil 43, 44, 53, 55
  overall ecological situation 60–62
  palladium 48
  permafrost 57–58
  phytomass 39
  platinum 48
  population density 62
  primary succession 39
  radioactivity 56–57
  selenium 48
  socio-economic issues 62–63
  soil contamination 55–57
  soil cover 41–42
  sulphur dioxide 49–52, 53
  temperature inversion 35
  timber 44
  transportation 63
  vegetation 37–41
  vulnerability of landscapes, factors
      contributing to 42–43
  water pollution 53–55
  see also tundra
arid soils, features 164
Armand, D.L. 13
ash 122, 191
aspen 122
Astrakhan 208
Astrakhan gas condensate plant 205, **206**, 209
Atlantic Gulf Stream 142
Atyrau oblast 237
Azerbaijan
  air pollution 226
  Caspian oil deposits, ownership 238, 239

  discharge of polluted waste waters into
      Caspian Sea **227**
  flooding due to rise in sea level 237
  flooding of oil extracting sites 236
  offshore oil reserves 222, 223
  oil pollution 226, 230
  oil production 1990–97 223, **224**
  water pollution 226
Azores high-pressure system 142

Babaev, A.G. 236
Baikal, Lake 3–4, 66–97
  age 70
  air pollution 86
    intra-regional air transfer 86
  air temperature 71
  aquatic life 89–90
    epishura 89
    golomyanka fish 90
    nerpa seal 89–90
    omul fish 89, 90
    phytoplankton 89
    sturgeon 90
    zoobenthos 89
    zooplankton 89
  atmospheric stagnation 71
  Baikal-Amur Mainline 84–85
  Baikalsk pulp and paper mill 79–80, 83–84,
      95, 96
  benzopyrene 86
  and Buddhism 67
  Buryatia 81, 83
  chlorides 83
  climate 70–71
  copper 88
  cryogenic processes 72, 73
  damming 85
  DDT 88
  decomposition of organic matter 72
  dioxins 91
  disappearance of small rivers 87
  discharge of waste waters within basin 81, **82**
  drinking water quality 94
  dyphenyl polychlorines 91
  economic development 79–80
  as emerging ocean 70
  fauna and flora 67
  Federal Law 96–97
  Federal programme, financial sources **95**
  fertility of land 87–88
  filtration by sponges and shrimp 76
  fish species 76
  fish stocks 76
  forests
    effect of air pollution 87
    role 74
  geographical location and relief 68–70
  health issues 94
  hydrological network 71–73
  International Baikal Centre for Ecological
      Studies 95
  interrelationship between anthropogenic
      factors and physical environment **69**
  ion composition of water 72

logging operations 85, 87, 94
national parks 79
nature reserves 79
nitrates 83
oil products 81, 84
organic life 76
overall ecological situation 91, **92–93**
oxygen content 72
period of total water exchange 72
permafrost 72–73, 74
phenols 88
poaching 90
present ecological situation **78**
prospects for future 94–97
protected areas **78**, 79
seismic activity 70, 91
Selenga catchment 81, 83
Selenginsk pulp and cardboard complex
    (SPCC) 80, 81
size 66
socio-economic issues 94
soil contamination 87
soil cover 74–76
soil erosion 87
solar radiance duration 71
sulphates 83, 84
temperature inversions 70–71
thematic atlases 7
thermokrast 73
timber rafting 94
timber transportation 85
tourism 88
Trans-Siberian railway 73
Ulan-Ude 81, 83
UNESCO heritage site 95
vegetation cover 73–74
    altitudinal zonality 73–74
vulnerability of landscapes 77–79
water fowl 90
water pollution 81–85
water quality 88–89
water temperature 72
wildlife 88
wind and water erosion 87
woodcutting 87
Baikal seal *see* nerpa seal
Baikal-Amur Mainline 80, 84–85
Baikalo-Lenskiy reserve 79
Baikalsk, air pollution 86
Baikalsk pulp and paper mill 79–80, 83–84, 86,
    95, 96
Baikalskiy reserve 79
Baku 236
Bananova, V.A. 169, 177
Barguzin 79
Barguzin river 71
Barguzin sable 79
Barguzinskiy reserve 79
bear 88
beech 142, 143
Belarus
    ability to cope with problems 281

Chernoybl accident contaminated areas
    127, **128**, 133, **134**
    agricultural lands **136**
    economic implications 136
    population settled in **135**
    time required to recuperate 138
    soil 122
    see also Chernoybl
Belaya river 189
Belopukhova, E.B. *et al* (1976) 41
Beluga sturgeon 221–222, 232, 235
benzapyrene 107
benzene 105
benzol 104Table
benzopyrene 86
    soil pollution 208–209
Bingham, S. 183
biomes 17
biosphere 13
birch 38, 39, 112, 122
Black lands *see* Kalmykia
Black Sea 162
Black Triangle 3, 4, 140–158
    air pollution 147–148
    climate 142
    forests
        broad-leaved deciduous 142–143
        coniferous 143
        deterioration 151–153
    future prospects 154–158
    health issues 154
    heavy metals 150
    historical perspective and economic
        development 144–146
    hydrological network 142
    location 141–142
    modification of landscapes by man 143
    nitrogen dioxide 149–150
    overall ecological situation 153–154
    socio-economic issues 154
    soil pollution 149–151
    soils 142
    sulphur deposition 155
    sulphur dioxide 149–150, 155–156
    summary of reasons for environmental
        situation 146
    temperature inversions 142
    vegetation 142–143
    vulnerability of landscape 143–144
    water pollution 148–149
Blais, J.M. *et al* (1999) 49–50, 55
Bliznyuk, A.I. 177
blood diseases, Lake Baikal 94
Blyava river 228
Boehmer-Christiansen, S. 274
Bogolyubov, S. 28
bogs 122, 124
Bohemia, North 141, 154
    see also Black Triangle
Bovanenkovo 44
Bovanenkovo gas condensate deposit 65
Bradshaw, M.J. 32, 64
Bratsk 50, 86, 91, 94

breast milk, dioxin levels 213, 214
Bridges, O. and J. 84, 85, 106, 109, 228
broad-leaved deciduous forests 12
    Black Triangle 142–143
    climatic, soil and vegetation features **15**
    geographical characteristics **18**
Brokgauz, F.A. and Efron, I.A. 235
bronchial asthma 63, 64
bronchitis 63, 64, 154
Brueggemann, E. and Spindler, G. 155
Bryansk 196, 209
Bryukhanov and Zaveryaeva 53
Buddhism 67
Budyko, M. 13, 16, 17
bulrush 122
Buryatia 81, 83

cadmium 117, **150**, 205
caesium–137 127, **128**, 131, 133
camel 169
cancers 63, 94, 135
carbon monoxide **104**, 105, 106
cars, pollution 205
Carter, F.W. and Turnock, D. 145, 147, 148, 154
Caspian region 4
Caspian Sea 5, 161, 162, 186, 217–241
    air pollution 229
    bacterial pollution 226
    catchment area 220
    climate 220, 237
    dimensions 217
    discharge of polluted waste waters by
        littoral countries **227**
    environmental problems faced by littoral
        states **225**
    eutrophication of Volga delta 228–229
    fish
        decline in catch and reserves 232–235
        number of species 221
        see also sturgeon *below*
    future prospects 239–241
    historical perspective and economic
        development 222–224
    hydrology 220
    Kara-Bogaz-Gol Bay 220–221
    landscapes 221
    legal status 219, 238
    level fluctuations 218, **219**
    oil deposits, ownership 238–239
    oil pollution 226, 229–232
    oil production 1990–97 223, **224**
    oil reserves 217, **218**, 223
    overall ecological situation 236
    phenol 229, 230
    physical geographical setting 219–222
    poaching 226, 234
    political divisions **218**
    political issues 237–239
    resettlement of people 237
    rivers flowing into 220
    salinity of water 221
    sea level rise 236, 237
        predicted effects of 239–240
    socio-economic issues 237–239
    sturgeon 221–222, 230
        Beluga 221–222, 232
        decline in catch and reserves 234–235
        health 233
        poaching 226, 234
        reproduction problems 232–233
        reserves 217, 234–235
        Russian 222, 232
        Sevryuga (star) 222
        spawning grounds 232
        Stellate 232
        threats 222, 223
    surface area 217
    vegetation 163
    water pollution
        from river flow 226–229
        main sources 226–227
Caspian seal 221, 233
catastrophes, definition 25
caviar 221, 235
    see also sturgeon
cellulose 79–80, 83–84
Central Asia, ability to cope with problems
        280–281
Central Europe
    ability to cope with problems 283
use of term 140
Centre for International Projects 240
Chapayevka river 203
Chapayevsk 202–203, 213–214
Chapayevsk syndrome 214
Cheboksary reservoir 204, 208
Cheleken 236
Cheremkhovo 86, 91
Cherepovets **50**, 202, 203, 205
Chernoybl 4, 28, 56, 120–139, 209
    afforestation measures 122
    amount of radiation released into
        environment 127
    areas contaminated 120
        with caesium–137 133, **134**
        population settled in **135**
    bogs 122, 124
    causes of accident 124–127
    climate 121–122
    composition of radioactive fallout 127
    costs of dealing with consequences 138
    date of accident 124
    distribution of contaminated food 137–138
    ecological consequences and recent changes
        127–134
    economic implications 136
    forests 122
    future prospects 137–139
    global effect of accident 126
    health issues 134–136
    intensity of radiation **132**
    location 121
    overall ecological situation 133–134
    Poles'ye 121, 122

political consequences 136–137
sarcophagus security 138
soil cover 123
    impact on 131–133
    thematic atlases 7
    vegetation 122, 124, 129–131
    vulnerability of landscapes 123–124
chernozem 192
chestnut soils 112, 164, 192, 247
chlorides, water pollution 53, 83, 228
chlororganic compounds, soil pollution 203
chromium pollution **150**, 228
Circum-Aral region *see* Priaralye
clay polygonal soils 249
climate
    changes
        forecast 65
        and reservoirs 199
    impact of urban activities 114
    Northern Eurasia **14–15**, 17
coal industry 43, 144–145
cobalt 44, 48, 49–51
collectivization of lands 167
Colls, J. 143, 155
complete recovery, definition 23
coniferous forests
    air pollution 39, **40**, 151–152
    climatic, soil and vegetation features **14**
    impact of radioactivity 124
    see also pine
copper
    air pollution 49–51
    soil pollution 113, **150**, 209
    water pollution 53, 55, 88, 203, 204, 228
cotton grasses 38
cotton industry
    Aral Sea region 244, 250–252, 267
    production **252**
    under Bolsheviks 251
    under Tsars 251
creeping environmental phenomena 25, 244
crisis stage 23
critical situation
    criteria for evaluation **29**
    description **27**
cryogenic processes, Lake Baikal 72, 73
cryogenic soil formation, main features 41
cryogeosystems 36
Cullen, R. 234
Czech Republic 141
    acid rain **149**, 150
    air pollution, forest damage **151**
    emission of sulphur dioxide and nitrogen
        dioxide 147, **148**
    heavy metal pollution **150**
    life expectancy 154
    lignite production 156
    see also Black Triangle

dachas 102, 103, 112
Dagestan 182, 237
Danilov-Danilyan, V.I. *et al* (1993) 282
data, reliability 6–7

DDT 88
deciduous forests 12
    air pollution 59
    climatic, soil and vegetation features **15**, 17
    geographical characteristics **18**
deglaciation reagents 102, 112
degradation 20–21
desertification
    Aral Sea region 255, **256**
    definition 172–173
    overgrazing 181
    Priaralye 258–260
    Volga basin 208
    wind erosion 175, 181
    see also Kalmykia
deserts 192
    climatic, soil and vegetation features **15**
    geographical characteristics **18**
    plant biodiversity 248
    vegetation 247–248
diamonds, Arctic North 43
dioxins
    breast milk 213, 214
    soil pollution 203
    water pollution 91, 203
disturbance, definition 22
Dnieper river 135
Dobris Assessment 12
Doklad 197, 199, 201, 203, 207, 208, 209, 212, 214
Dokuchaev, V. 13
Downing, D. 267, 272
Dregne, H. *et al* (1991) 161
drinking water consumption 111
drinking water quality
    Aral Sea region 261, 267
    Lake Baikal 94
    Moscow 119
    Russia 278
    Volga basin 213
dry steppe, geographical characteristics **18**
dust pollution 52
dyphenyl polychlorines 91
Dzerzhinsk 203

Eastern Europe, use of term 140
ecological catastrophe
    criteria for evaluation **29**
    description **27**
ecological crisis
    criteria for evaluation **29**
    definition 26
    description **27**
ecological disaster zones, definition 25
ecological maps 7
ecosystem, concept of, definition 13
Elbe river 142
Elektrostal 16
elk 88
elm 122, 191
endocrine system illnesses 213–214
energy resources, ineffective use 145
environment, definitions 9–10
environmental awareness 119

environmental catastrophe
  description **27**
  zones, definition 28
environmental concerns 275
environmental crisis 9
  definition 26
  description **27**
environmental degeneration 5, 20–30
  classification of stages 23–26, **27**
  interrlationships between human and
    natural factors **10**, 11
ephemeral plants 163, 174, 247, 248
epishura 89
Erzgebirge mountains 155
ethnic conflicts, water issues 271
European fir 122
European Union 283
evaluation of ecological situation, criteria 28, **29**
evolution 20, 21

feather grasses 163, 192
fertility of land 87–88
fertilizers 44, 261
Feshbach, M. and Friendly, A. 267, 278
field maple 122
Finlayson, C.M. *et al* (1993) 189, 228, 232
fir *see* coniferous
fish
  spawning grounds, and hydroelectricity 210
  stocks, decline 209–211
  toxic concentration of oil 233–234
  see also individual species
Fish Mafia 234
fluoride emissions 94
Fodor, I. 154, 158
Fondahl, G.A. 63, 64
forested steppe 191
  biodiversity 191
  climatic, soil and vegetation features **15**
  geographical characteristics **18**
  phytomass 191
  soil type 191
forested tundra 33, 38
  climatic, soil and vegetation features **14**
  geographical characteristics **18**
forests
  air pollution damage 151–153
  deterioration 59, 151–153
  fires 87
  see also coniferous forests; broad-leaved
    deciduous forests
formaldehyde **104**, 105, 205
free market economy 4
Fullenbach, J. 147, 280
functional recovery, definition 23
Furman, A.E. and Livanova, G.S. 20, 23

Galaziy, G. 87
gas industry
  air pollution 53
  Arctic North 43, 44, 53, 54–55
  water pollution 54–55
  Yamal-West gas pipeline 44, 48

gas reserves, Turkmenistan 223
gastritis 63
genetic defects 64, 213
geographical characteristics of major landscapes
  zones in northern Eurasia 17, **18**
geographical zone 5, 9, 11–20
  definition 13
  periodic law 17
geosystem, concept of, definition 13
Germany 274
  acid rain **149**, 150
  emission of sulphur dioxide and nitrogen
    dioxide 147, **148**
  heavy metal pollution **150**
  reunification 156
  water pollution 148
Glantz, M.H. 25, 244
Glazovskiy, N.F. 261
Glazovskiy, N.F. *et al* (1991) 25, 26, 28
global atmospheric circulation 13
Glogow 153
gold 43, 44, 48
golomyanka fish 76, 90
Golubchikov, Yu.N. 38, 42, 55, 62, 64
Golubev, G.N.
Golytsyn, G.S.
Gorkiy reservoir 204
Gorshkov, S.P. 13, 60
Goudie, A. 13
Grachev, *et al* (1989) 90
grain crops 265, **266**, 267
grass steppes 192
grasses 122, 174
grayling 76
greenhouse effect 53
Grigoriev, A.A. and Budyko, M.I. 17
Gvozdetskiy, N.A and Mikhailov, N.I. 34, 248
gypsophyte plants 247

Halkos, G.E. 153–154
halophyte plants 163, 174, 247, 248
Hannan, T. and O'Hara, S. 261, 271
Harasaveiskoye gas condensate deposit 65
hazel nut 191
health care 64, 278
heart problems 154
heavy metals
  presence in lichens and mosses 58
  soil pollution 56, 113, 150, 209
hepatitis, viral 268
hornbeam 122
hydrocarbons 54, 90
hydroelectricity 194, 199, 210

Ilke river 228
improvement in environment, causes of slow
  progress 275–280
infant mortality 268
information, reliability 6–7
Institute of Geography of Russian Academy of
  Sciences 25
international aid, Aral Sea region 271–272
International Baikal Centre for Ecological
  Studies 95

intestinal illnesses 214
iodine isotopes 127
Iral river 220
Iran
   Caspian oil deposits, ownership 238, 239
   flooding due to rise in sea level 237
   oil reserves 223
Irkutsk 86, 87, 91
Irkutsk hyrdroelectric plant 85
iron **150**
iron compounds 228
iron ores 44
irrigation 169, 178, 232, 249
   see also Aral Sea, dessication
Ishida, N. *et al* (1995) 267
Ivanov, Y.A. *et al* (1997) 124, 131, 133, 137
Ivanovo oblast 195–196, 209

Jerginskiy reserve 79
Jones, A. 149
Juhasz, J. *et al* (1993) 284
juniper 38

Kalininskaya NPP 139
Kalmykia 4, 28, 159–184, 237, 240
   air pollution 178
   animal pressures on pastures 166, **167**, 168–169
   area 159
   Black lands 161, 162, 163, 173, 175, **176**, 182
   Black lands' nature reserve 178
   climate 162
      sukhovey winds 162
   Dagestan 182
   degeneration of pastoral vegetation 173–175
   desertification 160, 172, 179, **180**, 181, 182
   future prospects 182–184
   geographical relief 159, 161–162
   health issues 181–182
   historical perspective and agricultural development 165–172
   hydrological system 162–163
   Kizlyar pastures 182
   Kuma-Manych valley 161
   location 159
   main rivers 162
   Manych valley 164
   meat quality 181
   overall ecological situation 179–181
   people 159
   present ecological situation **160**
   saiga population 177–178
   Sarpa lowland 161
   shifting sands 175, **176**
   socio-economic implications 181
   soil salinization 173, 176–177
   soils 163–164
   state of ecological emergency 160
   vegetation 163
   vulnerability of landscape 164–165
   water quality 178
   waterlogging 173, 176–177

wind erosion 175
Yergeni upland 161, 164
Kama river 189
Kamyshev, A.P. 44, 58
Kara-Bogaz-Gol Bay 220–221
Karachaganak oil and gas condensate complex 230
Karakalpakstan 261, 263, 267, 268
Karakum Canal 251, 255, 261, 267
Karakum desert 245, 247–248
Kargalin hydraulic power system 232
Karimsakov, O. 263, 264
Kaspiyskoye oil and gas field 240
Kazakstan 241
   Caspian oil deposits, ownership 238, 239
   cotton production **252**
   discharge of polluted waste waters into Caspian Sea **227**
   flooding due to rise in sea level 237, 240
   flooding of oil deposits 236
   grain crops 265, **267**
   infant mortality 268
   oil pollution 226, 230
   oil production 1990–97 223, **224**
   oil reserves 223
   per capita GDP **270**
   per capita water withdrawal **254**, 255
   soil salinization 261, 264
   water withdrawal 271
Kennedy, P. 156
Khalmg Tangch *see* Kalmykia
Khesina, A.Ya. *et al* (1996) 106, 107
Khibiny mountains 39, 44
Khorezm oblast 267
Khrushchev, N. 279
Kiev 135
Kirovsk 44
Klarer, J. *et al* (1994) 141
Klarer, J. and Francis, P. 157
Klin 16
Klyaz'ma river 101
Kola Peninsula
   air pollution 50, 51, 52
   climate 34
   degradation of natural vegetation and pastures 58–59
   ecological situation 60
   economic development and ecological change 44, **46–47**, 48
   future prospects 65
   geographical location and relief 33
   health issues 63
   hydrology 35–36
   soil contamination 56–57
   soil cover 41
   transboundary pollution, sulphur dioxide 53
   vegetation 37–38, 38–39
   water pollution 53
Kol'skaya nuclear power plant 57
Kotlyakov, V.M. and Agranat, G.A. 33, 44
Kovdor 44
Krainiy Sever 33

Krakow 153
Krakum canal 271
Krasovskaya, T.M. 38
Krasovskaya, T.M. and Yevseev, A.V. 33, 39
Kremenetsky, C. *et al* (1997) 35
Kryuchkov, V.V. 62
Kudelsky, A.V. *et al* (1996)
Kuibyshev reservoir 199, 203, 204, 208
Kuksa, V.I. 229–230, 233, 234
Kuma river 162, 220
Kura river 220
Kyrgyzstan
    cotton production **252**
    fertilizer use 261
    grain crops **267**
    infant mortality 268
    per capita GDP **270**
    per capita water withdrawal **254**, 255
    soil salinization 261, 264
    water issues, conflict with Uzbekistan 271
Kyzlkum desert 245

lakes, acidification 58
land tenure 196
landscape
    definition 11–12
    study of 12
landscape zone, definition 13
landslides 113
larch 38, 74, 143
Laskorin, B. 79
latitudinal zonality 19
'Law on the Protection of the Environment' 26
Lawrence, E. *et al* (1998) 11
lead
    in children's hair 117
    pollution **150**, 157, 209
Lebedev, A.T. *et al* (1998) 90
Legnica 153
Leipzig 154
Lerman, Z. *et al* (1996) 271
Levy, M.A. 147
lichens 38
    presence of heavy metals 58
    thermal resistance 41
life expectancy 154
lignite production 144, 145, 146, 156
Lipetsk **50**, 205, 213
London, drinking water consumption 111
Lower Don river 161
Lower Silesia 153
Lubin, N. 261, 263, 268
lung cancer 64
Luzin, G.P. *et al* (1994) 44
L'vov, G.N. 105, 109, 111, 112, 114, 119
Lychagina, N.Yu. *et al* (1998) 204
Lyuri, D.I. 25–26, 284–285

McCauley, M. 265
Makarova, O.A. *et al* (1997) 37, 38
Makarova, T.D. 56
Mangyshlak peninsula 230
Mankovska, B. 143, 152

Manser,R. 141, 153, 273–274
Manych river 162
maple 112, 122, 191
Markert, B. *et al* (1996) 150
Marples, D.B. 136, 138
Massal'skiy, V.I. 250
Matthews, J.A. and Saiko, T.A. 157, 278
Matzner, E. and Murach, D. 150, 151
Mayak 139
meat quality 181
mechanized agriculture 168
Medvedev, Zh.A. 120, 124, 125, 126, 129, 133,
    134, 137, 138
Mekhyiev, A.Sh and Gul A.K. 227, 230, 236
Mellor, R.E.H. 140
mercury, water pollution 55
metallurgical industries
    Arctic North
        air pollution 49–52
        degradation of natural vegetation and
            pastures 58
        sulphur dioxide emissions 49–52
meteorological conditions, and air pollution
    102
methane 53
Micklin, P. 271
Milanova, E.V. and Kushlin, A.V. 19
mining 43, 52, 144–145
Miniprirody RF 28
Minvodkhoz 255
mixed forests 190–191
    biodiversity 191
    climatic, soil and vegetation features **14**, 17
    geographical characteristics **18**
    phytomass 191
    soil type 191
Mnatsakanian, R.A. 187, 229
moisture, global distribution 13
Monchegorsk 53, 56, 58, 60
Monchegorsk industrial complex 59
Moravia 153
Moscow
    administrative divisions **99**
    air pollution **104**, 105
    airborne pollutants **50**
    area 98
    climate, effect on air pollution 102
    distribution of industrial centres **99**, 105
    drinking water quality 111, 119
    environmental awareness 119
    extent of industrial enterprises 100
    financial crisis (1998) 103
    green belt 4, 101–102
    health and demographic issues 117–118
    history of urban development 102–104
    main polluters 105
    orbital motorway (MKAD) 98
    overall ecological situation 114–116
    population 98
    population density 4, 100
    pressure on lithosphere 113
    prospects for the future 118–119
    recycling 119

soil contamination 112–113
transport pollution 118
tree planting programme 119
vegetation cover changes 111–112
waste, accumulation of solid 114
waste load 113–114
water, per capita consumption 109, 110
water consumption 4, 119
water pollution 107–109
    ground waters 111
water supply problems 109–111
see also Moscow region
Moscow City Ecological Profile 103, 111
Moscow NPZ 105
Moscow region 98–119
    air pollution 104–107
    climate 101, 102, 114
    degradation of urban and suburban
        landscape 114
    health and demographic issues 117–118
    landslides 113
    location 100
    overall ecological situation 114–116
    physical geographical setting 100–102
    population changes 1975–98 **113**
    pressure on lithosphere 113
    prospects for the future 118–119
    relief 101
    rivers 101
    seismic activity 113
    soil contamination 112–113
    soils 101
    transport pollution 106–107
    urban encroachment 103–104
    vegetation cover 101–102
Moskovskiy Gorodskoi 101
Moskva river 101
Moskvoretsko-Okskaya moraine 101
mosses 38
    presence of heavy metals 58
    thermal resistance 41
Mote, V. 85
mountain regions, vertical zonality 19
mountains, altitudinal zonality 39
Murat, A.B. 251
Murghab river 244
Murmansk 34, 44, 51, 53, 54, 57, 60, 63
Murzayev, E. 260
Muynak 264, 265
Myach, L.T. 49, 53, 65
Myalo, E.G. and Volodina, I.A. 174
Mytishchi 117

N-nitrosamines (carcinogenic), air pollution,
    Moscow 106
Nara river 119
natural zones
    definition 13
    northern Eurasia 20, **21**
Nefedova, T. 145
Nefedova, T. and Treivish, A. 145, 276
Nekrasov, N. 279

Nentsi people 44, 64
nerpa seal 76, 77, 89–90
nervous illnesses 213
newborn abnormalities 94
nickel
    air pollution 49–51
    Arctic North 43, 44, 48, 49–51, 53, 55
    Black Triangle **150**
    Moscow 117
    in sulphide ores as source of sulphur
        dioxide pollution 48
    water pollution 53, 55
Nikel 53, 58, 60
NIMMW 50
nitrates, water pollution, Lake Baikal 83
nitrogen, water pollution 228
nitrogen dioxide
    air pollution **104**, 105, 106
    emissions in various countries 147, **148**
    soil pollution, Black Triangle 149–150
Noginsk 116
Noril'lag 48
Norilsk 33–34
    air pollution, metallurgical industries 49–51
    airborne pollutants **50**
    climate 34–35
    copper 48
    degradation of natural vegetation and
        pastures 60
    ecological situation 60
    economic development and ecological
        change **46–47**, 48
    future prospects 65
    health issues 64
    hydrology 36
    nickel 48
    soil contamination 60
    soil cover 42
    vegetation 38, 39
    water pollution 55
    see also Arctic North
Norilsk Nickel 48, 50, **51**, 53
Norway spruce 143
Novikova, N.M. 260
Nowicki, M. 148
nuclear stations, staff strikes 139
nuclear testing 56–76
Nyuduai 54

oak 122, 142, 143, 191
oak groves 190
Ob river 35, 36, 54, 55
Oder river 142
Odra river 149
O'Hara, S.L. and Hannan, T. 258, 261, 271
oil
    air pollution 53
    Arctic North 43, 44, 53, 54–55
    offshore production 5
    reserves 223
        offshore 222, 223
    soil contamination 55

oil *continued*
   toxic concentration for aquatic marine life
      233–234
   water pollution 54–55
      Arctic North 54–55
      Caspian Sea 226, 229–232
      Lake Baikal 81, 84
      Moscow 111
      reservoirs 204
      Volga basin **201**, 202
Oka river 101, 119, 189
Olcott, M.B. 252
Olenegorsk 44
omul fish 89, 90
oncological illness 63, 94, 135
Orekhovo-Zuevo 116
organic cyanides 106
organochlorines 90
Orlovskiy, N.S. 251, 255, 261
overgrazing 172, 175, 181
Ozturk, M. *et al* (1997) 237

palladium 43, 48
Paris, drinking water consumption 111
pastoral vegetation, trampling 174
pear trees 191
Pechenga region 44
Pechenganickel 44, 51, 52, 54, 58
Pennsylvania, nuclear accident at Three Mile
      Island 134
Perel'man, A.I. 12
Perm oblast 211
permafrost 31, 35, 36, 41, 65, 72–73, 74
pesticides 88, 209
phenols
   air pollution **104**, 106
   contamination of water fowl 90
   water pollution 88, 229, 230
phytomass, major zones in northern Eurasia
      17, **18**
phytoplankton 89, 233–234
pine 38, 73, 76, 122, 143
plant biodiversity, major zones in northern
      Eurasia 17, **18**
platinum 43, 48
plutonium–239 138
poaching 88, 90, 226, 234
Podolsk 16
Poelzer, G. 63
Poland 153
   acid rain **149**, 150
   coal industry 144–145, 156
   economic revival 156
   emission of sulphur dioxide and nitrogen
      dioxide 147, **148**
   heavy metal pollution **150**
   life expectancy 154
   Silesian district 141
      see also Black Triangle
   water pollution 148, 149
   water quality monitoring service 158
polar belt
   climatic, soil and vegetation features **14**
   geographical characteristics **18**

polar day 34
polar desert
   climatic, soil and vegetation features **14**
   geographical characteristics **18**
polar night 34
Poles'ye 121, 122
pollutant pays principle 278
polychlorine diphenyl 202
poplar 112
Popov, S.N. 49
precipitation, removal of atmospheric
      pollutants 35
Priaralye
   area covered 242–243
   desertification processes 258–260
   health effects 267
   overall ecological situation 262
   population 244
Pribaikalskiy 79
primary succession duration 17, **18**
production, decline 195–196
Prokhorov, B.B. 62, 63, 64
Protasov, V.F. and Molchanov, A.V. 28
protected areas 30
Pryde, P.R. 30, 80, 84
psammophyte vegetation 247
Ptichnikov, A.V. 259
Pyasino Lake 60

radiation balance, major zones in northern
      Eurasia **18**
radiation index of dryness for each zone
      14–15, 16
radioactive contamination 4
   Arctic North 56–57, 60
   impact on coniferous species 124
   soil 56–57
   sources 139
   vegetation and pastures 58–59
   see also Chernoybl
radioactive waste disposal 57
radionuclides, migration in soil 124
Rafikov, A.A. 259, 260
railways, Baikal-Amur Mainline 80
rangeland pastoralism 248
Rasputin, V. 96
recycling 113, 119
redkolesye 38
reed 122
regional ability to cope with problems
      280–283
reindeer 43, 44, 59, 88
   pastures 59–60
reindeer moss 38
reservoirs
   and climatic change 199
   construction and flooding of fertile land 207
   erosion of banks 208
   pollution 203–204
   self-pollution 203–204
resilience 22, 23, **24**
resistance, definition 22
resistance threshold 25

respiratory diseases 63, 64, 94, 213, 214
Reznikov, N.I. 169, 172, 174, 179, 182, 183
rice growing, irrigation canals 178
river deltas, vegetation 248–249
Roginko, A.Yu. 59, 60
Romanova, E.P. 142–143
Rosgidromet 83, 86, 88
Rossiyskiy Statist... (1994–99) 214
Rozanov, B.G. 161
rush 122
Russia
    ability to cope with problems 281–283
    Caspian oil deposits, ownership 238, 239
    discharge of polluted waste waters into
        Caspian Sea **227**
    drinking water quality 278
    emission of sulphur dioxide and nitrogen
        dioxide 147, **148**
    environmental concerns 275
    oil pollution 230
    oil production 1990–97 223, **224**
    per capita GDP 270Fig
    per capita water withdrawal **254**, 255
    population in areas contaminated by
        Chernoybl **135**
Rybinsk reservoir 203

Sagers, M.J. 55
saiga 165, **166**, 177–178
Saiko, T.A. 51, 237, 240, 252, 265, 271, 275
Saiko, T.A. and Zonn, I.S. 164, 166
Sakata, T. 272
Sakwa, R. 234
saline soils 17, 247, 259
Sallnow, J.A. 63
Sallnow, J.A. and Saiko, T. 278
salmon 76
salt, use during winter 102, 112
salt storms 260
saltwort 174, 247
Samara **50**, 205, 213
Samur river 220
sand acacia 247
sand storms 260
Sandjiev, B.C. 162
sands, shifting 175, **176**
Saratov 205, 208, 213
Saratov reservoir 203, 208
satisfactory state
    criteria for evaluation **29**
    description **27**
Savchenko, V.K. 124, 131
saxaul 247
seals
    Caspian 221, 233
    nerpa 76, 77, 89–90
secondary succession duration 17, **18**, 19
sedges 38
seismic activity 70, 91, 113
Selenga 94
Selenga river 71, 90, 91
Selenginsk pulp and cardboard complex
    (SPCC) 80, 81

selenium 48
semi-deserts 38, 192
    climate and soil **15**, 17
    geographical characteristics **18**
    vegetation **15**, 17, 163
Sergiev, V.P. *et al* (1993) 268
serozems 249
Severonickel complex 44, 51, 52, 54, 58
Shahgedanova, M. *et al* (1999) 102, 105
Shaw, D.J.B. and Oldfield, J. 6
Shcherbak, Yu.N. 126, 134, 136, 138
sheep, effect of sharp hooves 168
sheep's fescue 163
shifting sands 175
Shlikhunov, V.M. 240
shrimp 76
shrub tundra subzone 38
Sib-Aral canal 268
Siberia *see* Baikal, Lake
Siberian cedar 74
Siberian stone pine 74
silver 48
silver fir 143
Simagin, Yu.A. 103, 118
slough grass 163
Slovakia
    coniferous forest 143
    emission of sulphur dioxide and nitrogen
        dioxide **148**
    heavy metal pollution **150**
Smolensk-Moscow 101
snow, retention of pollutants 58
Sochava, V.B. 22
Sodruzhestvo (1998b) 195
soil erosion
    Lake Baikal 87
    Volga basin 207–208
soil pollution
    Arctic North 55–57, 60
    cause by air pollution 208
    chlororganic compounds 203
    dioxins 203
    heavy metals, Black Triangle 150
    Lake Baikal 87
    Moscow and region 112–113
    Volga basin 208–209
soils
    Black Triangle 142
    chernozem 192
    chestnut 192, 247
    clay polygonal 249
    major zones in northern Eurasia **14–15**, 17
    migration of radionuclides 124
    salinization 173, 176–177, 208
    secondary salinization 261
    solonchaks 164, 192, 247, 249
    solonets 192
Sokhondinskiy reserve 79
Sokhraneniye (1997) 19, 30, 59, 193
solar radiation 13
solid waste accumulation, Volga basin 209
Solomatin, V.I. 36
solonchak soils 164, 192, 247, 249

solonets 192, 249
Sostoyaniye (1990) 26, 160, 203
Sos'vinskiy nature reserve 55
Soviet Union, former, area of 11
spawning grounds, and hydroelectricity 210
spindle-tree 191
sponges 76
statistical yearbooks 48
steppes 19, 191–192
  biodiversity 192
  climate and soil **15**, 17
  geographical characteristics **18**
  phytomass 192
  soils 163–164
  vegetation **15**, 17, 163
Stewart, J.M. 161
storms, sand and salt 260
strained situation
  criteria for evaluation **29**
  description **27**
strict nature reserves 30
strontium 52, 133
sturgeon 5, 76, 210, 221–222
  Beluga 221–222, 232, 235
  decline in catch and reserves 234–235
  Lake Baikal 90
  poaching 234
  Russian 222, 232
  Sevryuga (star) 222
  Stellate 232
subarctic belt, geographical characteristics **18**
subarctic tundra, geographical characteristics **18**
subpolar belt, climatic, soil and vegetation
     features **14**
Sudeten Mountains 142
sukhovey winds 162, 175, 246
Sulak river 220
sulphates, water pollution 53, 83, 84, 228
sulphur compounds, soil pollution 208
sulphur deposition 155
sulphur dioxide
  air pollution 49–52, **104**, 209
  emissions in various countries 147, **148**
  pollution, nickel in sulphide ores 48
  reduction in emissions 155–156
  soil pollution 149–150
  transboundary pollution 53
Sulphur Triangle *see* Black Triangle
Sultangazin, U.M. and Tsukatani, T. 230
Svitoch, A.A.
Sweden, emission of sulphur dioxide and
     nitrogen dioxide **148**
Syrdarya river 242, 244, 245, 246, 255, 269

taiga 33, 68, 73, 187
  biodiversity 190
  climatic, soil and vegetation features **14**, 17
  dark coniferous 73, 84
  geographical characteristics **18**
  light coniferous 73, 84
  phytomass 190
  typical landscape **75**

Tajikistan
  cotton production **252**
  fertilizer use 261
  grain crops **267**
  infant mortality 268
  per capita GDP **270**
  per capita water withdrawal **254**, 255
takyrs 249
Talyzin, N. 79
technical accidents 203
Tedjen river 244
temperate belt
  climatic, soil and vegetation features **14–15**
  geographical characteristics **18**
temperature inversion 35, 70–71, 142
Tenghiz oilfield 230
Terek river 220, 228, 232
TETS–23 105
thematic atlases 7
thermokarst lakes 58, 73
timber 44
  transportation 85
Tishkov, A.A. 19, 39
Tolyatti 205
tourism 88
Trans-Siberian railway 73
transboundary pollution 53, 86, 274–275
transport, shift from public to private 157
transport pollution, Moscow 106–107, 118
trees, stabilization of radiological situation 122
Trofimov, I.A. 164, 169, 172, 174, 181
Tsarist Russian agricultural policies 250
tuberculosis 64, 268
tugai 248
tumours 213
tundra 31, 33, 275–276
  biological diversity 39
  climatic, soil and vegetation features **14**, 17
  decomposition of organic matter 42
  destruction of vegetation and soil cover
     59–60
  flowering plants 38
  geographical characteristics **18**
  Kola Peninsula 37–38
  lichens 38
  mechanical disturbance 36
  mosses 38
  shrub subzone 38
  soils, microbiological activity 42
  typical landscape **37**
  zone 19, 30
    shift in boundary 62
  see also Arctic North
Tunkinskiy 79
Turkmenbashi 236
Turkmenistan
  average monthly earnings 263
  Caspian oil deposits, ownership 238, 239
  cotton production **252**
  discharge of polluted waste waters into
     Caspian Sea **227**
  fertilizer use 261
  flooding due to rise in sea level 240